Standard Form

$Ax + By + C = 0$

Slope-Intercept Form

Every first-degree equation of the form $Ax + By + C = 0$, $B \neq 0$, can be written in the form

$$y = mx + b$$

where b is the y-intercept and m is the slope. This form is called the *slope-intercept form* of the equation of a line.

Slope Formula

The slope, m, of a nonvertical line passing through $P_1(x_1, y_1)$ and $P_2(x_2, y_2)$ is given by

$$m = \frac{\text{RISE}}{\text{RUN}} = \frac{y_2 - y_1}{x_2 - x_1}$$

Horizontal and Vertical Lines

Horizontal lines have the form

$$y = \text{constant}$$

and have zero slope.

Vertical lines have the form

$$x = \text{constant}$$

and have no slope.

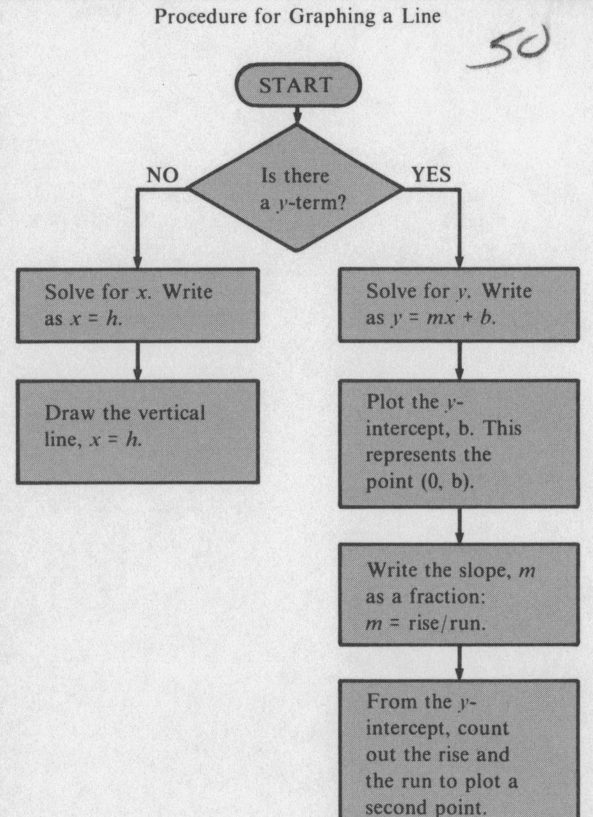

Systems of Linear Equations

Graph		Algebraic Equations	Solution
y, x	represent two *intersecting lines* a *single point* of intersection	You obtain a single value for each of the variables.	State the point of intersection as an *ordered pair*.
y, x	represent two *parallel lines* *no point* of intersection	The variable drops out of the equations, and you obtain an equation that is false, like 2 = 3.	Describe the system as *inconsistent*.
y, x	represent the *same line* *unlimited number* of solutions	The variable drops out of the equations, and you obtain an equation that is true, like 2 = 2.	Describe the system as *dependent*.

Beginning Algebra for College Students THIRD EDITION

I am the very pattern of a modern Major-General

. . . . .

I'm very well acquainted, too, with matters mathematical,
I understand equations, both simple and quadratical;
About binomial theorem I'm teeming with a lot o' news,
With interesting facts about the square of the hypotenuse.
I'm very good at integral and differential calculus,
I know the scientific names of beings animalculous.
In short, in matters vegetable, animal and mineral,
I am the very model of a modern Major-General.

Pirates of Penzance
by Gilbert and Sullivan

On the Cover: *A computer chip, photographed through a research telescope. Phillip Harrington began exploring these minute worlds as a diversion from his occupation as a world-traveling photojournalist and has become a specialist in photomicrography.*

The Smith and Boyle Precalculus Series

Beginning Algebra for College Students, Third Edition
KARL J. SMITH AND PATRICK J. BOYLE

Intermediate Algebra for College Students, Second Edition
KARL J. SMITH AND PATRICK J. BOYLE

Study Guide for Intermediate Algebra for College Students, Second Edition
KARL J. SMITH AND PATRICK J. BOYLE

College Algebra, Second Edition
KARL J. SMITH AND PATRICK J. BOYLE

Other Brooks/Cole Titles by the Same Authors

Essentials of Trigonometry
KARL J. SMITH

Trigonometry for College Students, Third Edition
KARL J. SMITH

Precalculus Mathematics: A Functional Approach, Second Edition
KARL J. SMITH

The Nature of Mathematics, Fourth Edition
KARL J. SMITH

Mathematics: Its Power and Utility
KARL J. SMITH

Basic Mathematics for College Students
KARL J. SMITH

Beginning Algebra for College Students THIRD EDITION

Karl J. Smith
Patrick J. Boyle

Santa Rosa Junior College

Brooks/Cole Publishing Company
Monterey, California

Consulting Editor: Robert J. Wisner

Brooks/Cole Publishing Company
A Division of Wadsworth, Inc.

Printed in the United States of America
10 9 8 7 6 5 4 3 2

Library of Congress Cataloging in Publication Data

Smith, Karl J.
Beginning algebra for college students.

Includes index.
1. Algebra. I. Boyle, Patrick J., [date]
II. Title.
QA152.2.S57 1984 512.9 83-14277
ISBN 0-534-02779-2

Subject Editor: *Craig F. Barth*
Production Editor: *Suzanne Ewing, Joan Marsh*
Manuscript Editor: *Jonas Weisel*
Permissions Editor: *Carline Haga*
Cover and Interior Design: *Jamie Sue Brooks*
Cover Photo: © *Phillip Harrington, Courtesy of Western Electric*
Art Coordinator: *Rebecca Tait*
Interior Illustration: *Carl Brown*
Typesetting: *Progressive Typographers, Emigsville, Pennsylvania*

To our wives,
Linda and Theresa

Preface

This book provides an introduction to algebra for college students. There are many reasons why such a course of study is needed at this level. Perhaps the student never mastered algebra in high school or perhaps avoided it entirely. Perhaps the student once felt no reason to "go on in math," but now finds an understanding of algebra is essential for further study in a chosen field. College students who enroll in beginning algebra have varied interests, abilities, and goals, but they do seem to share a distinct fear of mathematics. So we have designed this book to appeal to a variety of individuals and to allay that fear of math.

Those of you familiar with the second edition will notice a significant difference in format. The design has been changed to make it easier to follow the exposition. Important results are boxed, and important terms are indicated by boldface type. The order of topics has been changed so that polynomials, factoring, and rational expressions now precede graphing and systems. Many topics and discussions in this edition have been simplified; at the same time, more drill problems have been added. Word problems and applications, which remain an important part of the text, have been rewritten to simplify the process of obtaining a solution. Problem types have been introduced to simplify the selection of the proper method by the student.

We have assumed that students using this book have almost no mathematical background. We have structured the sections so that the methods and reasoning are followed by several examples; the examples are then followed by many problems. There is an example for each type of problem represented in the problem sets. Although we emphasize the *how* rather than the *why,* we do feel that algebra should not be presented as a "bag of tricks" that must be accepted without development. Therefore, we explain the *why* of algebra at those points where the reader has been given sufficient information to understand the concept.

Many schools are no longer requiring their students to follow a beginning algebra course with a geometry course. For that reason, geometrical terms and geometrical ideas are presented throughout the text whenever appropriate to reinforce the algebraic concept. For those schools that do not wish to cover geometrical ideas, the sections dealing with geometry are clearly marked optional and can be skipped with no loss of continuity.

We would like to thank the reviewers of this edition: Carol Achs of Mesa Community College, Carole Bauer of Triton College, Emily Dickinson of the University of Arkansas at Little Rock, Roy M. Gifford of Ferris State College, Robert G. Hammond of Utah State University, Roger Judd of Chemeketa Community College, Steven Marsden of Glendale Community College, and James Vineyard of Laney College.

And we would like to thank the staff at Brooks/Cole, especially Craig Barth, Jamie Brooks, Sue Ewing, and Joan Marsh. And, finally, we thank our wives and families for their patience and understanding.

Karl J. Smith
Patrick J. Boyle

Contents

* Optional section

* Optional section

* Optional section

Beginning Algebra for College Students THIRD EDITION

1

Numbers and Statements

The branches of arithmetic are
ambition,
distraction,
uglification,
and derision.

The Mock Turtle
in *Alice in Wonderland*
by Lewis Carroll

1.1 Introduction

This generation has seen an explosion of mathematical devices. The hand-held calculator was closely followed by personal computers. The widespread use of computers has contributed to many diverse developments: from utility bills to motor tune-ups, automated bank tellers to hotel reservations, and NASA Mission Control to Pac Man. It should not be surprising to find the computer and its companion mathematics invading an ever widening spectrum of fields. Your field of interest may be directly responsible for your current study of algebra. Your classmates will likely have diverse interests, abilities, and goals. Whatever your purpose in taking this particular course, we hope you find the text attractive and the course that it outlines relevant to your needs.

As you prepare to begin this study, recall that mathematics requires commitment and some organization if you are to realize your potential. Do not plan on working in spurts and cramming at the last minute for exams. It just does not work! If you set aside specific, regularly scheduled times to work on the course, you will maximize your chance of success. Be especially careful to use the text effectively. Read the discussion in the text *before* studying the examples; then study the examples *before* attempting the problems in your assignment. You will make more effective use of your time if you avoid backtracking to fill holes left by skipping around; and more importantly, you will be more likely to understand *why* you are doing things in a certain way.

In arithmetic you learned about numbers and operations with numbers. Elementary algebra is also a study of numbers and their properties, but it begins where arithmetic leaves off. Algebra requires extensive use of symbols. However, one of the difficulties in introducing symbols is that the same symbol might be interpreted in more than one way. For example, if you enter 39138 in your calculator, these symbols have a clear numerical value. However, if you invert the calculator and look at the display upside down, the symbols now seem to spell the word *BEIGE*.

In algebra we must define symbols carefully so that there is no chance for ambiguity. Consequently, our definitions must be clear and precise. Letters of the alphabet are used as symbols for numbers whose value is unknown. A letter used this way is called a ***variable.*** A variable represents a number from a given set of numbers called the ***domain*** of the variable. To express the idea

"the sum of a number plus five,"

we might represent the unknown number by the letter n and write

$$n + 5$$

The domain here is the set of all numbers, because there is no reason to restrict the possible replacements for n. However, if the domain of n is $\{2, 3, 7\}$, then the variable expression $n + 5$ represents

$2 + 5$ (that is, 7), if $n = 2$

$3 + 5$ (that is, 8), if $n = 3$

$7 + 5$ (that is, 12), if $n = 7$

It cannot represent

$$6 + 5$$

because the replacement $n = 6$ is not in the domain.

If n represents the number of stars in the cosmos, we would restrict n to the domain

$$\{1, 2, 3, 4, \ldots\}^*$$

because it would not make sense to let $n = \frac{1}{2}$. We call this set of numbers the set of ***counting numbers*** or ***natural numbers.***

If the domain has only one member, we say that the symbol is a ***constant.*** That is, a letter, symbol, or numeral that designates only one number during a particular discussion is called a constant.

Examples of Constants

1. Any numeral, such as 4, 7, 13
2. c, where c has the domain $D = \{7\}$
3. π, where π is the ratio of the circumference of any circle to its diameter. It is a constant approximately equal to 3.1416. ■

We often want to investigate the relationship between quantities. For example,

Ms. Muffet is older than Mr. Horner;
Dr. Phil S. Phee is taller than Dr. U. Klidd;
Martha's grade is the same as Mary's grade.

If we are given any two quantities, either these quantities are equal or one of these quantities is less than the other. Over time, symbols have been developed to represent these ideas.

We are familiar with the ***equality*** symbol "=." This symbol is used whenever the quantities are exactly the same. If two quantities are not exactly the same, we use the symbol "≠," meaning "not equal to." For example,

$$5 = 5 \text{ and } 5 \neq 10$$

The symbol "<" is used to express the relationship ***"is less than."*** For example, "seven is less than ten" is expressed

$$7 < 10$$

The symbol ">" is used to express the relationship ***"is greater than."*** "Ten is greater than seven" is expressed

$$10 > 7$$

If a number x is not greater than seven, we symbolize this relationship by using a slash:

$$x \not> 7$$

---

* Braces are used to enclose a set of elements; the three dots are used to indicate that some elements are not shown.

This is the same as saying that x is less than seven or x is equal to seven. These symbols are combined by writing

$$x \leq 7$$

The relationship

$$y \nless 10$$

is read "y is not less than ten" and can also be written

$$y \geq 10$$

meaning y is greater than ten or y is equal to ten.

Table 1.1
Summary of Relation Symbols

Symbol	Verbal Statement	Example
$=$	is equal to	$5 = 5$
$\neq$	is not equal to	$5 \neq 7$
$<$	is less than	$5 < 7$
$\leq$	is less than or equal to	$5 \leq 7$
$>$	is greater than	$5 > 3$
$\geq$	is greater than or equal to	$5 \geq 5$

Notice that the symbols $<$ and $>$ always point toward the smaller number. The symbols $<$, $\leq$, $>$, and $\geq$ are referred to as ***inequality symbols.*** Even though $\neq$ is not an equality symbol, it is usually not included when speaking of inequality symbols.

Examples of Relation Symbols

4. $5 < 7$ means 5 is less than 7.
5. $9 > 2$ means 9 is greater than 2.
6. $15 \nleq 4$ means 15 is not less than or equal to 4.
7. $10 \leq 10$ means 10 is less than or equal to 10.

Remember, the symbol $\leq$ means less than *or* equal to; that is, $10 \leq 10$ is true if *one* of the following is true:

$$10 < 10$$

or

$$10 = 10$$

8. $x \ngtr y$ means x is not greater than y. ■

In elementary algebra we combine and manipulate numbers in various ways. We use symbols to represent these operations.

Addition We indicate addition by using the familiar "plus" symbol from arithmetic. The result obtained is called the ***sum,*** and the numbers being added are called ***terms.***

Examples *Write the word statements or phrases in symbols.*

Solutions

9. Ten is equal to four plus six. — 9. $10 = 4 + 6$
10. The sum of the terms thirteen and four is seventeen. — 10. $13 + 4 = 17$
11. The sum of the terms x and five. — 11. $x + 5$
12. The sum of y added to z. — 12. $y + z$ ■

Subtraction We indicate subtraction by using the familiar "minus" symbol from arithmetic. The result is called the ***difference.*** As in addition, the numbers involved are called ***terms.***

Examples *Write the word statements or phrases in symbols.*

Solutions

13. Ten minus four is six. — 13. $10 - 4 = 6$
14. Thirteen is the difference of four subtracted from seventeen. — 14. $13 = 17 - 4$
15. The number x subtracted from the number y. — 15. $y - x$
16. The term six subtracted from the number x. — 16. $x - 6$
17. The difference of the number y from the number z. — 17. $z - y$ ■

Multiplication In algebra we cannot always use the same symbols that we use in arithmetic. The "cross" symbol for multiplication, as in 13×4, might be confused with the variable symbol x, so we rarely use the cross in algebra. Instead we use

1. a *dot,* as in $13 \cdot 4$ or $x \cdot y$;

Notice that the multiplication dot is raised so that it is not confused with a decimal point or period.

2. *parentheses,* as in (13)(4), 13(4), or (13)4;

Notice that one or the other or both numbers can be enclosed in parentheses.

3. *juxtaposition,* which means placed side by side, as in xy or $3x$. We do not use juxtaposition when two numerals are used, because 34 symbolizes thirty-four, not 3 times 4. On the other hand, xy *always* represents x times y, *not* a two-digit number.

The result of multiplying two numbers is called the ***product,*** and the numbers being multiplied are called ***factors.***

Examples *The first instance is the preferred form in each case. Notice that juxtaposition cannot be used to indicate multiplication of two numerals.*

18. *Two numerals:* 5 times 3 can be written as $\mathbf{5 \cdot 3}$, 5×3, 5(3), (5)3, (5)(3), $(5) \cdot (3)$, $(5) \cdot 3$, and $5 \cdot (3)$.
19. *A numeral and a variable:* 5 times x can be written as $\mathbf{5x}$ (this is juxtaposition

and uses no operation symbol), $5 \cdot x$, $5(x)$, $(5)x$, $(5)(x)$, $(5)\cdot(x)$, $(5)\cdot x$, and $5\cdot(x)$.

20. *Two variables:* x times y can be written as $\boldsymbol{xy}$, $x\cdot y$, $x(y)$, $(x)y$, $(x)(y)$, $(x)\cdot(y)$, $(x)\cdot y$, and $x\cdot(y)$. ■

Examples *Write the word statements or phrases in symbols.*

Solutions

21. Fives times eight is forty. — **21.** $5\cdot 8 = 40$

22. The product of three and seven is twenty-one. — **22.** $3\cdot 7 = 21$

23. The product of x and eleven — **23.** $11x$

24. The product of the factors y and z — **24.** yz ■

Division In arithmetic we denote division by using the symbol "÷." In algebra we also use the fractional notation

$$\frac{x}{y}$$

to mean x divided by y; x is called the ***dividend,*** and y the ***divisor.*** The result of dividing two numbers is called the ***quotient.*** Division can be explained in terms of multiplication. We say that

$$\frac{15}{5} = 3$$

because $15 = 5\cdot 3$

Examples *Check to see if the following problems are done correctly.*

Solutions

25. $\frac{18}{3} = 6$ — **25.** $18 = 3\cdot 6$ correct

26. $\frac{24}{6} = 4$ — **26.** $24 = 6\cdot 4$ correct

27. $\frac{0}{5} = 0$ — **27.** $0 = 5\cdot 0$ correct

28. $\frac{5}{0} = 0$ — **28.** $5 = 0\cdot 0$ not correct ■

If Example 28 is not correct, is there a value n for which $\frac{5}{0} = n$ is correct? Now, $\frac{5}{0} = n$ means $5 = 0\cdot n$. There is no such value for n, because 0 times any number is 0 and we can never get the answer 5. We, therefore, do not divide by 0.

Examples *Write the word statements or phrases in symbols.*

Solutions

29. Twenty divided by five is four. — **29.** $\frac{20}{5} = 4$

30. Six is the quotient of twenty-four divided by four. **30.** $6 = \frac{24}{4}$

31. The number x divided by eight **31.** $\frac{x}{8}$

32. The quotient of two numbers, y by z **32.** $\frac{y}{z}$ ■

Examples *Choose a letter to represent the variable and write a mathematical expression to express the idea.*

33. The sum of seven and a number.

Sum indicates addition, so the statement is

(SEVEN) + (A NUMBER)

Now select some variable (your choice), say s = A NUMBER. Then the answer is

$7 + s$

Other answers are possible, such as $s + 7$ or a similar expression if a different variable is chosen. However, once you choose a variable for a particular problem, that variable cannot represent any other number in that problem.

34. The difference of a number subtracted from seven:

(SEVEN) − (A NUMBER)

Let d = A NUMBER; then the expression is

$7 - d$

35. The quotient of two numbers:

$$\frac{\text{A NUMBER}}{\text{ANOTHER NUMBER}}$$

If there is more than one unknown in a problem, and no relationship between those unknowns is given, then more than one variable may be needed. Let m = A NUMBER and n = ANOTHER NUMBER:

$$\frac{m}{n}$$

36. The product of two consecutive numbers:

(A NUMBER)(NEXT CONSECUTIVE NUMBER)

If there is more than one unknown in a problem, but a given relationship between those unknowns, then *do not choose more variables than you need* for the problem. In this problem, a consecutive number means one more than the first number:

(A NUMBER)(A NUMBER + 1)

Let x = A NUMBER; then the variable expression is

$x(x + 1)$ ■

Problem Set 1.1

1. *Classify each of the following as a constant or a variable. See Examples 1–3.*

   a. 23 **b.** 2.3 **c.** $\frac{2}{3}$ **d.** 2 **e.** 2π
   f. w where the domain of w is $\{2\}$.
   g. x where the domain of x is $\{\pi\}$.
   h. y where the domain of y is $\{2, 3\}$.
   i. z where the domain of z is $\{2, 3, \pi\}$.

Fill in the word(s) or symbols that make the statements in Problems 2–10 complete and correct.

2. A ______________ is a letter used as a symbol for a number from a given set, called the ______________.
3. The set of numbers 1, 2, 3, 4, . . . is called the set of ______________ numbers.
4. A symbol that designates one and only one number is called a ______________.
5. The constant π is approximately equal to ______________.
6. The symbol ______________ is read "less than or equal to."
7. The result of addition is called the ______________ and the numbers being added are called ______________.
8. The result of subtraction is called the ______________.
9. The result of multiplication is called the ______________ and the numbers being multiplied are called ______________.
10. The division $\frac{a}{b} = c$ can be written in terms of multiplication as ______________.

In Problems 11–20 insert the appropriate inequality symbol in place of the question mark. See Examples 4–8.

11. 3 ? 4 **12.** 4 ? 3 **13.** 5 ? 2 **14.** 2 ? 1 **15.** 0 ? 0
16. 9 ? 0 **17.** 0 ? 7 **18.** 7 ? 7 **19.** π ? 3 **20.** π ? 4

Change the division problems in Problems 21–30 to equivalent multiplication problems, and check to see if the division is correct. See Examples 25–28.

21. $\frac{30}{5} = 6$ **22.** $\frac{32}{4} = 8$ **23.** $\frac{54}{8} = 7$ **24.** $\frac{42}{7} = 6$ **25.** $\frac{52}{13} = 4$
26. $\frac{51}{17} = 3$ **27.** $\frac{128}{4} = 32$ **28.** $\frac{4}{0} = 4$ **29.** $\frac{5}{0} = 0$ **30.** $\frac{0}{6} = 0$

In Problems 31–40 write the word statements in symbols. See Examples 9–24 and 29–32.

31. Nine is equal to four plus five.
32. Nine minus four is equal to five.
33. Seventeen is the sum of nine and eight.
34. Eight is the difference of nine from seventeen.
35. Seven is equal to nine minus two.
36. Seven times nine is equal to sixty-three.
37. Seven is the quotient of sixty-three divided by nine.
38. The product of eight and seven is less than sixty-three.

39. Sixty-three is greater than five times twelve.
40. Sixty divided by zero is not equal to six.

In Problems 41–60 choose a letter to represent the variable and write a mathematical expression for the idea. See Examples 33–36.

41. A number plus one
42. Twice a number
43. Three times a number
44. A number added to four
45. Five minus a number
46. A number divided by six
47. Seven plus a number
48. Eight divided by a number
49. A number times nine
50. Ten times a number
51. The difference of a number from one
52. The product of a number and seven
53. The sum of eight and a number
54. The difference of nine subtracted from a number
55. The sum of a number and ten
56. Twice a number multiplied by three
57. Three plus twice a number
58. The product of three and twice a number
59. The sum of two numbers
60. The sum of two consecutive numbers

1.2 Operations and Conventions

In algebra we are concerned with both numerical and variable expressions. *A **numerical expression** is a phrase containing only constants and at least one defined operation.* Defined operations include addition, subtraction, multiplication, and nonzero division.

Examples *Which of the following are numerical expressions?*

	Solutions
1. $5 + 8$	**1.** Numerical expression
2. $6 \div 3$	**2.** Numerical expression
3. 4	**3.** Not a numerical expression (no operation symbol)
4. $5 \div 0$	**4.** Not a numerical expression (cannot divide by 0)
5. $0 \div 3$	**5.** Numerical expression
6. $3x + 4$	**6.** Not a numerical expression (contains a variable)
7. $5 + 2 \cdot 3$	**7.** Numerical expression ■

*A **variable expression** is a phrase containing one or more variables and at least one defined operation.* It may also contain constants. Example 6 in the preceding set of examples is a variable expression.

Say you want to simplify the expression given in Example 7. You might consider

$$5 + 2 \cdot 3 = 7 \cdot 3 \quad \textit{(First add } 5 + 2 = 7.) \\ = \mathbf{21}$$

or

$$5 + 2 \cdot 3 = 5 + 6 \quad \textit{(First multiply } 2 \cdot 3 = 6.) \\ = \mathbf{11}$$

Obviously this creates a difficulty because we want to get the same result each time. To overcome this difficulty we *agree to perform any multiplications or divisions before we perform additions and subtractions.* Thus, the correct value for $5 + 2 \cdot 3$ is 11. However, if we wish to do the addition first, we introduce parentheses to the expression:

$$(5 + 2) \cdot 3, \text{ or more simply, } (5 + 2)3$$

which means that *first* we perform the operation in parentheses.

If you have a calculator, try working Example 7 by pressing:

[5] [+] [2] [×] [3] [=]

If the calculator displays 11 as the answer, then your calculator uses *algebraic logic.* This means that it is preprogrammed to use the above order-of-operations agreement. If the answer is given as 21, then your calculator uses *arithmetic* or *serial logic.* It performs the operations in the order entered, in this example addition first. You will have to be certain that *you* use the order-of-operations convention even if your calculator does not. For instance, the preceding example may be entered as:

[2] [×] [3] [+] [5] [=]

with multiplication entered first. If parentheses are available, the problem may also be entered as:

[5] [+] [(] [2] [×] [3] [)] [=]

Summary: Order-of-Operations Convention

1. First, perform any operations enclosed in parentheses.
2. Next, perform multiplications and divisions by working from left to right.
3. Finally, perform additions and subtractions by working from left to right.

Examples *Completely simplify the numerical expression.*

	Solutions	
8. $3 + 4 \cdot 5$	**8.** $3 + 20 =$ **23**	*A numerical expression is completely simplified if it is represented by exactly one constant.*
9. $(3 + 4) \cdot 5$	**9.** $7 \cdot 5 =$ **35**	
10. $15 \div 5 + 2$	**10.** $3 + 2 =$ **5**	
11. $10 - 7 - 2$	**11.** $3 - 2 =$ **1**	
12. $10 - (7 - 2)$	**12.** $10 - 5 =$ **5**	

13. $10 + 2 \cdot 3 - 6 \div 3$
$= 10 + 6 - 2$ *First step, multiplication and division*
$= 16 - 2$ *Second step, addition and subtraction*
$= \mathbf{14}$

14. $6 + 4(3 + 2) - 12 \div 4$
$= 6 + 4(5) - 12 \div 4$ *First step, parentheses*
$= 6 + 20 - 3$ *Second step, multiplication and division*
$= 26 - 3$ *Third step, addition and subtraction*
$= \mathbf{23}$

15. $18 \div 6 \div 3$
$= 3 \div 3$ *First step, left to right*
$= \mathbf{1}$

16. $6 \cdot 6 \div 2$
$= 36 \div 2$ *First step, left to right*
$= \mathbf{18}$ ■

Some problems may have more than one set of parentheses. If we get confused about which pairs of parentheses go together, we can use other ***grouping symbols*** as well:

Grouping Symbols

Parentheses:	( )
Brackets:	[ ]
Braces:	{ }

The division symbol is also sometimes used as a grouping symbol. For example,

$$\frac{2 + 3}{5} \text{ means } (2 + 3) \div 5$$

When one pair of grouping symbols is contained inside another pair, you should start with the innermost pair.

Examples *Simplify each expression to find the value of the expression.*

17. $3[(4 + 3) - 2]$ *Remember that grouping symbols say "Do this first."*
$= 3[7 - 2]$ *Innermost parentheses first.*
$= 3 \cdot 5$ *Next set of parentheses is simplified.*
$= \mathbf{15}$

18. $(4 + 1)\{(5 - 2)[6 + (3 + 2)]\}$
$= (4 + 1)\{(5 - 2)[6 + 5]\}$ *Innermost parentheses first.*
$= (4 + 1)\{3[11]\}$ *Next set of parentheses is simplified according to the order-of-operations convention.*
$= 5\{33\}$
$= \mathbf{165}$

19. $(13 + 15) \div [(5 \cdot 2) + (2 \cdot 6 \div 3)]$
$= (13 + 15) \div [10 + (12 \div 3)]$
$= (13 + 15) \div [10 + 4]$
$= 28 \div 14$
$= \mathbf{2}$ ■

It is important when studying algebra to tie algebraic symbols to word phrases. The following examples ask you to translate word phrases into symbols.

Examples *Write the word phrases in symbols, then simplify.*

	Solutions
20. The product of four and three	**20.** $4 \cdot 3 = 12$
21. The sum of three and two	**21.** $3 + 2 = 5$
22. Four times three plus two (Remember the order-of-operations convention.)	**22.** $4 \cdot 3 + 2 = 12 + 2 = 14$
23. Four times the sum of three and two	**23.** $4(3 + 2) = 4 \cdot 5 = 20$ ■

Problem Set 1.2

Classify the expressions in Problems 1–14 as numerical expressions, variable expressions, or neither. See Examples 1–7.

1. $7 + 9$ **2.** $7 \cdot 3$ **3.** 15 **4.** 8.3
5. $15 \div 3$ **6.** $15 \div 0$ **7.** $0 \div 15$ **8.** $2x + 1$
9. $6x$ **10.** $2 + 7 \cdot 3$ **11.** $7 - 2x$ **12.** $12(3x + 1)$
13. $7 + 6 \cdot 3 \div 2$ **14.** $3 - 2 \cdot 5 + y$
15. In your own words state the order-of-operations convention.

Completely simplify the numerical expressions in Problems 16–30. See Examples 8–16.

16. $2 + 3 \cdot 4$
17. $35 - 4 \cdot 3$
18. $(2 + 3) \cdot 4$
19. $(35 - 4) \cdot 3$
20. $4 + 8 \div 2$
21. $(4 + 8) \div 2$
22. $4 + 2 \cdot 3 - 6 \div 2$
23. $(4 + 2) \cdot 3 - 6 \div 2$
24. $[(4 + 2) \cdot 3 - 6] \div 2$
25. $[4 + 2 \cdot 3 - 6] \div 2$
26. $15 + 3 \cdot 2 - 12 \div 3$
27. $25 + 24 \div 6 + 5 \cdot 7$
28. $5 + 2(3 + 12) - 5 \cdot 3$
29. $17 - 4(12 - 2 \cdot 6) + 3$
30. $[2 + 3(4 + 5)][6 + 2(3 \cdot 4 - 2)]$

Completely simplify the expressions in Problems 31–40 to find the value of each expression. Notice that the numbers and operations are the same and in the same order in each problem. The only difference is the symbols of grouping. See Examples 17–19.

31. $[(3 + 9) \div 3] \cdot 2 + [(2 \cdot 6) \div 3]$
32. $3 + [(9 \div 3) \cdot 2] + [(2 \cdot 6) \div 3]$
33. $[(3 + 9) \div (3 \cdot 2)] + [(2 \cdot 6) \div 3]$
34. $(3 + 9) \div [(3 \cdot 2 + 2 \cdot 6) \div 3]$
35. $(3 + 9 \div 3 \cdot 2 + 2) \cdot (6 \div 3)$
36. $[(3 + 9) \div 3 \cdot 2 + 2] \cdot (6 \div 3)$
37. $[(3 + 9 \div 3) \cdot 2] + (2 \cdot 6 \div 3)$
38. $[(3 + 9) \div (3 \cdot 2) + 2] \cdot 6 \div 3$
39. $3 + (9 \div 3) \cdot (2 + 2 \cdot 6 \div 3)$
40. $3 + \{[(9 \div 3) \cdot (2 + 2)] \cdot (6 \div 3)\}$

In Problems 41–50 write the word phrases in symbols. Simplify where possible and choose a letter to represent the variable where necessary. See Examples 20–23.

41. The sum of three and the product of two and one
42. The product of three and the sum of two and one
43. The sum of three and the quotient of two and one

44. The quotient of three divided by the sum of two and one
45. Eight times the sum of nine and ten
46. Eight times nine plus ten
47. The sum of a number and three
48. The product of a number and ten
49. The sum of ten times a number and three
50. Ten times the sum of a number and three

1.3 Exponents and Powers

Recall that numbers related by multiplication are called factors. For example, given $5 \cdot 4 = 20$, 5 and 4 are factors of 20. If one factor is repeated several times, as in $5 \cdot 5 \cdot 5 \cdot 5$, a simpler notation called *power notation* is used. The number being multiplied is called the ***base,*** and the number of times it is used as a factor is called the ***exponent.*** The exponent is written smaller and placed above and to the right of the base as in 5^4. This is read "five to the fourth power."

Definition of Exponent

$$b^n \text{ means } \underbrace{b \cdot b \cdot b \cdot \cdots \cdot b}_{n \text{ factors}}$$

b is called the *base* and n is called the *exponent.* This expression is called "b to the nth power," or simply "b to the nth."

For historical reasons, special terminology is used with exponents of 2 or 3.

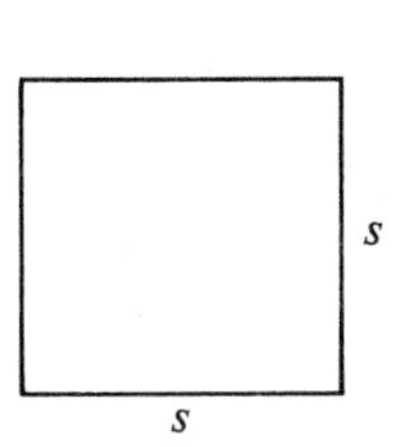

The area of a *square* is $A = s \cdot s = s^2$. Thus, when we see an exponent "2," we usually call it "squared."

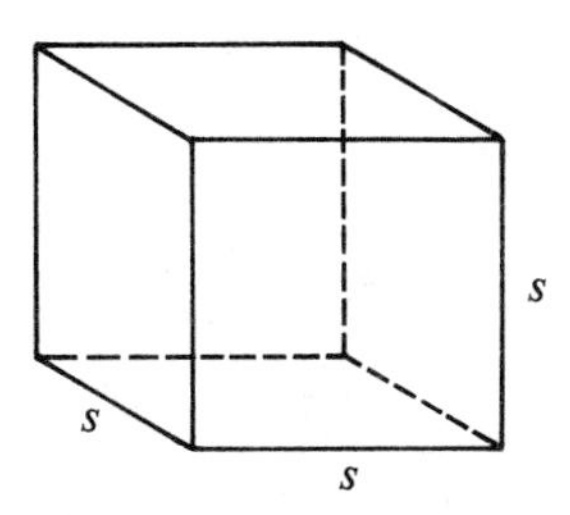

The volume of a *cube* is $V = s \cdot s \cdot s = s^3$. Thus, when we see an exponent "3," we usually call it "cubed."

In order to write a factorization of a number using exponents, construct a "factor tree," which eventually consists of a product of numbers that cannot be further factored. Such numbers are called ***prime numbers.***

Definition of Prime Numbers

A *prime number* is a counting number that has exactly two *distinct* factors. These factors are 1 and the number itself.

The primes less than 50 are: 2, 3, 5, 7, 11, 13, 17, 19, 23, 29, 31, 37, 41, 43, and 47. Do you see why 1 is not prime?

Example 1. Find the prime factors of 72, and write them in exponent form.

Solution Use a factor tree.

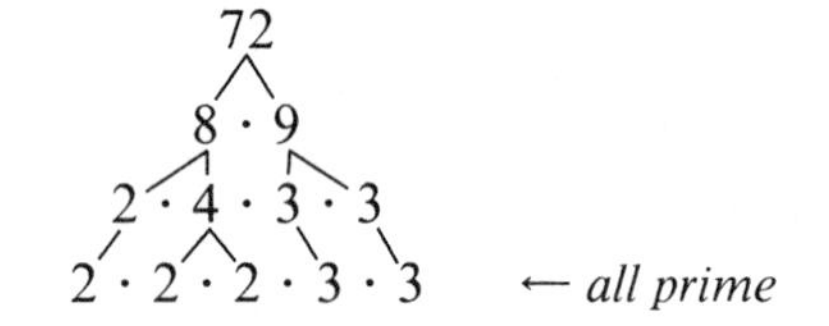

Step 1 Find *any* two factors of 72:

Step 2 Find *any* factors of each:

Step 3 Continue this process until all the numbers are prime. If a number is prime, simply recopy it in the next line of the factor tree.

Therefore, the prime factorization of 72 is

$2 \cdot 2 \cdot 2 \cdot 3 \cdot 3$

Use exponents to write $72 = 2^3 \cdot 3^2$. ■

When building your factor tree, you can find the factors in any order. If you continue the process until you have all prime factors, the result is the same except the order in which you have listed those factors. This process is illustrated by the following examples.

Examples *Write the prime factorization using exponents.*

2.

50
5 · 10
5 · 2 · 5

or $2 \cdot 5^2$

Alternate factor tree

50
2 · 25
2 · 5 · 5

or $2 \cdot 5^2$

3.

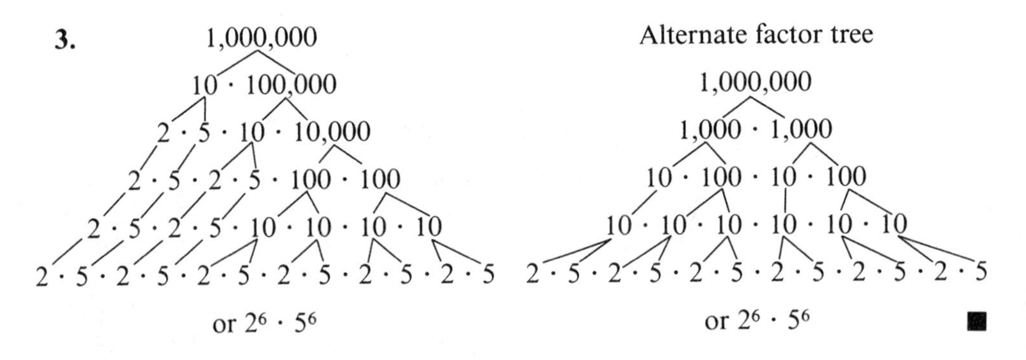

■

The reverse procedure from factoring is to change an expression in exponent form to nonexponent form by using multiplication.

Examples *Write each expression in nonexponent form, and evaluate where possible.*

4. $3^4 = 3 \cdot 3 \cdot 3 \cdot 3$
$= 81$

5. $4^3 = 4 \cdot 4 \cdot 4$
$= 64$

6. $x^2 = xx$

7. $y^5 = yyyyy$

8. $4(2)^3 = 4(2)(2)(2)$
$= 32$

9. $4x^3 = 4xxx$

10. $(4x)^3 = (4x)(4x)(4x)$
$= 4 \cdot 4 \cdot 4 \cdot xxx$
$= 64xxx$

11. $2(xy)^2 = 2(xy)(xy)$
$= 2(xxyy)$
$= 2xxyy$

12. $(2xy)^2 = (2xy)(2xy)$
$= 2 \cdot 2 \cdot xyxy$
$= 4xxyy$

■

Examples 8, 9, and 11 show that we must raise to a power before multiplying. Examples 9, 10, and 12 illustrate that parentheses still take precedence over the other operations.

Order-of-Operations Convention

1. First, perform any operations enclosed by parentheses.
2. Next, perform any operations involving raising to a power.
3. Next, perform multiplications and divisions by working from left to right.
4. Finally, perform additions and subtractions by working from left to right.

Examples *Completely simplify the following.*

13. $2 \cdot 5^2 = 2 \cdot 25$
$= 50$

14. $3^2 \cdot 3^4 = (3 \cdot 3) \cdot (3 \cdot 3 \cdot 3 \cdot 3)$
$= 9 \cdot 81$
$= 729$

15. $2^3(3 + 4)^2 - 25 = 2^3(7)^2 - 25$
$= 8(49) - 25$
$= 392 - 25$
$= 367$

■

Notice in Example 14 that $3^2 \cdot 3^4 = 729$ and $3^6 = 729$, thus $3^2 \cdot 3^4 = 3^6$. This suggests a general result. The product of two powers with the same base is a power with that base and an exponent that is the sum of the exponents.

First Law of Exponents

$$b^n \cdot b^m = b^{n+m}.$$

To multiply two numbers with like bases, add the exponents.

This is the first of five very important properties of exponents, which will be discussed completely in Section 4.3.

Examples *Write the solutions in exponent form, applying the First Law of Exponents whenever possible.*

16. $2^2 \cdot 2^3 = 2^{2+3} = 2^5$

17. $x^3x^5 = x^{3+5} = x^8$

18. $aa^3 = a^1a^3 = a^{1+3} = a^4$

19. a^2b^3 — *Cannot be simplified further. Because the bases are not the same, you cannot add exponents.*

20. $8 \cdot 16 = 2^3 \cdot 2^4$ — *First, write each factor in exponent form with a prime base.*
$= 2^{3+4}$ — *Second, apply the First Law of Exponents.*
$= 2^7$ ■

Examples *Write the word phrases in symbols and simplify, if possible.*

21. The sum of two cubed and four squared

$$2^3 + 4^2 = 8 + 16 = 24$$

22. The cube of the sum of three and seven

$$(3 + 7)^3 = 10^3 = 1000$$

23. The sum of the cubes of three and seven

$$3^3 + 7^3 = 27 + 343 = 370$$

24. The sum of the cubes of a number and two

$$n^3 + 2^3 = n^3 + 8$$ ■

Problem Set 1.3

In Problems 1–20 find the prime factorization using exponents. See Examples 1–3.

1. 8	**2.** 27	**3.** 16	**4.** 81
5. 32	**6.** 54	**7.** 36	**8.** 24
9. 72	**10.** 60	**11.** 625	**12.** 346
13. 3150	**14.** 3300	**15.** 8910	**16.** 1000
17. 4900	**18.** 3310	**19.** 1,000,000	**20.** 100,000,000

Write out Problems 21–40 in nonexponent form, and evaluate, when possible. See Examples 4–12.

21. 2^5	**22.** 4^3	**23.** $2^2 \cdot 3^3$	**24.** $2^3 \cdot 3^2$	**25.** x^4
26. y^7	**27.** t^6	**28.** s^5	**29.** $2x^2$	**30.** $3y^3$
31. $8a^2$	**32.** $9b^3$	**33.** $5x^2$	**34.** $(5x)^2$	**35.** $10a^2b^2$
36. $(10ab)^2$	**37.** $3^3 \cdot 2^3$	**38.** $(3 \cdot 2)^3$	**39.** $4^2 \cdot 3^2$	**40.** $(4 \cdot 3)^2$

Use the order-of-operations convention to simplify completely Problems 41–54. See Examples 13–15.

41. $3^2 - 2^3$	**42.** $3^3 - 5^2$
43. $(5 \cdot 3)^2 + 5$	**44.** $100 + 5^2 \cdot 3^2$

45. $3^2(2+3)^2-10$
46. $(2+1)^2 5^2-10^2$
47. $(3-2)^2+(3^2-2^2)$
48. $(5+4)^2-(5^2+4^2)$
49. $3^4-3^3-3^2-3$
50. $2^5-2^4-2^3-2^2-2$
51. $(5-4)^2+7(1^2+1^3)$
52. $2(4+3)^2-(4-3)^2$
53. $2^3+3(3+4)^2-7$
54. $3^2+4(2+3)^2-5$

Write Problems 55–70 in exponential form and use the First Law of Exponents where applicable. See Examples 16–20.

55. $aaaaaa$
56. $bbbbbbb$
57. a^2a^3
58. b^4b^3
59. $x^3x^4y^2y^3$
60. $x^2x^3y^3y^3$
61. $xxyy^3z$
62. x^2xy^3x
63. $8\cdot4$
64. $27\cdot9$
65. $25x^2x^3$
66. $49yy$
67. $a^2a^3a^5b^2b^3$
68. $w^2w^3w^5z^4z^3$
69. $(x+y)^2(x+y)^3$
70. $(a+b)^3(a+b)^4$

Write the word phrases in Problems 71–80 in symbols and simplify, if possible. See Examples 21–23.

71. The sum of three squared and five squared
72. The square of the sum of three and five
73. The square of three increased by the cube of two
74. The cube of three decreased by the square of two
75. The square of a number plus five
76. Six times the cube of a number
77. The sum of the squares of seven and a number
78. The square of the sum of seven and a number
79. The difference of the cubes of two numbers.
80. The difference of two numbers times their sum

1.4 Evaluation of Formulas

The equality symbol is used when two quantities are exactly the same. That is, if we write

$$x = a$$

we mean that x and a name the same number; x may be replaced by a in any expression and the value of that expression will remain unchanged. To ***evaluate an expression*** means to replace the variable with the number from the domain of the variable and then to simplify the resulting expression. A variable is replaced by the same value each time it occurs in an expression.

Examples *Evaluate the following expressions where $x = 1$ and $y = 2$.*

1. $$\begin{aligned} 5x+9 &= 5(\mathbf{1})+9 \\ &= 5+9 \\ &= 14 \end{aligned}$$

2. $$\begin{aligned} y^3+2y^2-7 &= (\mathbf{2})^3+2(\mathbf{2})^2-7 \\ &= 8+2\cdot4-7 \\ &= 8+8-7 \\ &= 9 \end{aligned}$$

■

Examples *Expressions with more than one variable can also be evaluated. Let $x = 1$, $y = 2$, and $z = 6$.*

3. $3z^2 - (x + y) = 3(\mathbf{6})^2 - (\mathbf{1} + \mathbf{2})$
$= 3(36) - (3)$
$= 108 - 3$
$= 105$

Generate a numerical expression from a variable expression by replacing each variable by its numerical value, then simplify.

4. $(6x - 3y)\cdot z + 4xy$
$= (6\cdot 1 - 3\cdot 2)\cdot 6 + 4\cdot 1\cdot 2$
$= (6 - 6)\cdot 6 + 4\cdot 1\cdot 2$
$= 0 + 8$
$= 8$

A variable is replaced by the same number every time that variable occurs. Do not forget to insert the dot to indicate multiplication of numbers. ■

Recall that the domain of a variable is the set of all possible replacements for the variable. Consider an example where the domain has several values.

Example 5. What numerical values does the variable expression $[4(x + 3)] \div 2$ generate if the domain is $D = \{2, 13, 4\}$?

Solution If $x = 2$, then $[4(x + 3)] \div 2 = [4(2 + 3)] \div 2$
$= [4\cdot 5] \div 2$
$= 20 \div 2$
$= 10$

If $x = 13$, then $[4(x + 3)] \div 2 = [4(13 + 3)] \div 2$
$= [4\cdot 16] \div 2$
$= 64 \div 2$
$= 32$

If $x = 4$, then $[4(x + 3)] \div 2 = [4(4 + 3)] \div 2$
$= [4\cdot 7] \div 2$
$= 28 \div 2$
$= 14$ ■

A variable expression with a particular interpretation is called a ***formula.*** That is, a formula is a symbolic statement telling us what operations are performed on certain numbers to obtain a particular result. Table 1.2 describes some frequently used formulas.

Table 1.2 Formulas

Name	Description	Symbols
1. Perimeter of a rectangle	The perimeter of a rectangle is twice the length added to twice the width.	$P = 2l + 2w$
2. Area of a rectangle	The area of a rectangle is the product of the length and the width.	$A = lw$
3. Perimeter of a square	The perimeter of a square is four times the length of one side.	$P = 4s$
4. Area of a square	The area of a square is the length of one side times the length of one side.	$A = s^2$

Table 1.2 Continued

Name	Description	Symbols
5. Circumference of a circle	The circumference of a circle is pi times the diameter.	$C = \pi D$
6. Area of a circle	The area of a circle is pi times the radius squared.	$A = \pi r^2$
7. Simple interest	The interest is the principal times the rate times the length of time the money is invested.	$I = prt$
8. Business profit	The profit is the selling price less the cost.	$P = S - C$
9. Uniform motion	The distance traveled is the rate times the time.	$d = rt$

Example 6. A billiard table is 4 ft by 8 ft. Find the perimeter.

Solution Using Formula 1, $w = 4$ and $l = 8$. Thus,

$$\begin{aligned} P &= 2l + 2w \\ &= 2 \cdot 8 + 2 \cdot 4 \\ &= 16 + 8 \\ &= 24 \end{aligned}$$

so the perimeter is 24 ft. ■

Example 7. If the radius of a phonograph record is 15 cm, what is the circumference?

Solution The diameter is twice the radius, so $D = 30$. Thus, from Formula 5,

$$C = \pi \cdot 30$$

so the circumference is 30π cm. You can use an approximate value for π. Recall that $\pi \approx 3.14$ or $\pi \approx 3.1416$, where the symbol "$\approx$" means "is approximately equal to."

$$\begin{aligned} C &\approx 3.14 \cdot 30 \\ &= 94.2 \end{aligned}$$

and the circumference is about 94 cm. ■

Example 8. If you deposit \$5000 at 9% simple interest, how much money will you have at the end of five years?

Solution Use Formula 7,

$$I = prt$$

where $p = \$5000$, $r = .09$, and $t = 5$:

$$\begin{aligned} I &= 5000(0.09)(5) \\ &= 2250 \end{aligned}$$

The amount you would have in five years is \$7250. ■

Problem Set 1.4

Evaluate the expressions in Problems 1–10 for $x = 2$, $y = 5$, and $z = 1$. See Examples 1–4.

1. $2x + 3y$ **2.** $4y - 13z$ **3.** $x^2 + 3y$ **4.** $y^2 - z^2$

5. $\frac{x + y}{z}$ **6.** $\frac{y + z}{z}$ **7.** $x(y + z)$ **8.** $x^2 + xy - xz$

9. $(x + y)^2$ **10.** $z(x + y^2)$

Evaluate the expressions in Problems 11–20 for $a = 4$, $b = 3$, and $c = 7$. See Examples 1–4.

11. $4a - 5b$ **12.** $3c + 5a$ **13.** $b^2 + 2c$ **14.** $b^3 - 2c$

15. $\frac{a + b}{c}$ **16.** $\frac{2a + c}{b}$ **17.** $b(a + c)$ **18.** $a^2 - b^2 - c$

19. $(b + c)^2$ **20.** $b^2 + c^2$

21. If the domain is $D = \{3, 5, 7\}$, what numerical values does the variable expression

$$[2(x - 1)] \div 2$$

generate? See Example 5.

22. If the domain is $D = \{9, 99, 199\}$, what numerical values does the variable expression

$$[2(x + 1)] \div 10$$

generate? See Example 5.

Choose a formula from Table 1.2 and evaluate in order to answer the question in Problems 23–30. See Examples 6–8.

23. A field is 100 ft long and 75 ft wide. What length of fencing is necessary to enclose its perimeter?
24. How much carpeting is necessary to cover an area that is 6 yd wide and 12 yd long?
25. An automotive tire has a radius of 15 in. What is the circumference of the tire?
26. If the diameter of a phonograph record is 12 in., what is the area of one side of the record (before the hole is punched in the center)?
27. How much interest is earned if \$3000 is invested at 11% simple interest for five years?
28. If you invest \$2000 for 10 years and earn 14% simple interest, how much interest is earned?
29. A product is bought for \$17.35 and is sold for \$24.95. What profit is realized on the transaction?
30. An airplane maintains a constant rate of 670 mph for a 3-hour flight. How far does the plane travel?

In Problems 31–40 write a formula to express the given relation. See Table 1.2 for examples.

31. The area A of a parallelogram is the product of the base b and the height h.
32. The area A of a triangle is one-half the product of the base b and the height h.
33. The area A of a rhombus is one-half the product of the diagonals p and q.
34. The area A of a trapezoid is the product of one-half the height h and the sum of the bases a and b.
35. The volume V of a cube is the cube of the length s of a side.
36. The volume V of a rectangular solid is the product of the length l, the width w, and the height h.

37. The volume V of a circular cylinder is the product of pi, the square of the radius r, and the height h.

38. The volume V of a cone is one-third the product of pi, the radius r squared, and the height h.

39. The volume V of a sphere is the product of four-thirds pi times the cube of the radius r.

40. In a right triangle, the square of the hypotenuse c is equal to the sum of the squares of the sides a and b.

1.5 Properties of Natural Numbers

Recall that the set of numbers

$$\{1, 2, 3, \ldots\}$$

is called the set of ***counting numbers.*** The set is also known as the set of ***natural numbers*** and denoted by N, and we may write that

$$a \in N, b \in N, \text{ where } N = \{1, 2, 3, \ldots\}$$

to mean that a and b are elements of the set N.

The order-of-operations convention specifies the performance of addition and of multiplication from left to right. In symbols,

$$a + b + c = (a + b) + c; \qquad abc = (ab)c$$

Actually, we could have said that it does not matter whether we add or multiply from left to right or right to left. That is, it does not matter whether we begin by associating the a and b together or the b and c together. If we state this procedure formally, we call it the ***associative property.***

Associative Property

If $a \in N$, $b \in N$, and $c \in N$, then

Addition	*Multiplication*
$(a + b) + c = a + (b + c)$	$(ab)c = a(bc)$

Note that the associative property is actually two properties: the associative property of addition and the associative property of multiplication.

Another property of the natural numbers that we have been using is the property stating that when we add or multiply two natural numbers, it does not matter in which order we add or multiply. Because we can change the order back and forth as we please, we call the procedure the ***commutative property.***

Commutative Property

If $a \in N$ and $b \in N$, then

Addition	*Multiplication*
$a + b = b + a$	$ab = ba$

The commutative property is also two properties. The commutative property of addition states that the order in which the numbers are added does not change the sum, and the commutative property of multiplication states that the order in which the factors are multiplied has no effect on the product.

Examples *Classify each of the following as an example of the associative or the commutative property.*

	Solutions
1. $x + y = y + x$	Commutative property of addition
2. $xy = yx$	Commutative property of multiplication
3. $xy + w = w + xy$	Commutative property of addition
4. $xy + w = yx + w$	Commutative property of multiplication
5. $3(x + y) = (x + y)3$	Commutative property of multiplication
6. $(x + y) + z = x + (y + z)$	Associative property of addition
7. $(x + y) + z = z + (x + y)$	Commutative property of addition
8. $2(xy) = (2x)y$	Associative property of multiplication ■

To distinguish between the associative and commutative properties, remember the following:

1. When only the associative property is used, the numbers are grouped differently, but the order in which the elements appear from left to right is not changed.
2. When only the commutative property is used, the order in which the elements appear from left to right is changed, but the grouping of the numbers is not changed.

One of the most important properties of elementary algebra ties together the operations of multiplication and addition. When working with mixed operations, we classify an expression according to the last operation performed when simplifying that expression.

1. It is a *sum* if the last operation performed is addition;
2. It is a *difference* if the last operation performed is subtraction;
3. It is a *product* if the last operation performed is multiplication; and
4. It is a *quotient* if the last operation performed is division.

Remember parentheses and the order-of-operations convention.

Examples *Classify each expression as a sum, product, difference, or quotient.*

9. $12 - 3 \cdot 4$	**9.** a difference
10. $(9 - 6) \cdot 2$	**10.** a product
11. $2 + \frac{8}{2}$	**11.** a sum
12. $\frac{2 + 8}{2}$	**12.** a quotient
13. $2(3 + 4)$	**13.** a product
14. $2 \cdot 3 + 2 \cdot 4$	**14.** a sum ■

In order to understand another important property of the natural numbers, consider the following rather contrived example in order to discover a way to change certain products to sums and certain sums to products.

Example 15. Suppose you are selling tickets for a raffle, and the tickets cost two dollars each. You sell three tickets on Monday and four tickets on Tuesday. How much money did you collect?

Solution 1 You sold a total of $3 + 4 = 7$ tickets that cost 2 dollars each, so you collected $2 \cdot 7 = 14$ dollars. That is,

$$2 \cdot (3 + 4) = 14$$ ■

Solution 2 You collected $2 \cdot 3 = 6$ dollars on Monday and $2 \cdot 4 = 8$ dollars on Tuesday for a total of $6 + 8 = 14$ dollars. That is,

$$2 \cdot 3 + 2 \cdot 4 = 14$$

Because these equations are equal, we say:

$$2 \cdot (3 + 4) = 2 \cdot 3 + 2 \cdot 4$$ ■

Do you suppose this would be true if the tickets cost a dollars and you sold b tickets on Monday and c tickets on Tuesday? Then the equation would be

$$a \cdot (b + c) = a \cdot b + a \cdot c$$

Notice that the solution on the left is a product and the solution on the right is a sum. Because the a has been *distributed* to *each* of the numbers on the inside of the parentheses, we call this procedure the ***distributive property*** *for multiplication over addition.*

Distributive Property

If $a \in N$, $b \in N$, and $c \in N$, then

$$a(b + c) = ab + ac$$

and

$$ab + ac = a(b + c)$$

A geometric example of the distributive property is shown in Figure 1.1.

Recall that the area of a rectangle is found by $A = lw$.

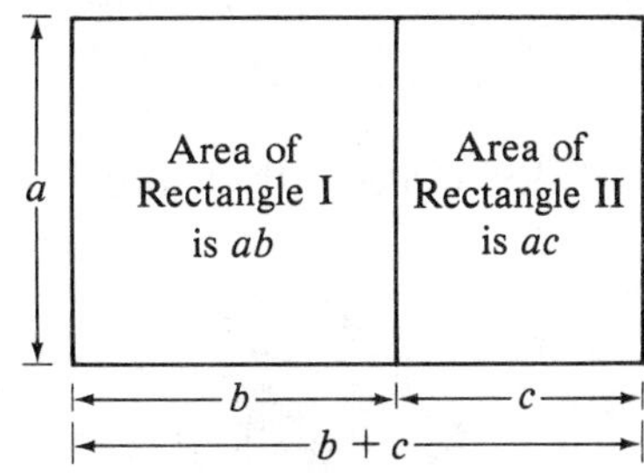

Area of large rectangle is $a(b + c)$.

Area of Rectangle I: ab
Area of Rectangle II: ac
Area of large rectangle is $ab + ac$.

Figure 1.1

Examples *Change each of the following to sums.*

16. $x(x + z) \quad = xx + xz = x^2 + xz$
17. $ab(c + d) \quad = abc + abd$
18. $(a + b)x \quad = ax + bx$
19. $4(x + 2) \quad = 4x + 4 \cdot 2$ or $4x + 8$ ■

In Examples 16–19 the distributive property was used to write

$$a(b + c) = ab + ac$$

This changed a product to a sum. In order to change a sum to a product, use the same distributive property considered as

$$ab + ac = a(b + c)$$

Examples *Change each of the following to products.*

20. $ax + ay \quad = a(x + y)$
21. $wxy + zxy \quad = (w + z)xy$
22. $5a + 5 \quad = 5a + 5 \cdot 1$
$\quad = 5(a + 1)$
23. $3x + 6 \quad = 3x + 3 \cdot 2$
$\quad = 3(x + 2)$ ■

The distributive property can be generalized to apply regardless of the number of terms in the parentheses, which allows us to distribute a number outside parentheses to any number of terms. For example,

$$x(a + b + c + d + e) = xa + xb + xc + xd + xe$$

and

$$\begin{aligned} 3x(x + y + z) &= 3xx + 3xy + 3xz \\ &= 3x^2 + 3xy + 3xz \end{aligned}$$

and

$$abx + abxy + abwz^2 = ab(x + xy + wz^2)$$

The distributive property changes sums and products. The following examples illustrate this in word statements and phrases.

Examples *Write the word statements or phrases in symbols.*

	Solution
24. The sum of four times a number and one	**24.** $4n + 1$
25. Five times the sum of a number and two	**25.** $5(n + 2)$
26. The sum of two times a number and six is equal to two times the sum of a number and three.	**26.** $2n + 6 = 2(n + 3)$

■

Problem Set 1.5

In Problems 1–20 classify each as an example of the associative property, the commutative property, or both. See Examples 1–8.

1. $a + 5 = 5 + a$
2. $5ax = 5xa$
3. $2 + (3 + 4) = (2 + 3) + 4$
4. $2 + (3 + 4) = 2 + (4 + 3)$
5. $2 + (3 + 4) = (3 + 4) + 2$
6. $2 + (3 + 4) = (2 + 4) + 3$
7. $5 + (3 + t) = (3 + t) + 5$
8. $5 + 3t = 3t + 5$
9. $w(a + b) = (a + b)w$
10. $w(a + b) = w(b + a)$
11. $(c + d) + f = c + (d + f)$
12. $(a + b) + c = c + (a + b)$
13. $(xy)(a + b) = x[y(a + b)]$
14. $(x + y) + (a + b) = x + [y + (a + b)]$
15. $(x + y)(2 + b) = (2 + b)(x + y)$
16. $(x + y)(2 + b) = (y + x)(b + 2)$
17. $(xy)z = x(yz)$
18. $(x + y) + z = x + (z + y)$
19. $(x + y)z = (y + x)z$
20. $xy + z = yx + z$

Classify each expression in Problems 21–40 as a sum, difference, product, or quotient. See Examples 9–14.

21. $3 + 4 \cdot 2$
22. $25 - 5 \cdot 2$
23. $6 \cdot 7 + 2$
24. $3 \cdot 7 - 10$
25. $7 \cdot 2 + 4$
26. $5 \cdot 8 - 9$
27. $2(3 + 4)$
28. $5(2 + 8)$
29. $6(7 + 1)$
30. $2 \cdot 3 + 2 \cdot 4$
31. $5 \cdot 2 + 5 \cdot 8$
32. $6 \cdot 7 + 6 \cdot 1$
33. $12 + \frac{8}{4}$
34. $\frac{12 + 4}{4}$
35. $\frac{12}{4} + \frac{8}{4}$
36. $\frac{15 + 25}{5}$
37. $\frac{15}{5} + \frac{25}{5}$
38. $15 + \frac{25}{5}$
39. $2[3 + 4(6 - 2)] - 3(4 + 6)$
40. $5(3 + 3) + 6[14 - 6(2 + 3)]$

In Problems 41–56 use the distributive property to change each of the given expressions to a sum. See Examples 16–19.

41. $a(b + c)$
42. $s(t + u)$
43. $x(y + z)$
44. $5(x + 2)$
45. $7(y + 4)$
46. $3(z + 5)$
47. $2(3x + 4)$
48. $3(2y + 1)$
49. $5(3z + 4)$
50. $x(x + 3)$
51. $y(y + z)$
52. $z(2 + z)$
53. $2x(x + y + z)$
54. $4x(x^2 + y^2 + z^2)$
55. $5x^2(x^2 + 2x + 7)$
56. $3y(2y^3 + 4y^2 + 5y)$

In Problems 57–70 use the distributive property to change each of the following to a product. See Examples 20–23.

57. $ax + aw$
58. $bs + ba$
59. $cx + cy$
60. $abc + abd$
61. $uvw + suv$
62. $mn + 2pn$
63. $3x + 3y$
64. $3x + 6y$
65. $9x + 6y$
66. $3a + 3$
67. $14y + 7$
68. $25z + 5$
69. $x^2 + 5x$
70. $xy + y^2$

Write the word statement or phrase in Problems 71–80 in symbols. See Examples 24–26.

71. The sum of a number and one is equal to one added to the number.
72. Two added to a number is equal to the sum of the number and two.
73. Three times the sum of a number and four is equal to sixteen.
74. The product of three and the sum of a number and four is sixteen.
75. Five times the sum of a number and one is equal to the sum of five times the number and five.
76. The sum of six times a number and twelve is equal to six times the sum of the number and two.
77. A number times the sum of seven and the number is equal to zero.
78. The sum of a number squared and eight times the number is equal to six.
79. A number times the sum of nine and the number is equal to nine times the number plus the square of the number.
80. The square of a number plus eight times the number is equal to the number times the sum of eight and the number.

1.6 Review

When preparing for a chapter test, you should first study the chapter; then work these review problems. *Next,* compare your answers with those given in the back of the book. Missed questions indicate that more review is needed, and you can then refer to the section or sections indicated in parentheses following the question numbers.

Self-Test

[1.1] **1.** Insert the appropriate inequality symbol in place of the question mark.
a. $9 ? 7$ **b.** $8 ? 8$ **c.** $7 ? 6$ **d.** $6 ? 2\pi$

[1.1] **2.** Change the division problems to equivalent multiplication problems, and check to see if the division is correct.
a. $\frac{45}{9} = 5$ **b.** $\frac{35}{6} = 7$ **c.** $\frac{54}{8} = 7$ **d.** $\frac{5}{0} = 0$

[1.1, 1.2] **3.** Write the word statements in symbols.
a. The sum of twelve and three is fifteen.
b. Five is equal to nine minus four.
c. Eight is the quotient of forty and five.
d. The product of six and eight is less than seven squared.

[1.1] **4.** Choose a letter to represent the variable and write a mathematical expression to express the following ideas.
a. The sum of twice a number and 5.
b. The sum of two consecutive numbers.
c. Twice the difference of 5 subtracted from a number.
d. The difference of 5 subtracted from twice a number.

[1.2, 1.3] **5.** Simplify the given numerical expressions.
a. $5 + 2 \cdot 3$
b. $10 + 2 \cdot 3 - 12 \div 4$
c. $(10 + 2 \cdot 3 - 12) \div 4$
d. $5 + 3^4$

[1.3] **6.** **a.** Write $7.13 \cdot 10^8$ in nonexponential form and simplify.
b. In part a, the number 8 is called the ______________.
c. In part a, 7.13 is a ______________ of $7.13 \cdot 10^8$.
d. In part a, 10 is called the ______________.

[1.3] **7.** Write out the following expressions in exponent form.
a. 360 **b.** $2ssstt$ **c.** $(5^2xy^2)(5^3xy)$ **d.** $(9a^2)(3ab^3)$

[1.1, 1.2, 1.4] **8.** Evaluate the expressions for $x = 3$, $y = 2$, and $z = 4$, and then insert an appropriate inequality symbol in place of the question mark.
a. $x(y + z) \mathrel{?} xy + yz$
b. $(x + y)^2 \mathrel{?} x^2 + y^2$
c. $(x + y)(x + z) \mathrel{?} x^2 + xy + yz$
d. $3[2x + (y + 2z)] \mathrel{?} 4xz$

[1.5] **9.** Change each of the following to a sum by using the distributive property.
a. $5(x + y)$ **b.** $3(2a + 3b)$ **c.** $2x^2y(3xy + y^2)$ **d.** $(3a^2b + 1)c$

[1.5] **10.** Change each of the following to a product by using the distributive property.
a. $stu + stv$ **b.** $5x + 15y$ **c.** $2mn + 6m^2n$ **d.** $x^2y^2 + x^3y^3$

Review Problems

Completely simplify the numerical expressions in Problems 1–30.

1. $3 + 4 \cdot 5$
2. $(3 + 4)5$
3. $7 - 2 \cdot 3$
4. $9 + 6 \div 3$
5. $(9 + 6) \div 3$
6. $9 - 6 \div 3$
7. $2 + 6 \div 2 + 2 \cdot 3$
8. $(2 + 6) \div 2 + (2 \cdot 3)$
9. $(2 + 6) \div (2 + 2) \cdot 3$
10. $(2 + 6) \div [2 + 2 \cdot 3]$
11. $3 + 6 \cdot 2 \div 3 + 3 \cdot 2$
12. $(3 + 6)2 \div (3 + 3 \cdot 2)$
13. $(3 + 6)2 \div (3 + 3)2$
14. $3 + (6 \cdot 2 \div 3) + 3 \cdot 2$
15. $[3 + (6 \cdot 2 \div 3) + 3]2$
16. $3 + [(6 \cdot 2 \div 3) + 3]2$
17. $5 + 25 \div 5 \cdot 2 + 3 \cdot 5 \div 3$
18. $[(5 + 25) \div 5(2 + 3)5] \div 3$
19. $(5 + 25) \div 5 \cdot 2 + (3)(5) \div 3$
20. $(5 + 25) \div [5 \cdot 2 + (3 \cdot 5) \div 3]$
21. $[5 + 25 \div 5(2 + 3 \cdot 5)] \div 3$
22. $[(5 + 25) \div (5 \cdot 2)] + (3 \cdot 5) \div 3$
23. $7^2 - 5^2$
24. $5^3 - 3^3$
25. $3^2 - 2^3$
26. $3^3 - 5^2$
27. $(2 + 5)^2 - (2^2 + 5^2)$
28. $(5 - 2)^3 + (5^3 - 2^3)$
29. $3^2 + 2(2 + 5)^2 - 7$
30. $2^4 + 2(1 + 7)^2 - 2^3$

Simplify Problems 31–45, using exponential form.

31. $dddd$
32. $hhhhhhh$
33. kk^2k^3
34. m^3m^4
35. n^2n^6
36. $p^2p^3p^4p$
37. $3x^2x^4$
38. $5y^2y^3y$
39. $2zzz^2$
40. $a^2a^3b^4b^5$
41. $bb^2b^3c^4c^5$
42. $cc^2cd^3dd^2$
43. $xxx^2yyz^3z^2$
44. $xx^2x^3yy^2zz$
45. xy^2yz^3z

Evaluate the expressions in Problems 46–60 for $a = 1$, $b = 3$, and $c = 5$.

46. $2a + b$
47. $2b - c$
48. $5a - c$
49. $4a + 2b$
50. $a + 3c$
51. $4b - c$
52. $a(b + c)$
53. $b(c - a)$
54. $(a + b)c$
55. $\dfrac{4a + 2b}{c}$
56. $\dfrac{a + 4c}{b}$
57. $\dfrac{4b - c}{a}$
58. $(a + b)^2$
59. $a^2 + c^2$
60. $b^3 - c^2$

Evaluate the expressions in Problems 61–75 for $x = 2$, $y = 5$, and $z = 7$.

61. $3x - y$ **62.** $2y - z$ **63.** $5x - z$ **64.** $3x + 2z$
65. $5y - 3z$ **66.** $5x - 2y$ **67.** $x(y + z)$ **68.** $(x + y)z$
69. $(x + z)y$ **70.** $\frac{3x + 2z}{y}$ **71.** $\frac{5y - 3z}{x}$ **72.** $\frac{5x - 2y}{z}$
73. $(x + y)^2$ **74.** $x^2z - y^2$ **75.** $xy^2 - z^2$

In Problems 76–87 use the distributive property to write each expression as a sum.

76. $r(s + t)$ **77.** $w(x + y)$ **78.** $p(m + n)$ **79.** $3(s + 2)$
80. $5(x + 2)$ **81.** $7(m + 3)$ **82.** $2(3s + 2)$ **83.** $3(5x + 2)$
84. $5(7m + 3)$ **85.** $s(s + 2)$ **86.** $x(x + 2)$ **87.** $m(m + 3)$

In Problems 88–100 use the distributive property to write each expression as a product.

88. $ax + ay$ **89.** $bx + bz$ **90.** $cy + cz$ **91.** $xyz + xyw$
92. $abc + abe$ **93.** $mnp + mnq$ **94.** $5x + 5y$ **95.** $2x + 6y$
96. $3y + 6z$ **97.** $7x + 7$ **98.** $7x + 14$ **99.** $49x + 7$
100. $x^2 + 7x$

2

Integers

I had been to school . . .
and could easily say the
multiplication table
up to 6 x 7 = 35,
and I don't reckon I could ever
get any further than that
if I was to live forever.

Huckleberry Finn
in *The Adventures of Tom Sawyer*
by Mark Twain

2.1 The Number Line and Opposites

The set of natural numbers is not sufficient for all types of problems. In winter, for example, we read weather reports such as "Zero Weather and Powder at Squaw Valley," "−18 Lowest in Continental Forty-Eight," or "−10 and Ice Slow New York." The Fahrenheit scale sets the temperature at which water boils at 212°F and the temperature at which water freezes as 32°F. But temperatures can fall well below freezing. The metric temperature scale, Celsius, measures boiling at 100°C and freezing at 0°C. The two scales agree at only one point, which is 40 units below their zero points, −40°. The units are equally spaced on the entire scale of each so that 20° is the same distance from zero as −20°, only in the opposite direction.

A thermometer or a ruler provides a good example of how we are used to ordering numbers on a line. A ***number line*** is constructed by choosing and labeling any point as 0, the ***origin.*** Then the line is divided to the right into segments of equal length and labeled with the counting numbers as shown.

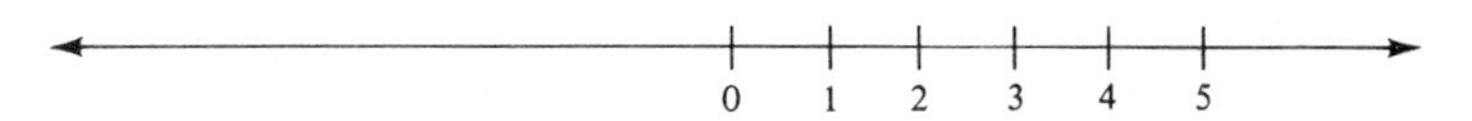

As these examples illustrate, we need to name the points on the number line to the left of zero or less than zero. The numbers to the right of zero or greater than zero are called ***positive,*** and those to the left or less than zero are called ***negative.*** Negative three (−3), for instance, is an equal distance (but in the opposite direction) from zero as positive three (+3). Such pairs of numbers are called ***opposites.*** Because of our reliance on symbols and graphs, we often call these numbers ***signed numbers*** or ***directed numbers.*** Zero is the division point between the positive and the negative numbers and hence is neither positive nor negative.

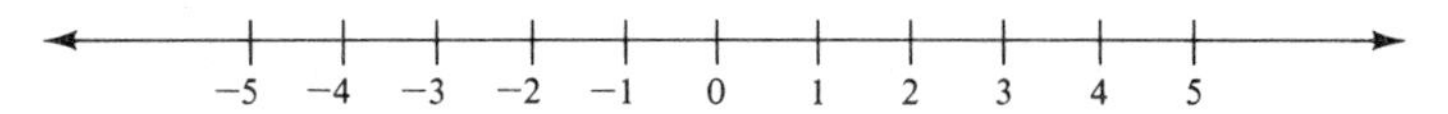

Our new and larger set of numbers, which is now made up of zero, the natural numbers, and the opposites of the natural numbers, is called the set of ***integers.*** We are able to list the integers, as we did the natural numbers: $\{\ldots -3, -2, -1, 0, 1, 2, 3, \ldots\}$.

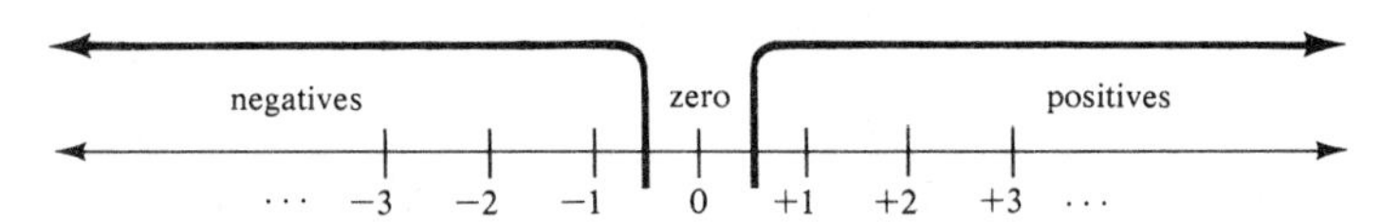

You probably already know many instances of opposites. A profit of ten dollars is the opposite of a loss of ten dollars. Two steps to the left is the opposite of two steps to the right. A charge of five dollars is the opposite of a payment of five dollars. A 6-yard gain is the opposite of a 6-yard loss. A clockwise movement of 30° is the opposite of a 30° movement counterclockwise.

Examples *Express each of the following as (a) an integer, (b) the opposite of the quantity in words, and (c) the opposite as an integer.*

Solutions

1. A 5° rise in temperature.
   a. $+5$
   b. A 5° drop in temperature
   c. -5
2. 3° below zero
   a. -3
   b. 3° above zero
   c. $+3$
3. 350 ft below sea level
   a. -350
   b. 350 ft above sea level
   c. $+350$
4. A profit of 200 dollars
   a. $+200$
   b. A loss of 200 dollars
   c. -200
5. Ten steps to the left
   a. -10
   b. Ten steps to the right
   c. $+10$ ■

There are three uses for the symbol "—". These are:

Minus used to indicate subtraction, an operation symbol;
Negative used to indicate those numbers to the left of the origin on a number line;
Opposite used to signify an equivalent distance from the origin, but in an opposite direction. This is a number that can either be positive or negative.

For example,

5 *minus* the *opposite* of *negative* 2
↕ ↕ ↕
$5 \quad - \quad [\quad - \quad (\quad - \quad 2)]$

illustrates three different uses for the symbol (—).

Examples *Indicate whether "minus," "negative," or "opposite" best describes the use of the "—" in each of Examples 6–9.*

Solutions

6. $(-3) + (+5)$ — 6. negative
7. $-(+3) + (+5)$ — 7. opposite
8. $(+3) - (+5)$ — 8. minus
9. $-x$ — 9. opposite ■

The symbol $-x$ is read *not* as "negative x" or "minus x" because x can be any number; $-x$ can be positive, negative, or zero. Read $-x$ as ***"the opposite of x"*** to save yourself a lot of confusion. Notice that a calculator has separate keys for subtraction [—] and opposite [+/—]. The [+/—] key changes the sign of a number to the opposite of its present sign. For example, $5 - (-2)$ would be entered as:

[5] [—] [2] [+/—] [=]

Because the calculator assumes all numbers entered are positive, a negative number is obtained by taking the opposite of a positive.

The opposite of an elevation above sea level is the same number of feet below sea level. Whether we have a profit of ten dollars $(+10)$ or a loss of ten dollars (-10), we have a change of ten (10). Whether we move five steps to the left (-5), to the right $(+5)$, down (-5), or up $(+5)$, we have moved five steps (5). When we talk about the numerical value of a quantity independent of direction or sign, we speak of its ***absolute value.***

Definition of Absolute Value

> The absolute value of n is its distance from zero. We use the symbol $|n|$ for absolute value.

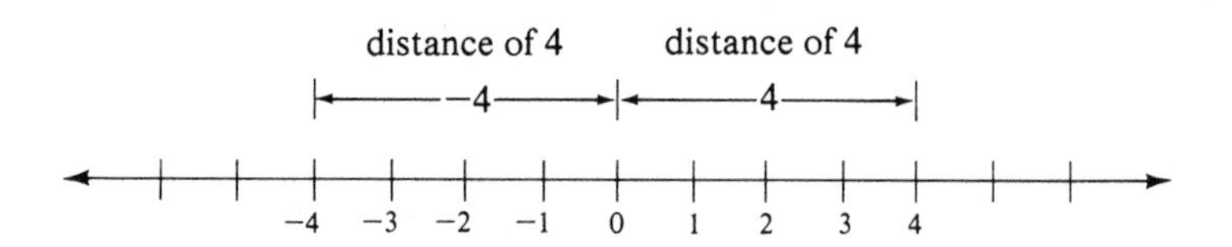

$|n|$ is the distance from n to 0 on the number line. The absolute value is always positive or zero; that is, nonnegative.

Examples *Find the value of each.*

Solutions

10. $-(+2) = ?$

10. The opposite of a positive is negative.

$$-(+2) = -2$$

11. $-(-3) = ?$

11. The opposite of a negative is positive.

$$-(-3) = +3$$

12. $|+7| = ?$

12. The absolute value is always nonnegative.

$$|+7| = +7$$

13. $|-5| = ?$

13. The absolute value is always nonnegative.

$$|-5| = +5$$

14. $-|-8| = ?$

14. $|-8| = +8$ and $-(+8) = -8$ ■

In the next section you will be asked to distinguish the larger of two numbers, as well as the larger absolute value. Examples 15–20 will give you some experience with this concept.

Examples *Given a pair of integers, first identify the larger integer, and secondly find the absolute value of each and give the larger of those two.*

	Numbers	Larger Number	Larger Absolute Value
15.	$6, 2$	6	6
16.	$-6, 2$	2	6
17.	$6, -2$	6	6
18.	$-6, -2$	-2	6
19.	$-8, 0$	0	8
20.	$-3, -\pi$	-3	π

■

Problem Set 2.1

In Problems 1–20 express as (a) a signed number, (b) the opposite of the quantity in words, and (c) the opposite as a signed number. See Examples 1–5.

1. A gain of six dollars
2. A loss of eight dollars
3. Seven steps to the left
4. Nine steps to the right
5. 2° above zero
6. 5° below zero
7. 12° latitude North
8. A deposit of 150 dollars
9. A 10-yard gain
10. Down two floors
11. A mountain 6200 ft high
12. A 3-yard loss
13. A $500 bonus
14. A $150 penalty
15. A five dollar increase in price
16. A desert 100 ft below sea level
17. 70° longitude East
18. A twenty-five dollar reduction in price
19. 50° latitude South
20. 100° longitude West

Indicate whether "minus," "opposite," or "negative" best describes the use of the "−" in each of Problems 21–30. See Examples 6–9.

21. $(+8)-(+5)$
22. $(+8)+(-5)$
23. $-(+8)+(+5)$
24. $(+8)-(+5)$
25. $-(+8)(+5)$
26. $-(8+5)$
27. $(-1)(+2)(+3)$
28. $2-x$
29. $-|x|$
30. $|-x|$

In Problems 31–40 give the integer that represents the final position, relative to the starting point.

31. Move five units to the left, then four units to the right, and finally three back to the left.
32. Move seven units to the right, then five units to the right, and finally three units to the left.
33. Move five units in a negative direction, eight units in a positive direction, and then three units in a negative direction.
34. Move nine units in a positive direction, twelve in a negative direction, and then five in a positive direction.
35. Move up two units, five units down, and then six units upward.
36. Move ten units down, then four units up, and then five units up.
37. Move ten to the left, twelve to the right, and then twenty to the left.
38. Move eleven to the right, thirty to the left, and then ten to the right.
39. Move twenty to the left, twelve farther to the left, and fifteen to the right.
40. Move sixty to the right, forty to the left, and then twenty farther to the left.

Find the integer value in Problems 41–50. See Examples 10–14.

41. $-(-1)$
42. $-(+2)$
43. $|-3|$
44. $|+4|$

45. $-|+5|$ **46.** $-|-6|$ **47.** $-|-8|$ **48.** $-(-8)$
49. $-(-|-8|)$ **50.** $-|-(-8)|$

Given each pair of integers in Problems 51–60, find the larger of integers. Then find the larger of the absolute values of the pair. See Examples 15–20.

51. 3, 4 **52.** 3, −4 **53.** −7, 5 **54.** −8, 9 **55.** −10, −11
56. −12, −20 **57.** −13, 4 **58.** −4, 13 **59.** 4, π **60.** −4, $-\pi$
61. Is $-x$ always negative? Explain.
62. When is $-x$ (a) positive? (b) negative? (c) zero?

2.2 Addition of Integers

Now that we have discussed negative numbers, what do we do with them? We add, subtract, multiply, and divide them. Balancing a checkbook, charting a football game, averaging temperatures, and keeping score in a card game all call for operations with signed numbers.

Consider a football game as a series of positives and negatives, or gains and losses, added together to give total or net yardage. Two consecutive gains, +3 and +5, produce a net gain of +8.

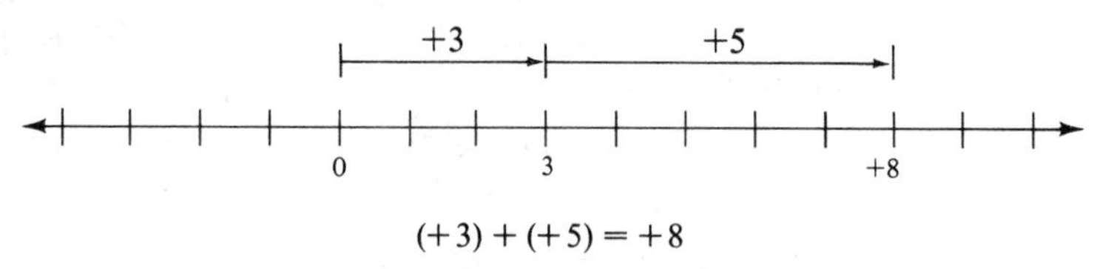

$(+3) + (+5) = +8$

Combining a gain and a loss, however, may be positive or negative. The amount of the gain and loss determines the net effect. If the gain is larger (in absolute value) than the loss, it is a net gain. If the loss is greater (in absolute value), then it is a net loss.

A 3-yard gain followed by a 5-yard loss.

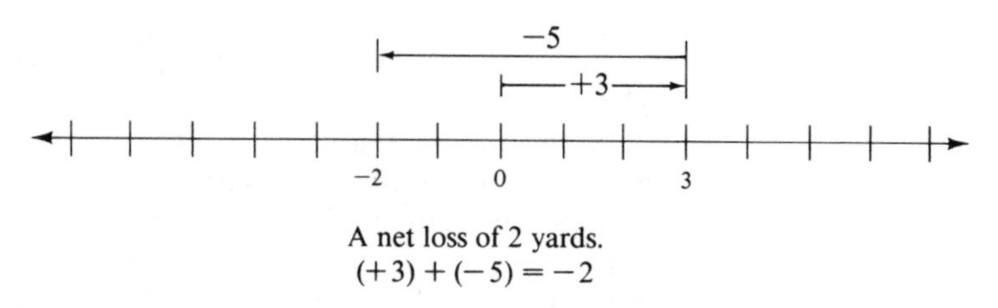

A net loss of 2 yards.
$(+3) + (-5) = -2$

A 3-yard loss followed by a 5-yard gain.

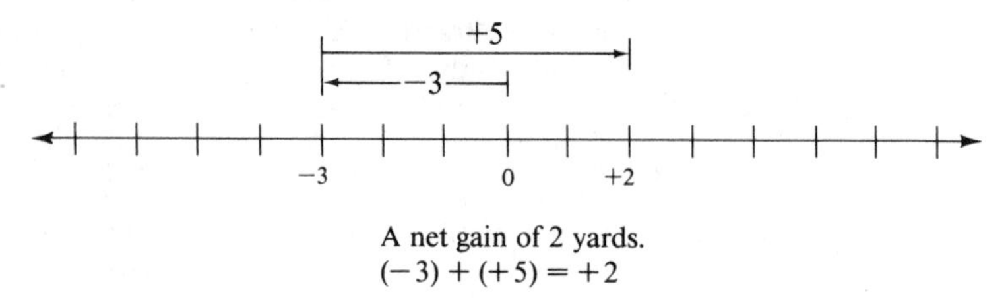

A net gain of 2 yards.
$(-3) + (+5) = +2$

Two consecutive losses will produce a net loss.
A 3-yard loss followed by a 5-yard loss.

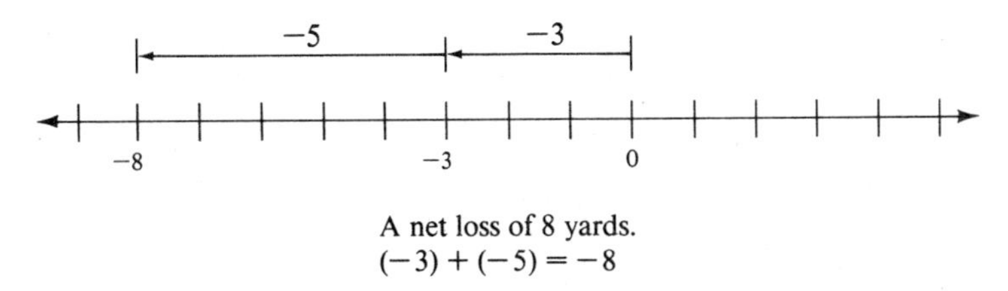

A net loss of 8 yards.
$(-3)+(-5)=-8$

A number line is a practical means of understanding operations with signed numbers. A thermometer is a number line that records increases and decreases in temperature. But you probably would not use a number line to balance your checkbook or keep score in a card game. You calculate with signed numbers, keeping in mind the principle illustrated by a number line. This principle is awkward to phrase as a simple word rule, because the lengths of the numbers on the line are their absolute values and the directions are their signs. With this in mind, we write two sets of rules that are essentially equivalent, one geometric and the other algebraic. The geometric rule describes a construction on a number line. The algebraic rule can be illustrated in a flowchart as shown in Figure 2.1.

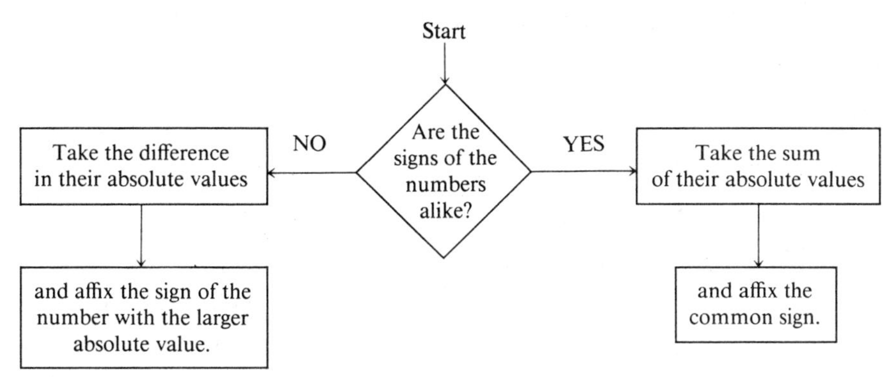

Figure 2.1
Flowchart for the addition of integers

Geometric

To add two numbers that have the same direction, take the sum of their lengths in that direction.

To add two numbers that have different directions, take the difference in their lengths in the direction of the number with the greater length.

Algebraic

To add two numbers with the same sign, take the sum of their absolute values and affix the common sign.

To add two numbers with different signs, take the difference in their absolute values and affix the sign of the number with the larger absolute value.

In practice you will probably not use either of these verbal rules. Rather, you will remember the principle illustrated by the gain/loss model on a number line.

Examples *Add each of the following on a number line.*

1. $(-7) + (-5)$

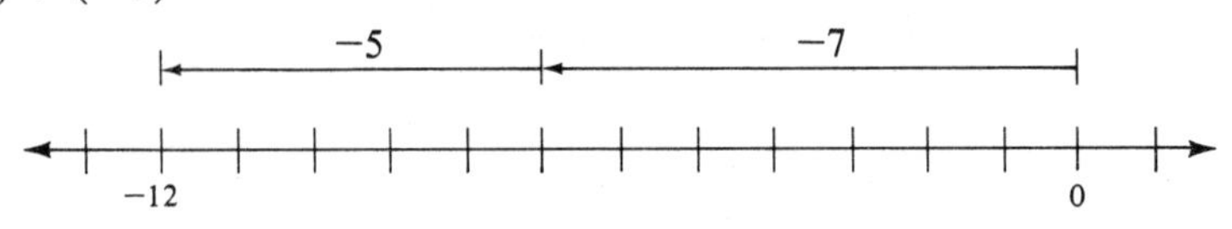

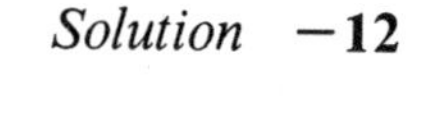

2. $(+2) + (-8)$

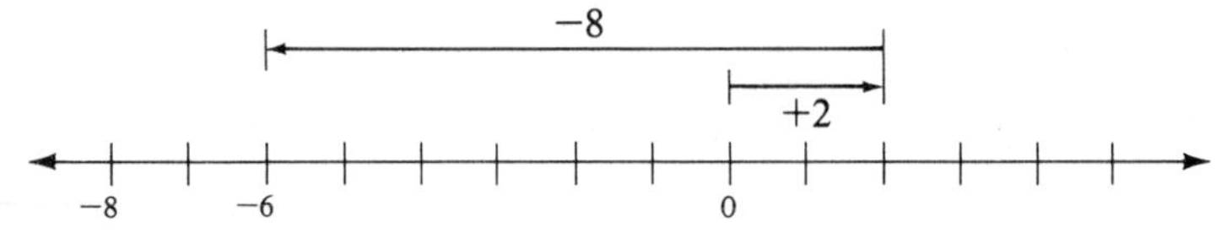

Solution **−6**

3. $(-3) + (+11)$

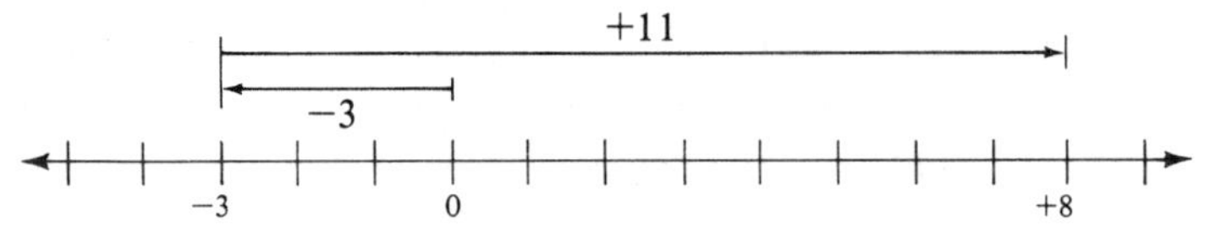

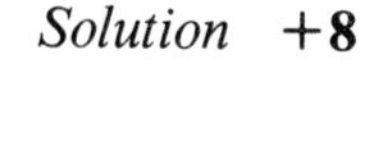

4. $(+33) + (-33)$

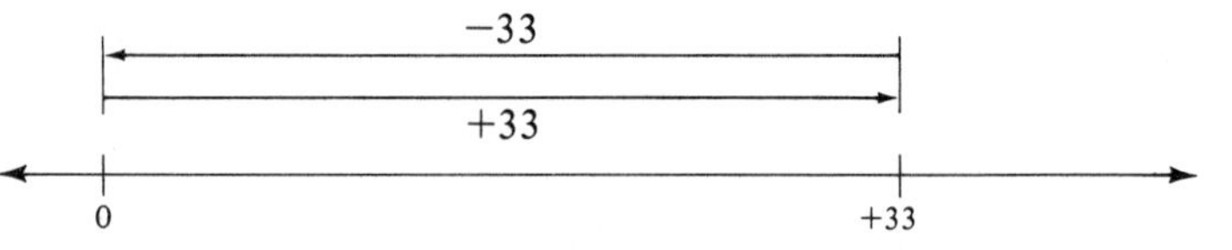

Solution **0** ■

Compare the process in the following examples of the algebraic rule with the flowchart in Figure 2.1. Note that most of what is described in the solutions would be performed mentally.

Find the following sums, algebraically.

Example **5.** $(+21) + (+14) = ?$

Solution The signs of the numbers are *alike and positive,* so take the sum $|21| + |14|$ and affix a *positive* sign.

$$\mathbf{(+21) + (+14) = +35}$$ ■

Example **6.** $12 + 7 = ?$

Solution The signs of the integers are understood:

$$\mathbf{(+12) + (+7) = +19}$$

or simply 19. ■

Example 7. $(-19) + (-15) = ?$

Solution The signs of the numbers are *alike and negative,* so take the sum $|-19| + |-15|$ and affix a *negative* sign.

$$(-19) + (-15) = -34$$ ■

Example 8. $(-7) + (+12) = ?$

Solution The signs of the numbers are different, so take the difference $|-7| - |12|$ and affix a positive sign because $|12| > |-7|$.

$$(-7) + (+12) = +5$$ ■

Example 9. $-8 + 5 = ?$

Solution The signs of the integers are understood:

$$(-8) + (+5) = -3$$ ■

Example 10. $(+4) + (-13) = ?$

Solution The signs of the numbers are different, so take the difference $|4| - |-13|$ and affix a negative sign because $|-13| > |4|$.

$$(+4) + (-13) = -9$$ ■

Example 11. $16 + (-9) = ?$

Solution The signs of the integers are understood:

$$(+16) + (-9) = 7$$ ■

Examples 12–20 are for your practice and are shown as you would write them. Make certain that you understand each of them.

Examples

12. $-15 + 8 = -7$
13. $14 + (-27) = -13$
14. $0 + (-15) = -15$
15. $-7 + 2 = -5$
16. $-9 + (-5) = -14$
17. $-5 + 8 = 3$
18. $-10 + (-4) + 3 = -11$
19. $6 + (-8) + 6 + 8 = 12$
20. $5 + (-7) + (-4) + 2 = -4$ ■

Problem Set 2.2

Use the number line in Problems 1–10. See Examples 1–4.

1. On the number line start at 0 and move +3 units. Then move +6 units. Then move −12 units. Where do you stop?
2. On the number line start at 0 and move −5 units. Then move +7 units. Then move −4 units. Where do you stop?

3. Start at 0, move -5 units, move $+2$ units, move -4 units, and then move $+3$ units. Where do you stop?
4. Start at 0, move $+5$ units, move -2 units, move $+4$ units, and then move -3 units. Where do you stop?
5. Start at 0, move -4 units, move -5 units, move $+7$ units, and then move -2 units. Where do you stop?
6. Start at 0, move -8 units, move $+3$ units, move $+4$ units, and then move -1 unit. Where do you stop?
7. Start at 0, move $+7$ units, move $+4$ units, move -9 units, and then move -5 units. Where do you stop?
8. Start at 0, move -8 units, move -6 units, move $+7$ units, and then move $+7$ units. Where do you stop?
9. Start at -5, move $+7$ units, move -3 units, and then move $+4$ units. Where do you stop?
10. Start at $+7$, move -9 units, move $+5$ units, move -6 units, and then move $+4$ units. Where do you stop?

Find the sums in Problems 11–90. See Examples 1–20.

11. $4 + 7$
12. $(-5) + (-6)$
13. $(+2) + (+3)$
14. $-8 + 6$
15. $9 + (-5)$
16. $4 + (-7)$
17. $-2 + 7$
18. $7 + (-2)$
19. $(-2) + (+7)$
20. $(+22) + (+25)$
21. $(+14) + 25$
22. $21 + 25$
23. $37 + 34$
24. $(+23) + (+17)$
25. $27 + (+33)$
26. $(-19) + (-12)$
27. $(-27) + (-13)$
28. $-23 + (-14)$
29. $-12 + (-8)$
30. $(-23) + (-23)$
31. $-9 + (-18)$
32. $(-3) + (-5)$
33. $(-8) + (-2)$
34. $(-5) + (-7)$
35. $(+9) + (-7)$
36. $(+7) + (-4)$
37. $7 + (-4)$
38. $(+3) + (-8)$
39. $(+2) + (-5)$
40. $(-5) + (+2)$
41. $14 + (-11)$
42. $13 + (-19)$
43. $19 + (-13)$
44. $(-8) + (+5)$
45. $(-9) + (+4)$
46. $(-7) + (+6)$
47. $(-3) + (+7)$
48. $(-2) + (+6)$
49. $(-1) + (+8)$
50. $-11 + 9$
51. $-13 + 13$
52. $-16 + 10$
53. $9 + (-11)$
54. $10 + (-16)$
55. $13 + (-13)$
56. $(-12) + 7$
57. $8 + (-14)$
58. $13 + (-9)$
59. $-7 + (-5)$
60. $-6 + (-7)$
61. $-9 + (-4)$
62. $3 + (-2) + (-1)$
63. $-5 + 2 + 3$
64. $-4 + (-3) + 2$
65. $7 + 4 + (-5)$
66. $3 + 6 + (-8)$
67. $5 + 8 + (-7)$
68. $-9 + 16 + (-11)$
69. $5 + (-19) + 15$
70. $-7 + (-12) + (-17)$
71. $(-3) + (-2) + (-1)$
72. $(-2) + (-5) + (-3)$
73. $(-4) + (-5) + (-3)$
74. $5 + (-4) + 12$
75. $-7 + 5 + (-4)$
76. $-7 + 5 + 4$
77. $(-6) + 9 + (-4)$
78. $12 + (-9) + (-8)$
79. $-9 + 13 + (-7)$
80. $33 + (-28) + 15$
81. $-42 + 32 + (-18)$
82. $-39 + (-23) + 19$
83. $-[12 + (-9)]$
84. $-[(-23) + 14]$
85. $-[-18 + 25]$
86. $-[|-3| + |5|]$
87. $[-|-3| + |-5|]$
88. $-|-3 + (-5)|$
89. $|-8| + (-8)$
90. $-(-18) + |-18|$

2.3 Subtraction of Integers

Determining the amount of change a customer should receive from a purchase certainly uses subtraction. If the purchase is \$1.65 and you give the clerk \$2.00, the operation is $2.00 - 1.65 = .35$, or 35¢ change. But that is *not* the way a clerk determines your change. Usually, the clerk hands you a dime and a quarter and says "that's \$1.65, \$1.75, and \$2.00." That is, the subtraction is done by adding: $1.65 + .10 + .25 = 2.00$.

Many children play "take away" games. "What is 7 take away 5?" Holding up seven fingers, the child simply counts off "7, 6, 5, 4, 3; 2 is 7 take away 5." To take away, the child does not subtract but simply counts backward.

On the number line, subtraction would be going back in the opposite direction instead of going ahead as in addition.

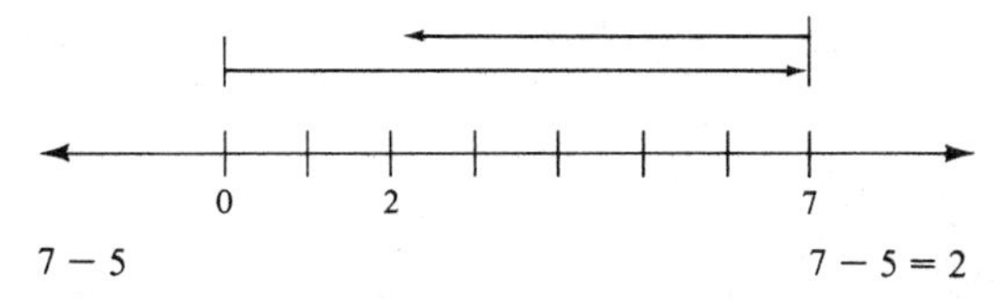

But what if we have negative numbers? Negative already indicates "going back" or to the left. Would subtracting a negative mean going to the right?

Look again at a sequence of problems and find the most logical solutions. Subtract each problem.

$5 - 5 = 0$
$5 - 4 = 1$
$5 - 3 = 2$
$5 - 2 = 3$
$5 - 1 = 4$
$5 - 0 = 5$
$5 - (-1) = ?$
$5 - (-2) = ?$
$5 - (-3) = ?$

It does seem that subtracting successively smaller numbers from 5 should give successively larger differences. On the number line $5 - (-3)$ would mean "going in the opposite direction from -3."

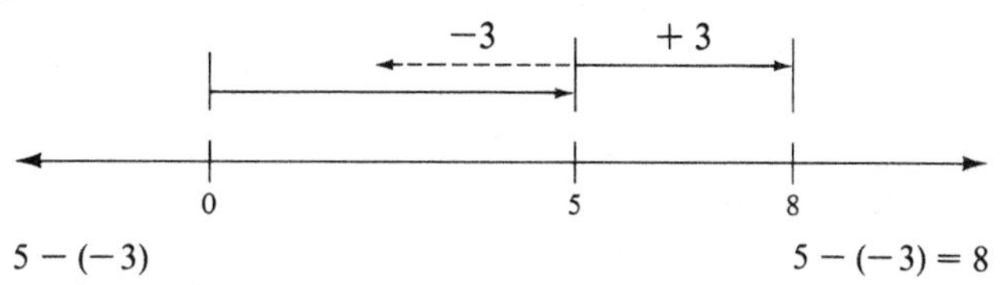

Wouldn't this be the same as $5 + 3$? Look at another sequence: subtracting smaller and smaller numbers. Subtract each of the following problems.

$-5 - 4 = -9$
$-5 - 3 = -8$
$-5 - 2 = -7$
$-5 - 1 = -6$
$-5 - 0 = -5$
$-5 - (-1) = ?$
$-5 - (-2) = ?$
$-5 - (-3) = ?$
$-5 - (-4) = ?$

It does look as if $-5-(-3)$ is going to have to be equal to -2. On the number line, to subtract -3, go 3 units in the opposite direction from -3.

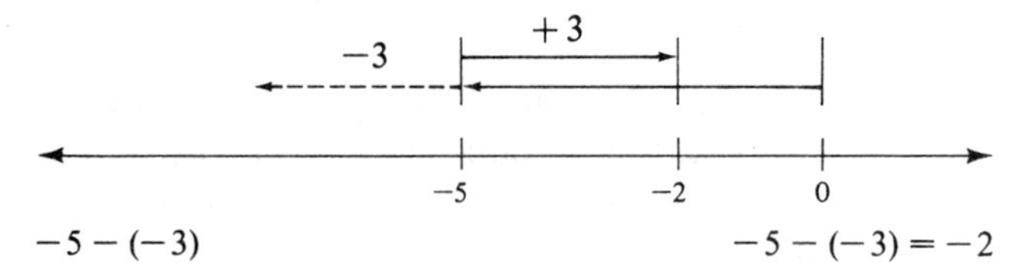

$-5-(-3)$ $\qquad$ $-5-(-3)=-2$

Guided by these results, we define ***subtraction*** as follows.

Definition of Subtraction

$$a-b=a+(-b)$$

To subtract, add the opposite of the number to be subtracted.

Examples

1. $3-2=3+(-2)$
   $=1$

2. $12-(-5)=12+(+5)$
   $=17$

3. $(-5)-8=(-5)+(-8)$
   $=-13$

4. $(-9)-(-15)=(-9)+(+15)$
   $=6$

5. $(-3)+(-5)-(-4)=(-3)+(-5)+(+4)$
   $=(-8)+(+4)$
   $=-4$

6. $75-14-38=75+(-14)+(-38)$
   $=75+(-52)$
   $=23$

7. Add the sum of negative three and positive two to the difference of negative four from positive seven.

$$\begin{aligned}[(-3)+(+2)]+[(+7)-(-4)] &= [(-3)+(+2)] + [(+7)+(+4)] \\ &= -1 + 11 \\ &= 10\end{aligned}$$

■

Problem Set 2.3

Perform the indicated operations, and state an answer in integer numbers for Problems 1–30. See Examples 1–4.

1. $(+3)-(+5)$ **2.** $(+7)-(+4)$ **3.** $(-2)-(+8)$
4. $(+6)-(-1)$ **5.** $(-8)-(-1)$ **6.** $(-5)-(+4)$
7. $(-7)-(+8)$ **8.** $(-9)-(+3)$ **9.** $(-2)-(+5)$
10. $(+6)-(-6)$ **11.** $(+9)-(-7)$ **12.** $(+8)-(-5)$
13. $0-(+2)$ **14.** $-7-0$ **15.** $0-(-3)$
16. $(+4)-(+4)$ **17.** $(-5)-(-5)$ **18.** $(+6)-(-6)$
19. $(+8)-(-7)$ **20.** $(-6)-(-8)$ **21.** $(-5)-(-4)$
22. $(+5)-(-9)$ **23.** $(+7)-(+2)$ **24.** $(+8)-(-3)$
25. $0-(-6)$ **26.** $(-1)-0$ **27.** $0-(+5)$
28. $6-(-8)$ **29.** $(-9)-(-2)$ **30.** $7-(-3)$

Perform the indicated operations and state an integer answer for Problems 31–60. See Examples 5 and 6.

31. $(+3)-(+2)+(+4)$
32. $(+7)+(+4)-(+5)$
33. $(+3)-(+2)-(+4)$
34. $(+7)-(+4)+(+5)$
35. $(+3)-(+2)+(-4)$
36. $(-7)-(+4)-(-5)$
37. $(-3)-(+2)+(-4)$
38. $(-7)+(-4)-(-5)$
39. $(-3)-(-2)-(+4)$
40. $(-7)-(-4)-(-5)$
41. $-5-(-6)+7$
42. $-9+8-(-7)$
43. $8-5-(-6)$
44. $4-7-5$
45. $-5-8-(-9)$
46. $-7-3-(-9)$
47. $-8-4-2$
48. $-9-3-6$
49. $11-14-(-19)$
50. $17-(-18)-15$
51. $-23-12-(-31)$
52. $-35+(-13)-17$
53. $34-(-21)-29$
54. $37+(-14)-23$
55. $56-(-23)-(-25)$
56. $-51-(-27)-(-29)$
57. $-93+(-17)-(-61)$
58. $-67-(-39)+(-32)$
59. $31-(-31)+(-31)$
60. $-37+(-37)-(-37)$

Write Problems 61–70 using symbols and find the value of the expressions, if possible. See Example 7.

61. Add the difference of seven from five to the difference of negative six from negative eight.
62. Subtract the sum of ten and negative eight from the sum of negative nine and positive six.
63. Subtract the difference of five from negative four from the sum of negative six and positive three.
64. Add the difference of negative nine from positive six to the sum of negative five and negative four.
65. Add the number negative two to the difference of eight from negative two.
66. Add the number negative five to the quantity five minus negative five.
67. Subtract the number negative six from the sum of negative three and negative six.
68. Subtract the number negative one from the difference of negative six from negative one.
69. Subtract the sum of the number negative three and one from negative three.
70. Subtract the difference of the number negative four from two from negative four.

2.4 Multiplication and Division of Integers

What is multiplication? We really should answer that question before considering *how* to multiply. What actually happens when we multiply? Well, what does three times four mean? "Twelve," you answer proudly. That *is* the product—the answer—but how did you get that value? What did you do to multiply? You probably memorized that answer long ago, using a multiplication table, but what if you could no longer use that table or you forgot that particular entry? You might have to think about what the operation really is.

Three times four, a young child would tell us, is "three fours," or $4+4+4$; it tells the number of fours you add. That is, we could rebuild the multiplication table from an addition table. The addition table could be constructed by counting—essentially all operations are a form of counting. If you count by twos, you are reciting the multiplication facts for two. Multiplication is repeated addition, and, because it is addition, we can display multiplication on the number line.

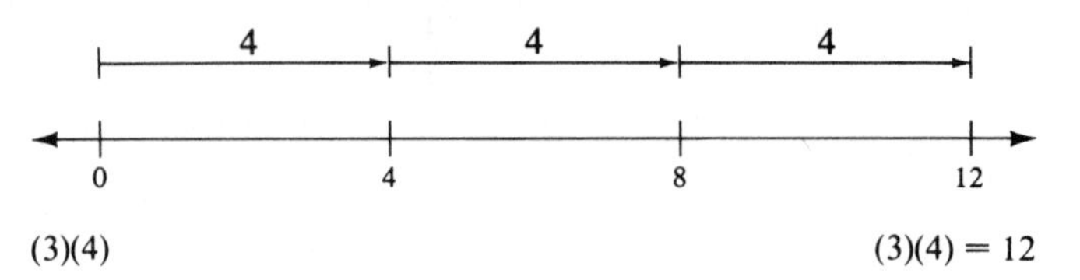

The product of two positive integers is positive.

Using this same idea, $3(-4) = (-4) + (-4) + (-4)$. From our work with addition, we know this result is negative and can be illustrated as shown.

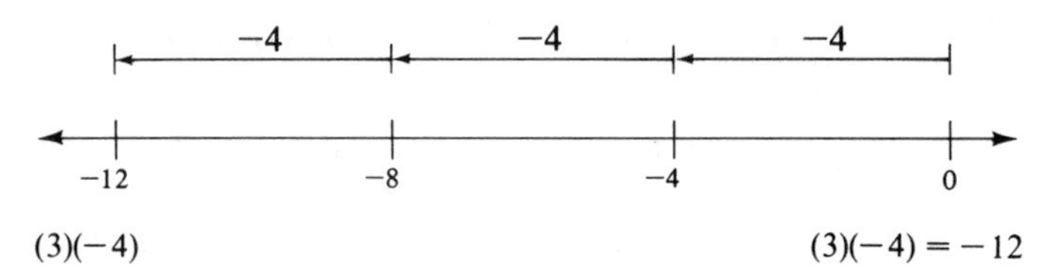

Moreover, we can see from a sequence of products that this result is reasonable and consistent with the arithmetic we have discussed to this point. Multiply in each of the following.

$$\begin{aligned} 4\cdot 3 &= 12\\ 3\cdot 3 &= 9\\ 2\cdot 3 &= 6\\ 1\cdot 3 &= 3\\ 0\cdot 3 &= 0\\ -1\cdot 3 &= ?\\ -2\cdot 3 &= ?\\ -3\cdot 3 &= ? \end{aligned}$$

The next number in the sequence should certainly be less than zero, because the sequence of answers is decreasing. $\{\ldots 12, 9, 6, 3, 0, -3, -6, -9, \ldots\}$ would seem to be the complete sequence and consistent with what we have seen on the number line. Multiply each of the following by continuing the preceding pattern.

$$\begin{aligned} -1\cdot 3 &= -3\\ -2\cdot 3 &= -6\\ -3\cdot 3 &= -9\\ -4\cdot 3 &= -12 \end{aligned}$$

Guided by these results, we can make two general statements. We will add to these statements and later summarize them into a general principle governing the multiplication of integers.

> The product of a positive and a negative number is negative.

> The product of zero and any integer is zero.

In Section 1.5 we mentioned the commutative, associative, and distributive properties of natural numbers. We would like these properties (in particular the commutative property) to apply to the integers as well, so that $(3)(-4) = (-4)(3) = -12$. Beginning with this assumption, consider another sequence of products. Multiply each of the following.

$$\begin{aligned}(3)(-4) &= -12\\ (2)(-4) &= -8\\ (1)(-4) &= -4\\ (0)(-4) &= 0\\ (-1)(-4) &= ?\\ (-2)(-4) &= ?\\ (-3)(-4) &= ?\\ (-4)(-4) &= ?\end{aligned}$$

It seems that the missing products of the sequence will have to be positive if the pattern $-12, -8, -4, 0, \ldots$ increasing by four is to continue. A general statement now seems plausible.

> The product of two negative numbers is positive.

To summarize what we have just established for multiplying integers, we look at the signs of the numbers. If the signs are different, the product is negative. If the signs are alike, the product is positive. Using a flowchart, such as Figure 2.2, we map our thinking process for multiplying two nonzero integers.

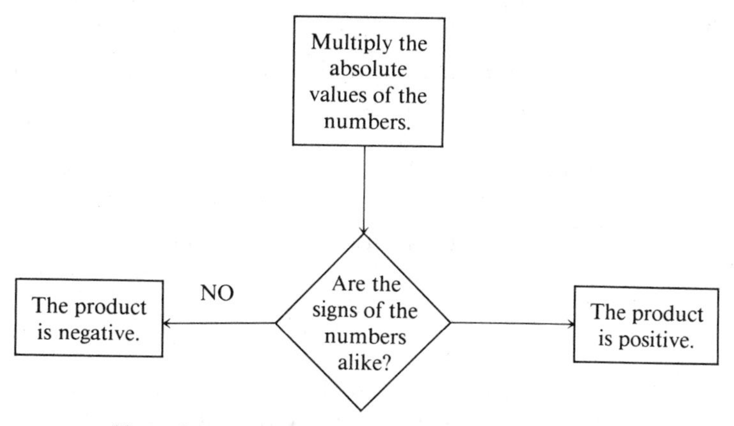

Figure 2.2
Flowchart for the multiplication of integers

Multiplication of Integers

> To multiply two nonzero integers, multiply (the absolute values of) the numbers and make the product positive if the signs (of the numbers) are alike, or negative if the signs are different. The product of zero and any integer is zero.

That is,

(positive)(positive) = positive
(positive)(negative) = negative
(negative)(positive) = negative
(negative)(negative) = positive
(any integer)(zero) = zero

Examples

1. $(-2)(+3)$
unlike signs
$(-2)(+3) = \mathbf{-6}$

2. $(-4)(-5)$
like signs
$(-4)(-5) = \mathbf{+20}$

3. $(-2)(-3)(-5) = [(-2)(-3)](-5)$ *Group two factors.*
$= (+6)(-5)$ *Find that product first.*
$= \mathbf{-30}$ ■

Before leaving multiplication, consider one more example. Certainly $(-4) = (4)(-1)$, as can be shown on the number line.

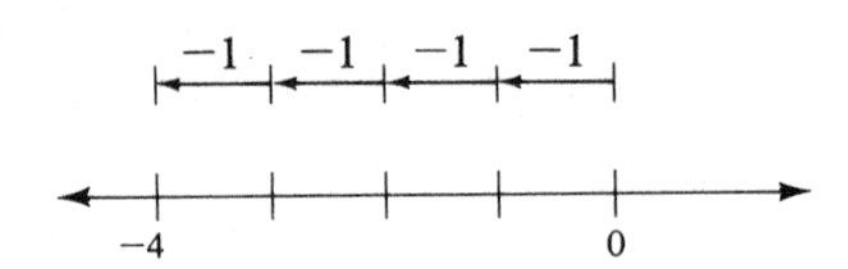

$(4)(-1) = (-1)(4) = -4$, and this would seem to apply for any integer n.

Opposite of an Integer

> $(-1)n = -n$
>
> Negative one times a number is the opposite of that number.

Now notice how this property can be applied in the previous example of $(3)(-4)$.

$$
\begin{aligned}
(3)(-4) &= (-4)(3) && \textit{commutative} \\
&= (-1)(4)(3) && (-1)(4) = -4 \\
&= (-1)(12) && \textit{associative} \\
&= -12 && (-1)(12) = -12
\end{aligned}
$$

Recall that division can be explained in terms of multiplication; that is, $10 \div 5 = 2$ because $10 = 5 \cdot 2$. Thus consider all the possibilities when dividing positive and negative numbers.

positive ÷ positive	$10 \div 5 = 2$	*because* $10 = 5 \cdot 2$
positive ÷ negative	$10 \div (-5) = -2$	*because* $10 = (-5)(-2)$
negative ÷ positive	$(-10) \div 5 = -2$	*because* $-10 = (5)(-2)$
negative ÷ negative	$(-10) \div (-5) = 2$	*because* $-10 = (-5)(2)$

Note that dividing numbers with like signs produces a positive quotient, and dividing numbers with unlike signs gives a negative quotient. A rule similar to that for multiplication of integers can be formulated for division.

Division of Integers

> To divide two integers, we divide (the absolute values of) the numbers and make the quotient positive if the signs (of the numbers) are alike, and negative if the signs are different.

Examples

4. $(-12) \div (-4)$
like signs
$(-12) \div (-4) = +3$

5. $(+35) \div (-7)$
unlike signs
$(+35) \div (-7) = -5$

6. $(-9)(+5) \div (-15) = (-45) \div (-15) = 3$ *Work left to right.*

7. $(-2)^4 = (-2)(-2)(-2)(-2) = (+4)(-2)(-2) = (-8)(-2) = +16$

8. $(-3)^3 = (-3)(-3)(-3) = (+9)(-3) = -27$

9.
$$\frac{(-5)(+4) - (-2)^3}{(-2)(+3)} = \frac{(-20) - (-8)}{-6}$$ *Multiply first.*
$$= \frac{(-20) + (+8)}{-6}$$ *Change to addition.*
$$= \frac{-12}{-6}$$ *Divide.*
$$= +2$$

10. Find the average of 10, -17, -39, 26, and -15.

$$\text{AVERAGE} = \frac{\text{SUM OF THE NUMBERS}}{\text{NUMBER OF NUMBERS}}$$
$$= \frac{10 + (-17) + (-39) + 26 + (-15)}{5}$$
$$= \frac{-35}{5}$$
$$= -7$$

The average of the numbers is -7. ■

Given that $a = -2$, $b = 3$, and $c = -5$, evaluate the expressions in Examples 11–13.

Examples

11. $a - bc = (-2) - (3)(-5)$
$= -2 - (-15)$
$= -2 + 15$
$= 13$

12. $a^3b + c^2 = (-2)^3(3) + (-5)^2$
$= (-8)(3) + (25)$
$= -24 + 25$
$= 1$

13. $\dfrac{b - c}{a} = \dfrac{3 - (-5)}{-2}$
$= \dfrac{8}{-2}$
$= -4$ ■

Problem Set 2.4

Find the value of each of the expressions in Problems 1–66. See Examples 1–9.

1. $(+6)(-8)$
2. $(-7)(+4)$
3. $(-5)(+3)$
4. $(-4)(-7)$
5. $(+9)(-9)$
6. $(-8)(-2)$
7. $(+7)(+5)$
8. $(+5)(0)$
9. $(0)(-7)$
10. $(+3)(+9)$
11. $(-8)(+5)$
12. $(4)(-6)$
13. $(-8)(+8)$
14. $(-6)(-9)$
15. $(-7)(5)$
16. $0 \cdot (-3)$
17. $(+7)(-6)$
18. $(-8)(-8)$
19. $(-9)(+8)$
20. $(+5)(-2)$
21. $(5)(+2)$
22. $(-1)(-7)(-6)$
23. $(+2)(+9)(-3)$
24. $(-4)(8)(-5)$
25. $(-4)(+2)(-1)$
26. $(+5)(-4)(-2)$
27. $(-2)(-1)(3)$
28. $(-4)(+2)(-1)$
29. $(+4)(-7)(+3)$
30. $(2)(-5)(-7)$
31. $(+42) \div (+7)$
32. $(+45) \div (-9)$
33. $48 \div (-12)$
34. $(-32) \div (+4)$
35. $(-60) \div (-3)$
36. $(-51) \div 17$
37. $\dfrac{-15}{-3}$
38. $\dfrac{+18}{-6}$
39. $\dfrac{-16}{4}$
40. $(-6)(-6) \div (+9)$
41. $(-20)(+3) \div (-5)$
42. $(-3)(-4) \div (-6)$
43. $2 \cdot (-14) \div (-4)$
44. $(-6) \cdot 4 \div 3$
45. $5 \cdot 8 \div (-10)$
46. -2^4
47. $(-2)^4$
48. -2^5
49. $(-2)^5$
50. $(-1)^8$
51. $(-1)^9$
52. $(-3)^2 - (-2)^3$
53. $(-2)^4 - (-3)^3$
54. $(-2)^5 - (-1)^3$
55. $(-5)^2 - (-3)(+8)$
56. $(-7)(-9) - (-2)^6$
57. $(-3)^3 - 2(-5)$
58. $\dfrac{8 + (-3) - (-7)}{-4}$
59. $\dfrac{-7 - (-2) - 9}{-2}$
60. $\dfrac{-3 + 9 - 6 - 8}{(-2)(-2)}$
61. $\dfrac{7 - 2 + 8 - 1}{(-1)(-3)}$
62. $\dfrac{(-7) - (-5)(-3)}{2 - (-9)}$
63. $\dfrac{(-5) - (-7)(-3)}{8 - (-5)}$
64. $\dfrac{(+4)(-8) + (-6)^2}{(-2)(-2)}$
65. $\dfrac{(-7)^2 + (-4)^3}{(-5)(-1)}$
66. $\dfrac{(-5)^2 + (-3)^3}{(-1)(-2)}$

67. Find the average of 2, −5, 4, −7, and −4.
68. Find the average of −7, −18, 4, −9, and 5.
69. Find the average of −3, −6, 5, 8, 0, −2, and 5.
70. Find the average of 21, 0, −34, 45, and −12.

Given that $u = -1$, $v = 2$, $w = -3$, and $x = -5$, evaluate each of the expressions in Problems 71–90. See Examples 11–13.

71. $u^2 + v^3$ **72.** $uvwx$ **73.** $u(v + w)$
74. $x - 2vw$ **75.** $u + v + w + x$ **76.** $uv + wx$
77. $u + vw + x$ **78.** $|x| + (w - v)$ **79.** $-(x - vw)$
80. $w^3 + x^2$ **81.** $(w + x)u$ **82.** $wx^2 + w^4$
83. vwx **84.** $x(v^3 - w^2)$ **85.** $\dfrac{x + u}{w}$
86. $\dfrac{v^3w}{x + u}$ **87.** $\dfrac{u - v}{v + x}$ **88.** $\dfrac{x + w}{u - x}$
89. $\dfrac{x^2 - v^4}{v + x}$ **90.** $\dfrac{u - w^2}{u - w}$

2.5 Simplification

We have devoted this chapter to simplifying numerical expressions involving integers. It is appropriate to turn our attention to simplifying variable expressions. For example,

$$(-2x)(5y)$$

can be rewritten as $(-2 \cdot 5)(xy)$ by using the commutative and associative properties. Thus

$$\begin{aligned}(-2x)(5y) &= (-2 \cdot 5)(xy)\\ &= -10xy\end{aligned}$$

Examples *Simplify.*

1. $\begin{aligned}(-5a)(-6b) &= (-5)(-6)ab\\ &= 30ab\end{aligned}$

2. $\begin{aligned}-3x^2 4y^3 &= (-3 \cdot 4)x^2y^3\\ &= -12x^2y^3\end{aligned}$ ■

In practice, most of the steps shown in Examples 1 and 2 will be performed mentally and appear more like Examples 3 and 4.

Examples **3.** $(-3x^2)(5y^3) = -15x^2y^3$ **4.** $(2a)(3b)(-4c) = -24abc$

5. $\begin{aligned}(-4z)(-z^3)(3z^2) &= (-4)(z)(-1)(z^3)(3)(z^2) && \textit{Property of opposites}\\ &= (-4)(-1)(3)zz^3z^2 && \textit{Reorder and regroup}\\ &= -12z^{1+3+2} && \textit{First Law of Exponents}\\ &= -12z^6\end{aligned}$

6. $(-2x)(3x^2)(-4x^5) = 24x^8$ *Mentally carry out the steps as shown in Example 5.*

The examples illustrate the operation of multiplication. In each case, the answer is still an indicated multiplication. Such expressions are called ***terms.***

Term

A term is a number, a variable, or the product of numbers and variables.

Examples *Tell which of the following are single terms.*

7. -5 is a term *(single number)*
8. 0 is a term *(single number)*
9. x is a term *(single variable)*
10. $5x$ is a term *(product of a number and a variable)*
11. $x + 2$ is not a term *(sums of numbers and variables are not single terms; they are actually two terms)*
12. $5abcdefghwxyz^3$ is a term *(product of numbers and variables)*
13. $\frac{3xy}{m}$ is not a term *(division by a variable)* ■

We will discuss terms and their operations more completely in Chapter 4, but in this section we introduce you to some terminology and some ways of writing certain expressions more simply. In a product of two or more factors, any collection of factors is said to be the ***coefficient*** of the rest of the factors. Thus, in $30ab$, $30a$ is the coefficient of b, $30b$ is the coefficient of a, and 30 is the coefficient of ab. The constant factor is called the ***numerical coefficient.*** If no constant factor is shown, as in

$$x^2y$$

we note that

$$x^2y = 1 \cdot x^2y$$

so the numerical coefficient of x^2y is 1. That is, *if no constant factor appears, the numerical coefficient is understood to be 1.*

Example **14.** Given $5x^3 + x^2 - 4x$; this algebraic expression consists of three terms. Indicate the numerical coefficient and exponent of the variable for each term.

Term	Numerical Coefficient	Exponent of the Variable
$5x^3$	5	3
x^2	**1**	2
$-4x$	-4	**1**

■

If two or more terms are exactly alike except for their numerical coefficients, we call them ***similar terms.***

Examples

15. $5x$, $-6x$, and πx are similar terms.
16. $2x^2$ and $2x$ are not similar terms (x is not exactly the same as x^2).
17. abc, abc^2, and ab^2c are not similar terms.
18. $-3a^2bc^3$, $-5a^2bc^3$, and $17a^2bc^3$ are similar terms. ■

The distributive property can be used to simplify expressions that contain similar terms. In the following examples, notice how the expressions are simplified by combining similar terms and by adding the numerical coefficients.

Examples *Simplify the following expressions.*

19. $3a + 7a = (3 + 7)$
$= 10a$

20. $8x - 3x = 8x + (-3)x$
$= [8 + (-3)]x$
$= 5x$

21. $-12y - (-9y) = -12y + 9y$
$= -3y$ *Note that* $-12 + 9 = -3$

22. $-3z - 6z = -3z + (-6)z$
$= -9z$ *Note that* $-3 + (-6) = -9$

23. $5a + 4b + 3a = (5a + 3a) + 4b$
$= 8a + 4b$ *Only two similar terms*

24. $4x - 5x + 2x = 4x + (-5)x + 2x$
$= x$ *Coefficient of x is 1*

25. $(2x + 3y) + (5x - 2y) = (2x + 5x) + [3y + (-2)y]$
$= 7x + y$

The middle steps in Examples 19–25 are commonly performed mentally and should not usually be shown as a part of your work.

26. $2(x - 3) + 4(1 - x) = 2x - 6 + 4 - 4x$
$= -2x - 2$

27. $2x(x - y) + 3y(x + 1) = 2x^2 - 2xy + 3xy + 3y$
$= 2x^2 + 5xy + 3y$

28. $3x(x + 2y) + 2y(y - 3) = 3x^2 + 6xy + 2y^2 - 6y$ *No similar terms* ■

Problem Set 2.5

1. Given $7x^4 + 6x^3 + x^2 + 5x$,
   a. what is the numerical coefficient of each term?
   b. what is the exponent of the variable of each term?
2. Given $a^5 + 3a^4 + 2a$,
   a. what is the numerical coefficient of each term?
   b. what is the exponent of the variable of each term?
3. Which of the following are similar terms? $3x$, $4xy$, $2x^2y$, $(5 \cdot 2)x$, $3xy^2$, $7xy$, $4y$, $6x^2$, $11x$
4. Which of the following are similar terms? $4a$, $6a^2b$, $5ab$, $4b$, $12a$, $3ab^2$, $12a^2b$, $7bc$

Simplify the expressions in Problems 5–20. See Examples 1–6.

5. **a.** $(7a)(3b)$ **b.** $(3a)(7b)$
6. **a.** $(2x)(4y)$ **b.** $(4x)(2y)$
7. **a.** $(9s)(-4t)$ **b.** $(-9s)(4t)$
8. **a.** $(2x)(-3y)$ **b.** $(-2x)(-3y)$
9. **a.** $x(4y)(3z)$ **b.** $x(-4y)(3z)$
10. **a.** $(4s)(-2t)(w)$ **b.** $(-4s)(2t)(-w)$
11. **a.** $(4x)(3y)^2$ **b.** $(4x)(3y^2)$
12. **a.** $(2x)(5y^2)$ **b.** $(2x)^2(5y)$
13. **a.** $(4m)(-2m)^2$ **b.** $(4n)(-2n)^3$
14. **a.** $(2x^2)(-3x)$ **b.** $(2x)^2(-3x)$

15. a. $y^3(-5y^2)(-4y)$ b. $y(-5y^2)(-4y^3)$
16. a. $-z^2(z^3)(6z)$ b. $z^3(-z)^2(6z)$
17. a. $-a(7a^2)(3b^3)$ b. $-b(7b^2)(7b^3)$
18. a. $r^2(-2s)(3r^3)$ b. $r^2(2r)(-3s^3)$
19. a. $-m^3(-9m^2)(-7m^2)$ b. $(-m)^3(-9m^3)(-7m)^2$
20. a. $x^4(-x^3)(-x^5)$ b. $x^5(-x^4)(-x^3)$

Combine similar terms in Problems 21–40. See Examples 19–22.

21. $2x + 5x$
22. $3y + 8y$
23. $6z + 9z$
24. $9z - 4z$
25. $7w - 6w$
26. $8u - 5u$
27. $-3a + 7a$
28. $-2b - (+5b)$
29. $-4c - 6c$
30. $-6h - (8h)$
31. $-5k - (-2k)$
32. $-7j - (-5j)$
33. $m - 4m$
34. $-n + n$
35. $-2p - 3p$
36. $-3x - (-7x)$
37. $-5y - (2y)$
38. $-4z - (-9z)$
39. $25a - (-37a)$
40. $49b - (-31b)$

In Problems 41–60 simplify by combining similar terms. See Examples 23 and 24.

41. $4x + 2x + 7x$
42. $2y + 3y + 8y$
43. $6a + 7a + (-4)a$
44. $9b + 2b + (-7)b$
45. $-2c + 3d + 4c$
46. $5e + 4f + (-6)f$
47. $3x + (-5)x + (-2)x$
48. $2y + (-9)y + 6y$
49. $-2a + (-5)a + 9a$
50. $-4b + 7b + (-5)b$
51. $2w - 3w + 7w$
52. $4x + 9x - 5x$
53. $y - 5y + y$
54. $z + 2z - 6z$
55. $3x - 4x - 7x$
56. $5y - 3y - 8y$
57. $7a - 4a - 3a$
58. $8b - 3b - 5b$
59. $12c - 9c + 5c - 8c$
60. $7d + 6d - 11d - 4d$

Simplify the expressions in Problems 61–70. See Examples 26–28.

61. $6(x - 5) + 3(x + 2)$
62. $5(x + 3) + 2(x - 3)$
63. $3(x - 2) + 4(x - 1)$
64. $5(y - 4) + 3(y - 1)$
65. $2(x^2 + x - 1) + 4(x^2 - 2x + 3)$
66. $2(a^2 - ab - b^2) + 4a(a - b)$
67. $4x^2(x - y + 2) + 3x(x^2 + xy - 1)$
68. $(6x^4 - 3x^3 + 2x - 1) + (5x^3 - 4x^2 + 3) + (5x^2 - 4x - 6)$
69. $(5r^3 - 4r^2s + 3rs^2 - 5) + (3r^3 + 2rs^2 - 5r^2s + 2)$
70. $(u^3 - 3u^2v + 3uv^2 + v^3) + (u^2v + 5uv^2 - v^3)$

2.6 Review

The dual nature of this chapter is important. First, the set of integers is developed, and the operations with integers are given special emphasis because of their continuing importance in every topic that follows. Second, integers are used with variable expressions. These expressions are evaluated for integer replacements as well as simplified using the properties of the integers.

Self-Test

[2.1] **1.** Express each of the following as a signed number.
a. 12° below zero. **b.** A fifty dollar gain. **c.** $-(+17)$
d. $-|-17|$ **e.** $-(-17)$

[2.1] **2.** Move five units to the left, then nine units to the right, and then six units to the left. What integer represents the final position relative to the starting point?

[2.2] **3.** Find the sum.
a. $5+(-7)$ **b.** $(-2)+(-6)$ **c.** $(-3)+|-4|$
d. $(-4)+(+7)+(-5)$ **e.** $|-6|+|-2|$

[2.3] **4.** Perform the indicated operations.
a. $5-(-7)$ **b.** $(-2)-(-6)$ **c.** $(-3)-|-4|$
d. $(-4)-(+7)-(-5)$ **e.** $2y-(-5)y-12y$

[2.3] **5.** Write each of the following using symbols, and simplify the expressions.
a. Subtract the sum of twelve and negative eight from the sum of negative eleven and positive nine.
b. Find the average of 9, -5, -11, 3, and -6.
c. Add the number n to the difference of eight from that number.

[2.4] **6.** Perform the indicated operations.
a. $5\cdot(-7)$ **b.** $(-2)(-6)$ **c.** $(-3)(-4)(+2)$
d. $(-4)(+7)\div(-2)$ **e.** $(-2)^3+(-3)^2$

[2.4] **7.** Find the value of each of the following given that $a=-2$, $b=3$, and $c=-5$.
a. $ab-c$ **b.** $a-b-c$ **c.** $a^3+b+|c|$
d. $\frac{a-c}{b}$ **e.** a^2bc+ab^2c

[2.5] **8.** Given x^3-2x^2+7x,
a. what is the numerical coefficient of each term?
b. what is the exponent of the variable of each term?

[2.5] **9.** Simplify the expressions.
a. $(5a)(-7b)$ **b.** $(-3x)(-8x^3)$ **c.** $(4y^2)(-2y)^2$
d. $(4y^2)(-2y^2)(-3y^3)$ **e.** $-b^2(-b^3)(2b^3)$

[2.5] **10.** Simplify the expressions.
a. $-2x+3y+4x$ **b.** $5y-6y-7y$ **c.** $z-8z+9z$
d. $2(a-3)+4(a+5)$ **e.** $a^2(a+2b)+b(b-2a^2)$

Review Problems

Perform the indicated operations and state an integer answer for Problems 1–75.

1. $(-13)+(-18)$
2. $(+14)+(-7)$
3. $(-14)+(+7)$
4. $(+27)+(-27)$
5. $(-29)+(-11)$
6. $(+18)+(-25)$
7. $(-13)+(+30)+(-17)$
8. $(+21)+(-9)+(-3)$
9. $(+18)-(+20)$
10. $(-9)-(-11)$
11. $(+13)-(-11)$
12. $(-17)-(-17)$
13. $(-17)-(+17)$
14. $(+17)-(-17)$
15. $8-11-3-14+9$
16. $-13+15-18+2-7$
17. $14+9-7+8-3$
18. $-1-17+14-9+11$
19. $(+13)(+3)$
20. $(-18)(-2)$
21. $(+12)(-4)$
22. $(+8)(+19)$

23. $(+14)(+7)$

24. $(-17)(-8)$

25. $(-3)(-18)$

26. $(-7)(+15)$

27. $(+6)(-26)$

28. $7(-4)(-1)$

29. $6(-3)(+2)$

30. $5(-2)(-3)$

31. $(-8)^2$

32. $(-2)^3$

33. $(-1)^5$

34. $\frac{-36}{-4}$

35. $\frac{-13}{+13}$

36. $\frac{18}{-3}$

37. $\frac{+651}{-21}$

38. $\frac{-527}{-17}$

39. $\frac{-161}{-7}$

40. $\frac{(-23)-(-5)}{-6}$

41. $\frac{(+8)-(-17)}{-5}$

42. $\frac{(-11)-(+17)}{-7}$

43. $\frac{(-15)+(+6)}{+3}$

44. $\frac{(+29)+(-8)}{-3}$

45. $\frac{(-17)-(+9)}{+2}$

46. $\frac{(-23)-(+13)+(-12)}{(-23)-(-15)}$

47. $\frac{(+39)-(-13)+(-22)}{(-23)+(-7)}$

48. $\frac{9-13+1-9-3}{-8-7}$

49. $\frac{17-8+4-13-18}{-6-3}$

50. $\frac{6+9+5+28-13+8}{20-(-5)}$

51. $\frac{12-7-9+15+9}{7-(-3)}$

52. $\frac{-34+10-11+8+3}{2-10}$

53. $\frac{-12-(-3)(-5)}{(-3)^2}$

54. $\frac{13-(+7)(-2)}{(-3)^2}$

55. $\frac{29-5(-2)(-2)}{7-(-2)}$

56. $\frac{23-2(-3)(+2)}{9-16}$

57. $\frac{(-2)(-5)-(+3)(-4)}{(-5)-6}$

58. $\frac{(-2)(-3)(+2)-(-13)}{(-8)+3}$

59. $\frac{(-9)-(-3)}{(-3)(-6)+2(-6)}$

60. $\frac{5(-25)-1}{5(-3)(-4)-(-3)}$

Evaluate the expressions in Problems 61–75 given that $a = -1$, $b = 3$, and $c = -2$.

61. $a+b+c$

62. $a-b+c$

63. $a+b-c$

64. $a-b(c)$

65. $ab-c$

66. $a(b-c)$

67. $a(b-ac)$

68. $(ab+c)c$

69. $b(ac-b)$

70. a^2-abc

71. a^3-bc^2

72. b^2+c^3

73. ab^2-c

74. a^2b+c

75. $(ab)^2-c$

Simplify each expression in Problems 76–85.

76. $(2x)(3x)$

77. $(-3y)(4y)$

78. $(-5z)(-z^2)$

79. $a(-a)(a^3)$

80. $-b(-b)^2(b^2)$

81. $(-c)(-c)(-c)^2$

82. $2x(-xy)(xy)$

83. $-3ab(5a)(-2b)$

84. $-mn(-2mn)(-3n)$

85. $-x(2xy)(7xy)$

Simplify by combining similar terms in the expressions in Problems 86–100.

86. $3x + 2x + 5x$

87. $3x - 2x + 5x$

88. $3x - 2x - 5x$

89. $4y - 7y - y$

90. $y - 2y - 3y$

91. $7y - 3y - 4y$

92. $5z + z(2 + z)$

93. $2(z - 1) - 5z$

94. $3z(1 - z) + 2z$

95. $2x + 3y - 8x$

96. $5y - 7z + 3y$

97. $2x - 3z - 4x$

98. $2(x - 3) + 4(2 - x)$

99. $5(y - 3) + 3(7 - y)$

100. $3(2 - z) + 5(z - 4)$

3

First-Degree Equations and Inequalities

"Can you do addition?"
The White Queen asked.
"What's one and one and one and
one and one and one and one and
one and one and one?"
"I don't know," said Alice.
"I lost count."
"She can't do addition,"
The Red Queen interrupted.

Through the Looking-Glass
by Lewis Carroll

3.1 Equations

In this chapter we will use algebra and symbols to help us solve some problems that otherwise might be very difficult. These problems usually have some known constants, some relationships, and some unknown values, and we must find the unknown values. As we saw in Chapter 1, *if two quantities are exactly the same, we use the equality symbol.* An ***equation*** is a statement of equality.

A value that makes an equation true is called a ***solution*** or a ***root.*** We also say that a root or a solution ***satisfies*** the equation. If some number is given and you want to determine if it is a root, simply replace the variable by that value and check to see if the resulting equation is true.

Determine whether the given value is a root of the equation.

Example 1. $5x + 3 = 13; x = 2$

Check: $5(\mathbf{2}) + 3 = 10 + 3$
$= 13$ ✓ ■

The given statement is true when x is replaced by 2; thus $x = 2$ is a root.

Example 2. $3(x + 1) + 4(x - 3) = 30; x = 3$

Check: $3(\mathbf{3} + 1) + 4(\mathbf{3} - 3) = 3(4) + 0$
$= 12 \neq 30$ ■

The given statement is false when x is replaced by 3; thus $x = 3$ is not a root.

Example 3. $6(x + 2) + 3(2 - 2x) = 18; x = 8$

Check: $6(\mathbf{8} + 2) + 3(2 - 2 \cdot \mathbf{8}) = 6(10) + 3(-14)$
$= 60 - 42$
$= 18$ ✓

Thus $x = 8$ is a root. ■

In this chapter we will develop some techniques to find the roots or solutions to certain types of equations and inequalities. Because most problems are stated in words, we need to express verbal sentences symbolically. It might be helpful to review Examples 29–36 of Section 1.1. At this point it is not necessary to know the number. (If you did know the number, you would not represent it with a variable.)

Express each of the following sentences symbolically.

Example 4. The sum of 5 and some number is 10.

Solution

5 + SOME NUMBER = 10
Let n represent SOME NUMBER
$5 + n = 10$ ■

Example 5. A person's intelligence quotient (IQ) is that person's mental age divided by chronological age, and that quality multiplied by 100.

Solution

$$\text{IQ} = \left(\frac{\text{MENTAL AGE}}{\text{CHRONOLOGICAL AGE}}\right)100$$

There may be more than one variable.
Let M = MENTAL AGE
C = CHRONOLOGICAL AGE

$$\text{IQ} = \left(\frac{M}{C}\right)100$$

■

Example 6. A man's son is 3 years younger than his daughter.

Solution

SON'S AGE + 3 = DAUGHTER'S AGE
Let s represent the SON'S AGE and d represent the DAUGHTER'S AGE.
Symbolically, $s + 3 = d$.
Notice that s and d represent numbers. It is not correct to say "Let s represent the son." A variable represents *a number.* ■

Example 7. The product of a number and 5 is 6 less than the number.

Solution

(A NUMBER)(5) is 6 less than (A NUMBER)
(A NUMBER)(5) + 6 = A NUMBER
Let x represent the NUMBER

$$(x)(5) + 6 = x$$

or

$$5x + 6 = x$$

■

Example 8. The product of a number and 5 is 6 less than another number.

Solution

(A NUMBER)(5) is 6 less than (ANOTHER NUMBER)
(A NUMBER)(5) + 6 = ANOTHER NUMBER
Let x represent the NUMBER and y represent ANOTHER NUMBER

$$5x + 6 = y$$

■

Notice from Examples 7 and 8 that different numbers require different variables, whereas one number in two places is represented by a single variable.

The equations in the examples are called ***linear*** or ***first-degree equations*** because the exponent on the variable is 1. We did not ask for the solutions of these problems, even though you could probably provide the replacements of the variables that satisfy the equations by inspection. In the next section we will consider a procedure for solving more difficult first-degree equations.

Problem Set 3.1

In Problems 1–24 determine whether the given value is a root of the equation. See Examples 1–3.

1. $x - 5 = 12; x = 7$
2. $y + 6 = 8; y = 2$
3. $z + 3 = 1; z = -2$
4. $x + 5 = 11; x = 6$
5. $y - 6 = 6; y = 0$
6. $z + 9 = 9; z = -1$
7. $2x + 1 = 7; x = 3$
8. $2y - 3 = 5; y = 2$
9. $3x - 4 = 8; x = 4$
10. $4x + 1 = 13; x = 3$
11. $5b + 3 = b + 7; b = 1$
12. $6a + 2 = a + 9; a = 2$
13. $7 - x = 9; x = -2$
14. $12 - y = 3; y = -9$
15. $4 - z = 6; z = -2$
16. $2c - 5 = 3c - 8; c = 3$
17. $2(x + 3) - 2x = 6; x = 5$
18. $3(d + 1) - 3d = 3; d = 7$
19. $3(2x + 1) + 2(3 - 3x) = 5; x = 7$
20. $2(x + 3) + 5(x + 3) = 7x + 12; x = 2$
21. $(6 - w) + (5 + w) = 3 - 2w; w = -7$
22. $(5 + y) + (3 + 2y) = 5 - 4y; y = -1$
23. $\dfrac{y - 3}{3} + 5 = 0; y = -2$
24. $\dfrac{x + 10}{2} + 2x = -3; x = 2$

Express each of the sentences in Problems 25–50 symbolically. See Examples 4–8.

25. The sum of 3 and some number is 11.
26. The sum of 7 and some number is 43.
27. The sum of -2 and some number is -6.
28. Three times a number is -7.
29. Four times a number is 72.
30. If a number is subtracted from 10, the difference is 3.
31. If six is subtracted from a number, the difference is 15.
32. The quotient of a number divided by 2 is -5.
33. If some number is divided by 7, the quotient is 12.
34. The product of five times a number minus three is 12.
35. The difference of five times a number minus three is 12.
36. Increasing a number by 12 is the same as four times the number.
37. Increasing a number by 12 is the same as four times another number.
38. If the product of 3 and a number is divided by 2, the result is 5 times another number.
39. If the product of 3 and a number is divided by 2, the result is five times that number.
40. Five times a number added to 7 is equal to three times a number subtracted from 4.
41. If the sum of a number and 7 is divided by 2, the result is 11 times the number.
42. The sum of the ages of a son and a daughter is 21.
43. A man's daughter's age is half his son's age.
44. The perimeter of a rectangular garden is 140 ft. (Use length and width as the variables.)
45. The width of a garden is 10 ft less than twice the length.
46. The length of a garden is 2 ft more than three times the width.
47. The circumference of a circle is three times the radius plus .7 ft.
48. A certain electrical current is equal to 110 divided by the resistance.
49. The dosage (in milligrams) of a certain medicine for a child is equal to the child's age plus 2, divided by the child's age times 50.
50. The volume of a sphere is equal to the product of $\frac{4}{3}$, π, and the square of the radius.

3.2 Addition Property of Equality

In this section we will introduce a procedure for solving equations. We begin with some simple examples that you can probably solve by inspection, but remember that we are building a *procedure* by which we can solve more complicated problems. Thus, the roots to equations such as $x = 5$, or $y + 2 = 10$ are obvious: they are 5 and 8, respectively. However, if you are given a more complicated equation, such as

$$5x + 3 - 4x + 5 = 6 + 9$$

the answer may not be so obvious. Therefore, the general procedure for solving an equation is to replace it with an ***equivalent equation*** that is easier to solve than the original equation. *Equivalent equations are equations with identical solutions.*

Example 1. The following equations are equivalent, because $x = 5$ is the only solution for each of them.

$$x + 4 = 9 \qquad 7x = 35 \qquad 2x - 3 = 7$$

How do we replace a given equation by a simpler equivalent equation? We set a goal. If the unknown variable is x, we wish to isolate this variable on one side of the equals sign:

Because equality means a relationship between quantities that are exactly the same, we can visualize an equation in terms of a two-pan balance scale.

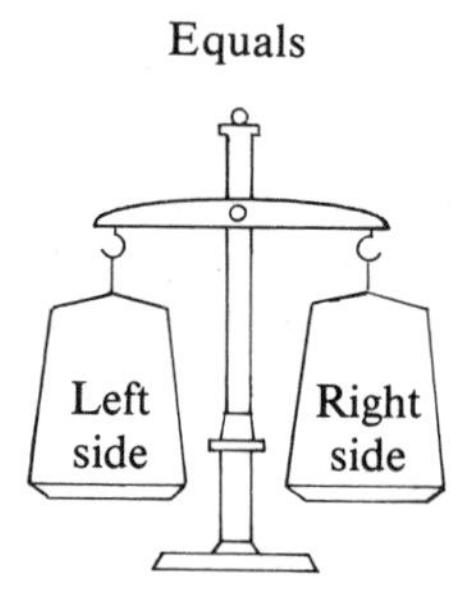

The scale is balanced if the left and right sides are equal.

The basic property we use in solving equations says that we can write an equivalent equation by *doing the same thing* (adding, subtracting, multiplying, or dividing by a nonzero number) to both sides of the equation. That is, the scale remains balanced as long as we do to one side of the scale exactly what we do to the other side. For example, if

$$x = 5$$

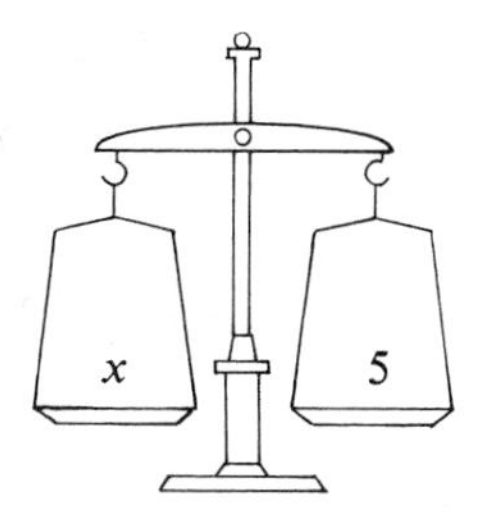

The scale is balanced if the left and right sides are equal.

then

$$x + 10 = 5 + 10$$

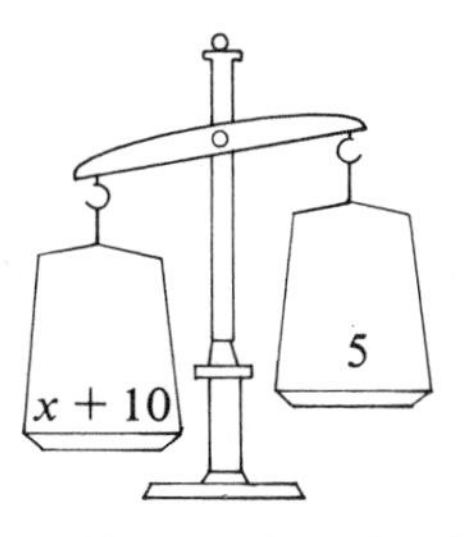

Add 10 to the left side.

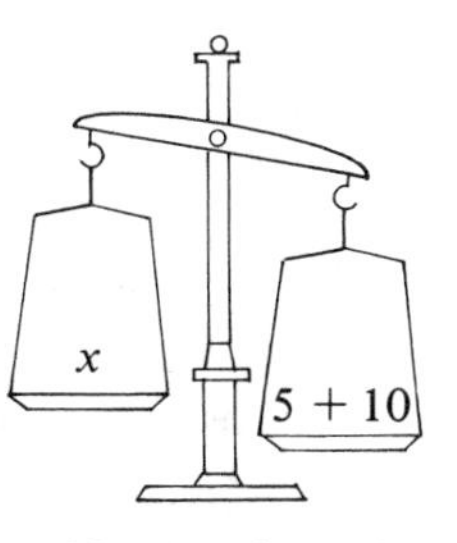

Add 10 to the right side.

Add 10 to both sides.

Also

$$x - 3 = 5 - 3$$

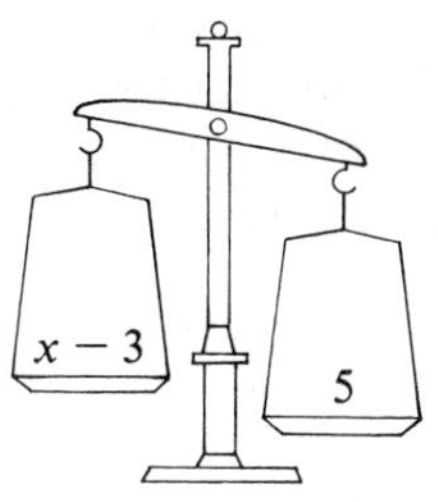

Subtract 3 from the left side.

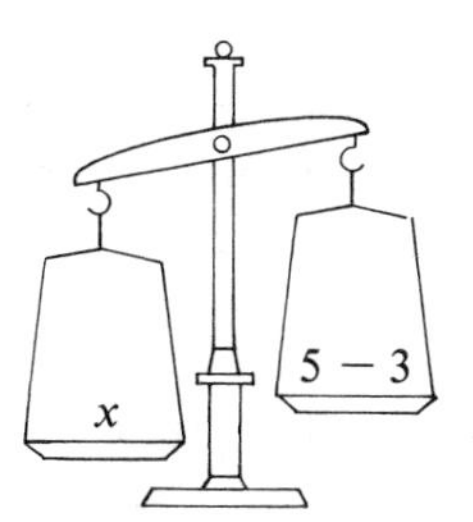

Subtract 3 from the right side.

Subtract 3 from both sides.

■

This operation can be summarized by the ***Addition Property of Equality.***

Addition Property of Equality

> If $a = b$, then $a + c = b + c$. Any number (positive or negative) can be added to both sides of an equation to obtain an equivalent equation.

Your ***goal*** in solving equations is to ***isolate the variable.*** You can do this by remembering that the sum of ***opposites*** is 0.

*Solve the following given equations. To **solve** an equation means to find all of its roots.*

Example 2. $x + 15 = 25$

Solution The *opposite* of 15 is -15, so add -15 to both sides.

$$x + 15 - \mathbf{15} = 25 - \mathbf{15}$$
$$x = 10$$
■

Example 3. $x - 36 = 42$

Solution The *opposite* of -36 is 36, so add 36 to both sides.

$$x - 36 + \mathbf{36} = 42 + \mathbf{36}$$
$$x = 78$$
■

Example 4. $-108 = y - 92$

Solution The variable may be isolated on either the left or the right side. Add 92 to both sides.

$$-108 + \mathbf{92} = y - 92 + \mathbf{92}$$
$$-16 = y$$
■

Example 5. $52 = 14 + x$

Solution

$$52 - \mathbf{14} = 14 + x - \mathbf{14} \qquad \textit{Add } -14 \textit{ to both sides.}$$
$$38 = x$$
■

The first step in learning to solve equations more difficult than those in this section is deciding which equality property to use and learning to do the arithmetic step in your head. It is important to do the arithmetic, not just put down the expected answer, but your work will be greatly simplified if it looks like Examples 6–8.

Solve the following equations.

Example 6. $x + 7 = -56$ *Copy the original equation.*
$x = -63$ *Mentally add -7 to both sides.* ■

Example 7. $y + 41 = 5$
$y = -36$ *Mentally add -41 to both sides.* ■

Example 8. $11 = z - 5$
$16 = z$ *Mentally add 5 to both sides.* ■

Sometimes it is necessary to combine similar terms before using the addition property. Consider the equation we stated earlier in this section.

$$5x + 3 - 4x + 5 = 6 + 9 \qquad \textit{Combine similar terms.}$$
$$x + 8 = 15 \qquad \textit{Add } -8 \textit{ to both sides.}$$
$$x = 7 \qquad \textit{The root is 7.}$$

Once you have a root, you might want to check your answer. Remember that a root is a value that satisfies the equation. If you substitute a root in the original equation, it should give a true statement. Check $x = 7$:

$$5x + 3 - 4x + 5 = 6 + 9$$ *This is the original equation.*

$$5(7) + 3 - 4(7) + 5 = 6 + 9$$

$$35 + 3 - 28 + 5 = 6 + 9$$ *Multiplication before addition and subtraction.*

$$15 = 15 \checkmark$$ *This is true, so $x = 7$ checks.*

Example 9. Solve and check.

$$6x + 3 - 5x + 7 = 11 + 2$$ *Combine similar terms.*

$$x + 10 = 13$$

$$x = 3$$ *The root is 3.*

Check: $6(3) + 3 - 5(3) + 7 = 11 + 2$

$$18 + 3 - 15 + 7 = 11 + 2$$

$$13 = 13 \checkmark$$

■

Problem Set 3.2

Find the root of the equations of Problems 1–32. See Examples 2–8.

1. $x + 5 = 11$
2. $y - 7 = 10$
3. $x - 10 = 6$
4. $y + 9 = 18$
5. $x - 14 = 3$
6. $x - 12 = 6$
7. $w + 23 = 5$
8. $t + 18 = 4$
9. $p - 18 = -52$
10. $x - 4 = -24$
11. $y - 7 = -50$
12. $y - 3 = -16$
13. $q - 5 = -23$
14. $5 + w = 8$
15. $12 + t = 8$
16. $7 + s = 2$
17. $8 + u = 23$
18. $6 = m - 6$
19. $5 = p + 19$
20. $3 = 5 + t$
21. $8 = 14 + u$
22. $43 + w = 0$
23. $112 + t = 0$
24. $15 + x = 8$
25. $11 + y = 5$
26. $4 + s = 1$
27. $13 + v = 6$
28. $3 = x + 5$
29. $-7 = y + 2$
30. $4 = s - 5$
31. $6 = k - 3$
32. $12 = m + 15$

Find the root of the equations in Problems 33–50. Example 9.

33. $6x + 2 - 5x = 12$
34. $-3q - 4 + 4q = 5$
35. $5y + 6 - 4y = 8$
36. $7t + 8 - 6t = -4$
37. $6x + 2 - 5x + 3 = 2(1 + 3)$
38. $4 - 3w + 8 + 4w = 6 + 11(1 + 1)$
39. $2y + 6 - y + 8 = 3(4 + 2)$
40. $5 - 6z + 4 + 7z = 5 + 6(8 - 7)$
41. $8 - 5y + 6 + 2y + 4y - 4 = 2 + 3 \cdot 4$
42. $4x + 6 - 3x = 5 \cdot 6 - 2$
43. $12 - 3z + 3 + 6z - 2z - 5 = 3 + 2 \cdot 5$
44. $6 - 8m + 5 - m + 3 + 10m = 6 + 2 \cdot 3$
45. $5t - 10 - 4t - 12 = 10 - 40$
46. $6z + 12 - 5z - 4 = -20$
47. $6 - 3x + 8 + 4x = 2 + 3 \cdot 4$
48. $5 - 2x - 4 + 3x = 1 + 2 \cdot 3$
49. $12 + 3x - 15 - 2x = 4 + 2 \cdot 5$
50. $7 + 8x - 10 - 7x = 3 + 2 \cdot 5$

3.3 Multiplication and Division Properties of Equality

Suppose we need to solve an equation in which the coefficient of the variable is not 1:

$$3x = 12 \qquad 5x = 10 \qquad \frac{1}{2}x = 4$$

The addition property is not sufficient for solving these equations. ***Our goal is still to isolate the variable,*** but now we need two additional properties to achieve this goal. Examples 1–4 will lead us to the first of those properties.

Examples *Simplify each of the following expressions.*

1. $\frac{3x}{3}$ *If some number, x, is multiplied by 3 and then the result is divided by 3, the result is the original number x. Thus,* $\frac{3x}{3} = x$

2. $\frac{5x}{5} = x$ **3.** $\frac{19y}{19} = y$ **4.** $\frac{-12q}{-12} = q$ ■

This process of simplifying expressions is used with the ***Division Property of Equality.***

Division Property of Equality

> If $a = b$, then $\frac{a}{c} = \frac{b}{c}$, $c \neq 0$. Both sides of an equation may be divided by any nonzero number to obtain an equivalent equation.

Find the roots of the given equations.

Example 5. $2x = 6$

Solution

$\frac{2x}{2} = \frac{6}{2}$ *Divide both sides by 2.*

$x = 3$ *Perform the indicated operation.* ■

Example 6. $-3x = 9$

Solution

$\frac{-3x}{-3} = \frac{9}{-3}$ *Divide both sides by* -3.

$x = -3$ *Perform the indicated arithmetic.* ■

If the varible has been divided by some number, say 2, as in

$$\frac{x}{2}$$

multiply both sides by 2 to isolate the variable. As before, you need to understand a certain simplification process as illustrated in Examples 7–10.

Examples *Simplify each expression.*

7. $\left(\frac{x}{2}\right)(2)$ *If some number, x, is divided by 2 and then the result is multiplied by 2, the result is the original number, x. Thus,* $\left(\frac{x}{2}\right)(2) = x$

8. $\left(\frac{x}{7}\right)(7) = x$ 9. $\left(\frac{y}{43}\right)(43) = y$ 10. $\left(\frac{z}{-5}\right)(-5) = z$ ■

To solve an equation in which some number is dividing the variable, you use the following property.

Multiplication Property of Equality

> If $a = b$, then $ac = bc$, $c \neq 0$. You can multiply both sides of an equation by the same nonzero number to obtain an equivalent equation.

Find the roots of the given equations.

Example **11.** $\frac{x}{2} = 6$

Solution

$$\left(\frac{x}{2}\right)\mathbf{2} = 6(\mathbf{2}) \qquad \textit{Multiply both sides by 2.}$$

$$x = 12 \qquad \textit{Perform the indicated operations.}$$ ■

Example **12.** $\frac{x}{-3} = \frac{1}{3}$

Solution

$$\left(\frac{x}{-3}\right)(\mathbf{-3}) = \frac{1}{3}(\mathbf{-3})$$

$$x = -1$$ ■

Sometimes you have an equation solved for the opposite of the variable. The procedure is to multiply both sides by -1.

Solve the given equations.

Example **13.** $-x = 12$

Solution

$$(-x)(\mathbf{-1}) = 12(\mathbf{-1})$$

$$x = -12$$ ■

Example **14.** $3 - x = 12$

Solution

$$3 - x - \mathbf{3} = 12 - \mathbf{3} \qquad \textit{Add } -3 \textit{ to both sides.}$$

$$-x = 9 \qquad \textit{Simplify.}$$

$$(-x)(\mathbf{-1}) = 9(\mathbf{-1}) \qquad \textit{Multiply both sides by } -1.$$

$$x = -9 \qquad \textit{Simplify.}$$ ■

Remember, the text examples show you *how* to solve an equation. *Your work* for Example 14 might look like the following:

$$\begin{aligned} 3 - x &= 12 \\ -x &= 9 \\ x &= -9 \end{aligned}$$

If an equation has mixed operations, it is generally best to isolate the single term involving the variable before multiplying or dividing both sides.

Find the roots of the given equations.

Example 15. $2x + 3 = 6$

Solution

$2x + 3 - \mathbf{3} = 6 - \mathbf{3}$ *Add -3 to both sides.*

$2x = 3$ *Perform the indicated arithmetic.*

$\frac{2x}{\mathbf{2}} = \frac{3}{\mathbf{2}}$ *Divide both sides by 2.*

$x = \frac{3}{2}$ *Perform the indicated arithmetic.* ■

Example 16. $\frac{x}{2} + 3 = 5$

Solution

$\frac{x}{2} = 2$ *Subtract 3 from both sides, and simplify.*

$x = 4$ *Multiply both sides by 2, and simplify.* ■

Problem Set 3.3

Simplify the expressions in Problems 1–9. See Examples 1–4 and 7–10.

1. $\frac{6x}{6}$
2. $\frac{-4t}{-4}$
3. $\frac{-8m}{-8}$
4. $\frac{125p}{125}$
5. $\left(\frac{y}{3}\right)3$
6. $\left(\frac{s}{-2}\right)(-2)$
7. $\left(\frac{z}{-14}\right)(-14)$
8. $\frac{-k}{-1}$
9. $\left(\frac{t}{-1}\right)(-1)$

Solve the equations in Problem 10–42. See Examples 5, 6, 11, and 12.

10. $4x = 12$
11. $3y = 15$
12. $7w = 14$
13. $5t = 30$
14. $2t = 50$
15. $4w = 52$
16. $-2x = 10$
17. $-y = -12$
18. $-z = 14$
19. $14 - x = 12$
20. $41 - y = -60$
21. $13 - y = 3$
22. $6 - x = 0$
23. $12 - y = 0$
24. $0 = 15 - m$
25. $-40 = x - 92$
26. $-137 = y - 49$
27. $0 = z - 112$
28. $\frac{x}{2} = 5$
29. $\frac{y}{3} = 4$
30. $\frac{z}{4} = 8$

31. $\frac{a}{-3} = 2$ **32.** $\frac{b}{-5} = 8$ **33.** $\frac{c}{-8} = 12$

34. $8 = \frac{p}{12}$ **35.** $-12 = \frac{k}{5}$ **36.** $-15 = \frac{t}{8}$

37. $11 = \frac{n}{-4}$ **38.** $16 = \frac{q}{-12}$ **39.** $\frac{-x}{3} = 10$

40. $\frac{x}{-3} = 10$ **41.** $\frac{-d}{-4} = -16$ **42.** $\frac{-p}{-2} = -6$

Solve the equations in Problems 43–60. See Examples 13 and 14.

43. $4x + 5 = 9$ **44.** $3x - 2 = 7$ **45.** $5x - 4 = 6$
46. $2x - 3 = 6$ **47.** $6x + 5 = 4$ **48.** $11x - 8 = 5$
49. $10 = 3x + 1$ **50.** $5 = 3 - 2x$ **51.** $6 = 10 - 2x$

52. $\frac{m}{2} + 1 = 4$ **53.** $\frac{n}{3} - 5 = 2$ **54.** $\frac{p}{6} + 4 = 1$

55. $\frac{z}{4} + 5 = 0$ **56.** $\frac{w}{2} + 5 = 0$ **57.** $\frac{u}{3} - 4 = 0$

58. $4 - 3x = -11$ **59.** $3 - 5x = -17$ **60.** $11 - 2x = 3$

3.4 Solving Equations

In this section we will summarize the procedures introduced earlier in this chapter. Our goal is to isolate the variable on one side of the equation. We do this by adding, subtracting, multiplying, or dividing both sides of the equation by the same number. If the variable appears on both sides of the equation, we use the addition property to obtain an equivalent equation that has all the variables on one side.

Example 1. Solve $5x + 2 = 4x - 7$.

Solution

$$5x + 2 - \mathbf{4x} = 4x - 7 - \mathbf{4x}$$
$$x + 2 = -7 \quad \textit{Now the variable is on the one side only.}$$
$$x + 2 - \mathbf{2} = -7 - \mathbf{2}$$
$$x = -9$$

■

Sometimes it is necessary to combine similar terms or use the distributive property to remove parentheses before carrying out the steps shown in Example 1.

Example 2. Solve $(-1)(x - 3) = -2(x + 9)$

Solution

$$(-1)(x - 3) = -2(x + 9)$$
$$-x + 3 = -2x - 18$$
Use the distributive property to remove the parentheses.
$$x + 3 = -18$$
Add 2x to both sides to isolate the variable on one side.
$$x = -21$$
Subtract 3 from both sides. ■

We can now summarize the procedure for solving simple first-degree equations.

Solution of First-Degree Equations

STEP 1.	Use the distributive property to eliminate parentheses in the problems. Then combine similar terms.
STEP 2.	Use the addition property of equality to obtain an expression with a single term involving the unknown on one side.
STEP 3.	Use the multiplication property of equality to isolate the variable on one side of the equation.

Solve the following equations. To ***solve*** *an equation means to find all possible roots.*

Example **3.** $3(m + 4) + 5 = 5(m - 1) - 2$

Solution

$3m + 12 + 5 = 5m - 5 - 2$ *Eliminate parentheses.*
$3m + 17 = 5m - 7$ *Combine similar terms.*
$24 = 2m$ *Use addition property.*
$12 = m$ *Use multiplication property.* ■

Example **4.** $2(t + 5) - 7 = 6 + 2(t + 1)$

Solution

$2t + 10 - 7 = 6 + 2t + 2$ *Eliminate parentheses.*
$2t + 3 = 8 + 2t$ *Combine similar terms and subtract 2t from both sides.*
$3 = 8$ *Simplify.*

Because there is no value of the variable for which $3 = 8$, we say that there is *no solution* or *no root* for the given problem. Do not get caught believing that all equations must have roots. ■

Example **5.** $5(s - 3) + 29 = 4 + 5(s + 2)$

Solution

$$5s - 15 + 29 = 4 + 5s + 10$$
$$5s + 14 = 14 + 5s$$
$$14 = 14$$

Because this is always true, regardless of the value of the variable, we say that *all values for the variable make this equation true.* ■

In most of the previous examples in this section we have included all of the steps explaining what is happening in the problem. Although this procedure shows you how to work the problem, your work will not look like our examples. The next two examples illustrate how your work might look when you are solving equations.

Solve the equations.

Example **6.** $5(y + 2) + 6 = 3(2y) + 5(3y)$

$$5y + 10 + 6 = 6y + 15y$$
$$5y + 16 = 21y$$
$$16 = 16y$$
$$1 = y$$

■

Example 7.
$$5x(7-3)+15=14x+x-5$$
$$20x+15=15x-5$$
$$5x=-20$$
$$x=-4$$
■

Problem Set 3.4

Solve the equations in Problems 1–61. See Examples 1–7.

1. $4(x+1)=4$
2. $3(x-2)=15$
3. $2(x-5)=30$
4. $4(n+1)+3=15$
5. $110(4)=40(m+4)$
6. $8(50-v)=5(40)$
7. $5x+50x=110$
8. $3x+x=20{,}000$
9. $3y+y=24$
10. $5p-610=-1090$
11. $5y-505=-100$
12. $150z=200(300)$
13. $2w+3w+4w=117$
14. $5m+6m+3m=42$
15. $12n-3n-10n=39$
16. $w+(w+50)=320$
17. $s+(s+1)=125$
18. $s+(s+1)+(s+2)=57$
19. $5u+10(u+12)=735$
20. $10t+8(t+3)=294$
21. $25(2u)+10u=840$
22. $5m+10(2m)+3(2m)=93$
23. $2n+2(3n-5)=54$
24. $5x+3(x-6)=6$
25. $2x+3=x+5$
26. $4y-7=3y+2$
27. $3w+4=4w-5$
28. $2t+1=3t+8$
29. $4m-11=3m+2$
30. $6p+4=7p-3$
31. $5n-2=6n+7$
32. $3k+7=4k-11$
33. $9c-21=10c-40$
34. $3v-5=v+1$
35. $3z+2=z+6$
36. $5t-8=3t+2$
37. $4x-3=2x+1$
38. $5y-6=2y+3$
39. $6z-2=3z-11$
40. $5s-3=7s+5$
41. $3t+4=5t-6$
42. $2u-3=4u-1$
43. $10z=6z+52$
44. $5(I+3)-6(I+5)=0$
45. $2(K-3)-2(K-12)=18$
46. $(-1)(z+2)=4-2z$
47. $(-1)(x-3)=10-2x$
48. $(-1)(y-4)=5-2y$
49. $6-3w=(-2)(w-3)$
50. $6-3d=(-4)(d+5)$
51. $6-5u=(-6)(u+1)$
52. $5(t-2)-4(t+3)=10-40$
53. $6(H-2)-4(H+3)=10-42$
54. $4(s+2)+7(s+1)=10s-7$
55. $3(t-4)-3(t-5)=t+3$
56. $6(z+2)=4+5(z-4)$
57. $3(v+1)-2(v-5)=6+v$
58. $5(g-3)+2(4-g)=3g-7$
59. $4(6C-81)=-3(4+5C)$
60. $5E+3(E+2)+(E+4)+17=0$
61. $5F+2(F+1)+(F+3)-5=0$

3.5 Literal Equations

In Section 1.4 we introduced the notion of a formula, and listed several of the more important ones with which you should be familiar. Because these formulas are made up of several letters, they are sometimes also called ***literal equations.***

Working with literal equations is important because you often need to solve formulas for different letters, not only in mathematics, but in business and science as well. However, it is also important to us at this time because it tests our understanding of the principles of the last two sections. When working with literal equations, we carry out the same steps that were explained in the last two sections while treating one letter as the unknown and the other letters as known quantities.

Example 1. Solve for the variable C.

$$P = S - C$$

Recall from Section 1.4, this is the formula for profit where $P =$ profit, $S =$ selling price, and $C =$ cost.

Solution Our goal is to isolate the C on one side.

$$P = S - C$$
$$P + \mathbf{C} = S - C + \mathbf{C} \qquad \textit{Add C to both sides.}$$
$$P + C = S$$
$$P + C - \mathbf{P} = S - \mathbf{P} \qquad \textit{Subtract P from both sides.}$$
$$C = S - P$$

■

Example 2. Solve $d = rt$ for t. This is the formula where $d =$ distance, $r =$ rate, and $t =$ time.

Solution

$$d = rt$$
$$\frac{d}{\mathbf{r}} = \frac{rt}{\mathbf{r}} \qquad \textit{Divide both sides by r.}$$
$$\frac{d}{r} = \mathbf{t}$$

■

Example 3. Solve for the capital letter:

$$v = k + \mathbf{G}t$$

Solution

$$v - k = k + \mathbf{G}t - k \qquad \textit{Subtract k from both sides.}$$
$$v - k = \mathbf{G}t$$
$$\frac{v-k}{t} = \frac{\mathbf{G}t}{t} \qquad \textit{Divide both sides by t.}$$
$$\frac{v-k}{t} = \mathbf{G}$$

■

A frequently encountered equation in algebra involves two variables x and y. You must be able to solve for y as illustrated by Examples 4–9.

Solve for y.

Example 4. $y - mx = b$

Solution

$$y = mx + b \qquad \textit{Add mx to both sides.}$$

NOTE: This could be written as $y = b + mx$, but it is customary to write the x-term first.

■

Example 5. $y - 5x = 6$

Solution $y = 5x + 6$ *Add 5x* to both sides. ■

Example 6. $3x + y = 8$

Solution $y = -3x + 8$ *Subtract 3x* from both sides.

NOTE: This could be written as $8 - 3x$, but we are following the custom described in Example 4. ■

Example 7. $6x + 3y = 12$

Solution

$3y = -6x + 12$ *Subtract 6x from both sides.*

$y = -2x + 4$ *Divide both sides by 3.* ■

Example 8. $2x - y = 3$

Solution

$-y = -2x + 3$ *Subtract 2x from both sides.*

$y = 2x - 3$ *Divide (or multiply) both sides by −1.* ■

Example 9. $6x - 2y = 4$

Solution

$-2y = -6x + 4$ *Subtract 6x from both sides.*

$y = 3x - 2$ *Divide both sides by −2.* ■

Literal equations can occur in many applications, as illustrated by Example 10. You will not be expected to derive the equation, but given the equation, you should be able to solve for the other variables.

Example 10. If you know that the circumference of the earth is about 24,000 mi at the equator, what is the radius? The formula for the circumference is $C = 2\pi r$. Solve for r and then answer the question.

Solution

$C = 2\pi r$ *Given.*

$\frac{C}{2\pi} = \frac{2\pi r}{2\pi}$ *Divide both sides by 2π.*

$\frac{C}{2\pi} = r$

If $C = 24{,}000$ then $r = \frac{24{,}000}{2\pi}$

$$r = \frac{12{,}000}{\pi}$$

The approximate answer can be found on a calculator:

[12000] [÷] [π] [=]

DISPLAY: 3819.718634

The answer is approximately 3800 mi. ■

Problem Set 3.5

In Problem 1–15 solve for the variable that is capitalized. See Examples 1–3.

1. $a + \mathbf{B} = c$
2. $c = d + \mathbf{E}$
3. $x + \mathbf{Y} = 7$
4. $x - \mathbf{Y} = 3$
5. $3x - \mathbf{Y} = 5$
6. $aw + \mathbf{S} + t$
7. $m = rt + \mathbf{P}$
8. $v = \mathbf{K} + gt$
9. $\frac{\mathbf{D}}{t} = r$
10. $\frac{d}{t} = \mathbf{R} + k$
11. $d = \mathbf{R}t$
12. $p = 2l + 2\mathbf{W}$
13. $c = 2\pi\mathbf{R}$
14. $i = \mathbf{P}rt$
15. $\mathbf{X} + 3a = 2y + a$

Solve for y in Problems 16–36. See Examples 4–9.

16. $y - 2x = 3$
17. $y - 6x = 5$
18. $y - 9x = 7$
19. $3x + y = 4$
20. $2x + y = 7$
21. $2x + y = -4$
22. $3x + y + 2 = 0$
23. $3x + y + 8 = 0$
24. $5x + y + 1 = 0$
25. $4x - y + 5 = 0$
26. $2x - y - 3 = 0$
27. $x - y - 5 = 0$
28. $9x + 3y + 6 = 0$
29. $6x + 3y - 12 = 0$
30. $4x + 2y + 8 = 0$
31. $8x - 4y - 12 = 0$
32. $2x - 2y - 100 = 0$
33. $6x - 3y - 21 = 0$
34. $9x - 3y - 6 = 0$
35. $4x - 2y - 8 = 0$
36. $10x - 5y + 15 = 0$

Solve the literal equations as directed in the applied Problems 37–47. See Example 10.

37. The formula for the circumference, C, of a circle with radius r is $C = 2\pi r$. Solve for r to find a formula for the radius.

38. The formula for the simple interest, I, of an investment of p dollars at r percent interest for t years is $I = prt$. Solve for p to find a formula for the principal.

39. The formula for the current, I, in an electrical circuit is $I = \frac{k}{R}$, where R is the resistance and k is some constant. Solve for R.

40. The formula for the current, I, in an electrical circuit is $I = \frac{kE}{R}$, where E is the electromotive force, R is the resistance, and k is some constant. Solve for E.

41. The formula $W = Fd$ relates the amount of work, W, by a force F moving through a distance d. Solve for F.

42. The pressure, P, exerted by a liquid at a given point is $P = kd$, where d is the depth at that point below the surface of a liquid and k is some constant. Solve for k.

43. The frequency of vibration, f, of air in an open pipe organ is $f = \frac{k}{l}$, where l is the length of pipe and k is some constant. Solve for l.

44. The centripetal force, C, of a body moving in a circular path at a constant speed is $C = \frac{k}{r}$, where r is the radius of the path and k is some constant. Solve for r.

45. The displacement, d, of a V-8 engine is $D = 4\pi b^2 s$, where s is the stroke and b is the bore of the engine. Solve for s.

46. The amount of heat, H, put out by an electrical appliance is $H = \pi tRI^2$, where R is the resistance, I is the current, and t is the time. Solve for R.

47. Suppose you fit a band tightly around the earth at the equator. You wish to raise the band so that it is uniformly supported 6 ft above the earth at the equator.

a. Guess how much extra length would have to be added to the band (not the supports) to do this.
b. Calculate the amount of extra material that would be needed.

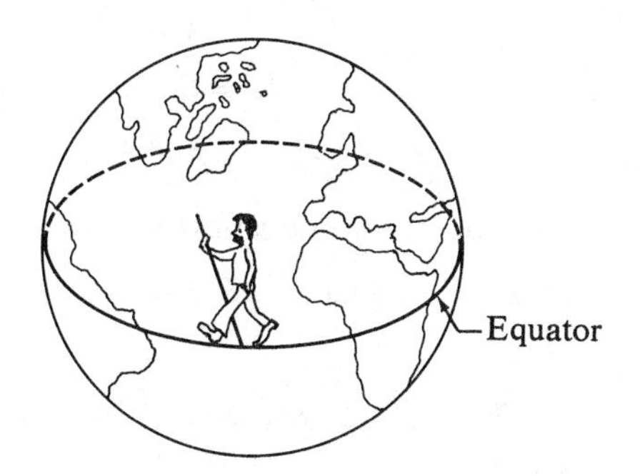

3.6 Word Problems

Our goal in algebra is to apply certain techniques to find solutions to a wide variety of problems. In previous sections we have been studying the techniques of solving equations. In this section we will introduce a *technique* for solving word problems. *Eventually* we want to be able to solve word problems to obtain *answers,* but when beginning, we must necessarily start with simple problems in which the answers are obvious or trivial. You must, therefore, keep in mind that in this section we are seeking not answers, but a strategy for attacking word problems.

Many students say, "I can do algebra, except for the word problems." If you cannot apply algebra to word problems, you cannot do algebra. It is therefore imperative that you make the necessary effort in working the verbal problems, not only in this section, but throughout the book. For that reason, we have aleady provided a considerable amount of practice in translating from words to algebra. This section should be a natural extension of the problems you have already solved.

The first type of problem involves number relationships.

Example 1. If you add 10 to twice a number, the result is 22. What is the number?

Solution *Step 1* Read the problem carefully. Make sure you know what is given and what is wanted.

Step 2 Write a verbal description of the problem using operation signs and an equal sign, but still using key words.

$$10 + 2(\text{A NUMBER}) = 22$$

Step 3 If there is a single unknown, choose a variable.

Let n = A NUMBER

Step 4 Replace the verbal phrase by the variable. This is called **substitution.**

$$10 + 2n = 22$$

Step 5 Solve the equation and check the solution in the original problem to see if it makes sense.

$$\begin{aligned} 10 + 2n &= 22 \\ 2n &= 12 \\ n &= 6 \end{aligned}$$

Check: add 10 to twice 6 and the result is 22.

Step 6 State the solution to a word problem in words.

The number is 6. ■

The second type of problem involves consecutive numbers. If n = A NUMBER, then

THE NEXT CONSECUTIVE NUMBER = $n + 1$, and
THE THIRD CONSECUTIVE NUMBER = $n + 2$.

Also, if E = an even number, then

$E + 2$ = the next even number

or if F = an odd numer, then

$F + 2$ = the next consecutive odd number.

Example **2.** Find three consecutive integers whose sum is 42.

Solution *Step 1* Read the problem.

Step 2 Write down a verbal description of the problem using operation signs and an equal sign.

(INTEGER) + (NEXT INTEGER) + (THIRD INTEGER) = 42

Step 3 Choose a variable.

Let g = INTEGER
$g + 1$ = NEXT INTEGER
$g + 2$ = THIRD INTEGER

Step 4 Substitute the variable into the verbal equation

$$g + (g + 1) + (g + 2) = 42$$

Step 5

$$\begin{aligned} g + (g + 1) + (g + 2) &= 42 \\ 3g + 3 &= 42 \\ 3g &= 39 \\ g &= 13 \end{aligned}$$

Check: $13 + 14 + 15 = 42$

Step 6 State the solution to a word problem in words.

The integers are 13, 14, and 15. ■

Notice that we are illustrating a *procedure* with these examples. Let us summarize this approach before considering further examples.

1. *Read the problem.* Note what it is all about. Focus on processes rather than numbers. You cannot work a problem you do not understand.
2. *Restate the problem.* Write down a verbal description of the problem using operations signs and an equal sign. Look for equality. If you cannot find equal quantities, you will never formulate an equation.

Symbol	Verbal Description
$=$	is equal to; equals; are equal to; is the same as; is; was; becomes; will be; results in
$+$	plus; the sum of; added to; more than; greater than; increased by
$-$	minus; the difference of; the difference between; is subtracted from; less than; smaller than; decreased by; is diminished by
$\cdot$	times; product; is multiplied by
$\div$	divided by; quotient of

3. *Choose a variable.* If there is a single unknown, choose a variable.
4. *Substitute.* Replace the verbal phrase by the variable.
5. *Solve the equation.* This is the easy step. Be sure your answer makes sense by checking it with the original question in the problem.
6. *State an answer.* There were no variables defined when you started, so $x = 3$ is not an answer. Pay attention to units of measure and other details of the problem. Remember to answer the question that was asked.

Applications involving area and perimeter are also common types of word problems. Recall from Section 1.4 that for a rectangle,

$$\text{AREA} = (\text{LENGTH})(\text{WIDTH}) \quad \text{and} \quad \text{PERIMETER} = 2(\text{LENGTH}) + 2(\text{WIDTH}).$$

Example 3. Find the dimensions of a rectangle with a perimeter of 16 ft if the length is 1 less than twice the width.

Solution *Step 1* Read the problem.
Step 2 $\text{PERIMETER} = 2(\text{LENGTH}) + 2(\text{WIDTH})$

Step 3 Let $\text{WIDTH} = w$

$$\left.\begin{aligned} \text{LENGTH} &= 2w - 1 \\ \text{PERIMETER} &= 16 \end{aligned}\right\}$$

This information comes directly from the problem.

Step 4 $16 = 2(2w - 1) + 2w$ *Substitute the values from Step 3 into Step 2.*

Step 5 $16 = 4w - 2 + 2w$

$16 = 6w - 2$

$18 = 6w$

$3 = w$

so $2w - 1 = 2(3) - 1$

$= 5$

Check: $P = 2(5) + 2(3)$

$= 10 + 6$

$= 16$

Step 6 ***The dimensions are 3 ft by 5 ft.*** ■

Many additional types of word problems will be considered later.

Problem Set 3.6

Solve Problems 1–30. Because you are practicing a procedure, *you must show all your work. See Examples 1–3.*

1. If you add 7 to twice a number, the result is 17. What is the number?
2. If you subtract 12 from twice a number, the result is 6. What is the number?
3. If you add 15 to twice a number, the result is 7. What is the number?
4. If you subtract 9 from three times a number, the result is 0. What is the number?
5. If you multiply a number by 5 and then subtract −10, the difference is −30. What is the number?
6. If you multiply a number by 4 and then subtract −4, the result is 0. What is the number?
7. The sum of a number and 24 is equal to four times the number. What is the number?
8. The sum of 6 and twice a number is equal to four times the number. What is the number?
9. If 12 is subtracted from twice a number, the difference is four times the number. What is the number?
10. If 6 is subtracted from three times a number, the difference is twice the number. What is the number?
11. Find two consecutive numbers whose sum is 117.
12. Find two consecutive even numbers whose sum is 94.
13. Find three consecutive numbers whose sum is 54.
14. The sum of three consecutive numbers is 105. What are the numbers?
15. The sum of four consecutive numbers is 74. What are the numbers?
16. The sum of two consecutive even numbers is 30. What is the larger number?
17. The sum of two consecutive odd numbers is 48. What is the smaller number?
18. Find two consecutive odd numbers whose sum is 424.
19. The sum of two consecutive even integers is 2 more than twice the smaller integer. What are the integers?
20. The sum of two consecutive even integers is 2 less than twice the smaller integer. What are the integers?
21. What is the length of a rectangular lot whose perimeter is 750 m and whose width is 75 m?
22. Find the dimensions of a rectangle with a perimeter of 54 cm if the length is 5 less than three times the width.

23. The perimeter of ΔABC is 117 in. Find the lengths of the sides.

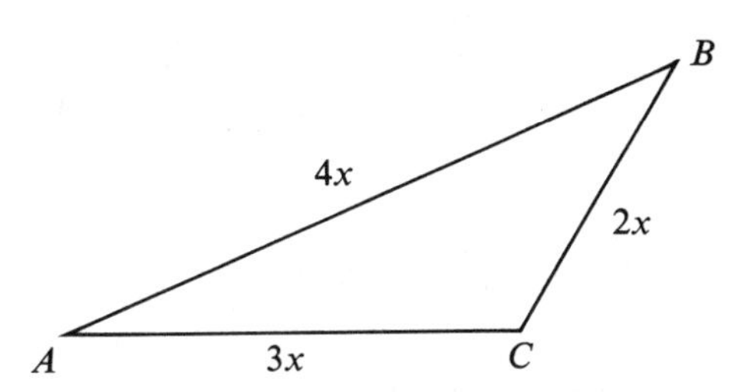

24. Mr. C. D. Money is making out his will and wants to divide some property between his two sons. He has one parcel that is 300 m by 200 m. He wishes to divide up another piece of land that is 150 m wide so that it has the same area as the first parcel. How long should the second parcel be?

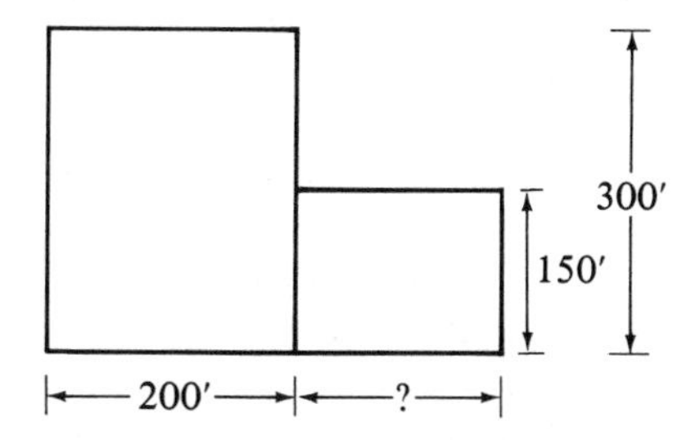

25. A rectangular field is 2 m narrower than it is long. What is the area if the width is 48 m?

26. A house and a lot are appraised at $112,200. If the house is worth five times the value of the lot, how much is the lot worth?

27. A cabinet shop produced two types of custom-made cabinets for a customer. If one cabinet cost four times the cost of the other, and the total price for one of each type of cabinet is $2075, how much does each cabinet cost?

28. What is the cost of installing a hardwood floor in a rectangular hallway whose length is 5 ft less than nine times its width? The perimeter is 70 ft and the cost is $10 per sq ft.

29. The largest angle of a triangle is twice the smallest angle, and the third angle is 20° less than the largest. How large is each angle? (NOTE: The sum of the measures of the angles in any triangle is 180°.)

30. The pressure (P) under an object depends on the amount of force applied (F) and the area under the object (A); the relationship is given by

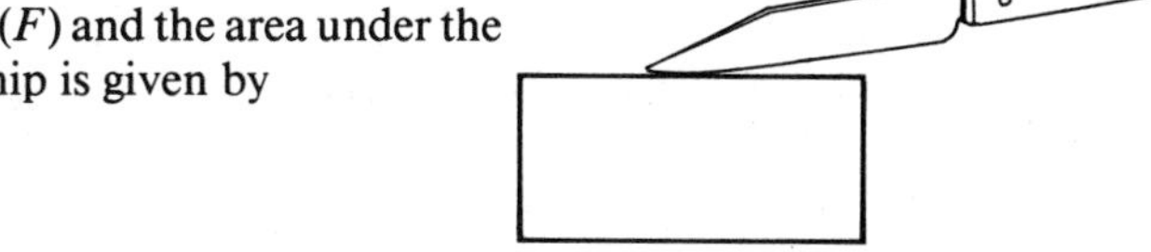

$$P = \frac{F}{A}$$

If the blade on a knife is 2 in. long and is sharpened to a width of .001 in., find the pressure under the blade where a force of 4 lb is applied to the blade.

In psychology, IQ is measured according to the formula

$$IQ = \frac{MA}{CA}(100)$$

where MA = mental age and CA = chronological age. Use this information in Problems 31–34.

31. If a child has a chronological age of 15 and a mental age of 18, what is the child's IQ?
32. If a child has a chronological age of 10 and a mental age of 12.8, what is the child's IQ?
33. If a 9-year-old has a superior intelligence (IQ of 132), what is the child's mental age?
34. If a 6-year-old has superior intelligence (IQ of 132) what is the child's mental age?

3.7 Solving Inequalities

The techniques of the previous sections can be applied to quantities that are not equal. If we are given any two numbers x and y, then obviously

$$\text{either } x = y \quad \text{or} \quad x \neq y$$

If $x \neq y$, then

$$\text{either } x < y \quad \text{or} \quad x > y$$

Comparison Property*

For any two numbers x and y, exactly one of the following is true.

1. $x = y$; x is equal to y (the same as)
2. $x > y$; x is greater than y (bigger than)
3. $x < y$; x is less than y (smaller than)

This means that if two quantities are not exactly equal, then we can relate them with a greater-than or a less-than symbol. The solution of

$$x < 3$$

has more than one value, and it becomes very impractical to write "The answers are 2, 1, −110, 0, $2\frac{1}{2}$, 2.99, ..." Instead, relate the answer to a number line, as shown in Figure 3.1. The fact that 3 is not included in the solution set is indicated by a circle at the point 3.

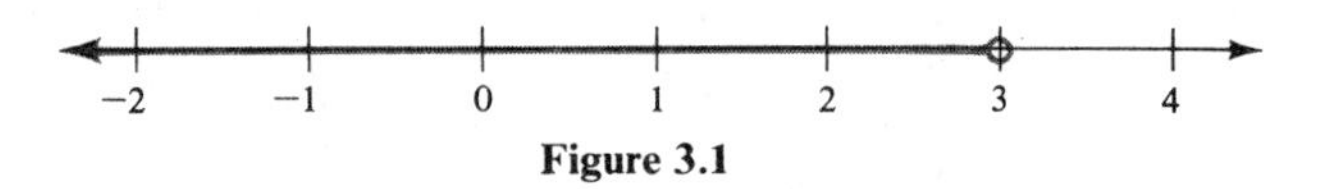

Figure 3.1

Graph the following solution sets.

Example **1.** $x \leq 3$

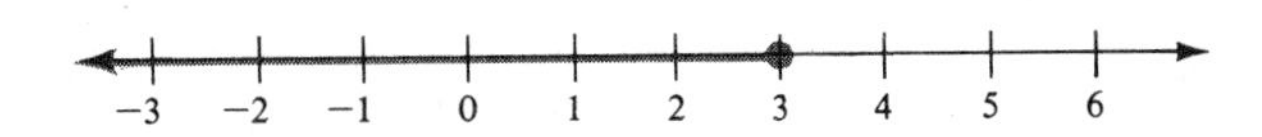

Notice that in this example the endpoint is included because with $x \leq 3$ it is possible that $x = 3$. This is shown as a solid dot on the number line. ■

* Sometimes this is called the Trichotomy Property.

Example 2. $x > 5$

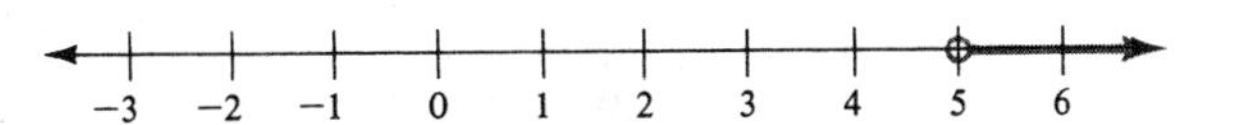

■

Example 3. $-2 \leq x$
Although you can graph this directly, you will have less chance of making a mistake when working with inequalities if you rewrite the inequalities so that the variable is on the left. That is, reverse the inequality to read $x \geq -2$. Notice that the order of the inequality is changed. This is because the symbol requires that the arrow always point to the smaller number. That is, if $2 \leq 5$, then $5 \geq 2$. Now graph $x \geq -2$ on the number line as shown.

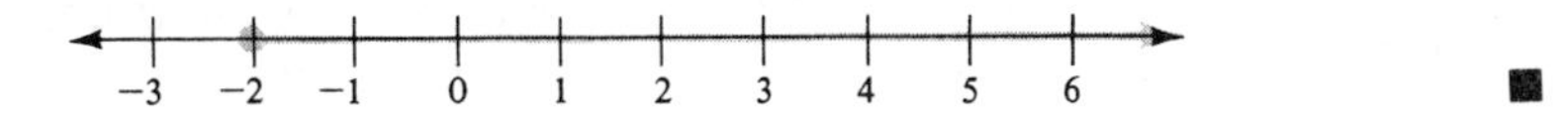

■

In terms of a number line, if $x < y$, then "x is to the left of y." Suppose the coordinates x and y are plotted as shown in Figure 3.2.

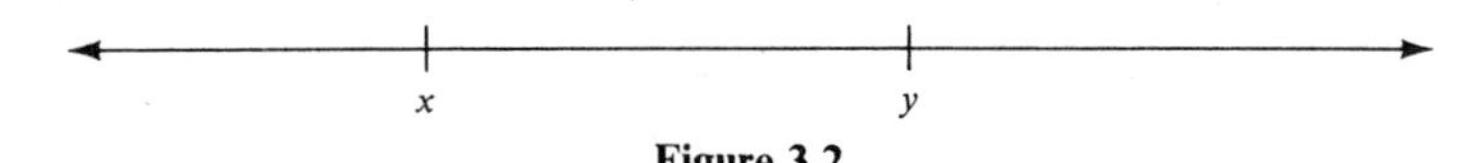

Figure 3.2

If you add 2 to both x and y, you obtain $x + 2$ and $y + 2$. From Figure 3.3, you see that $x + 2 < y + 2$.

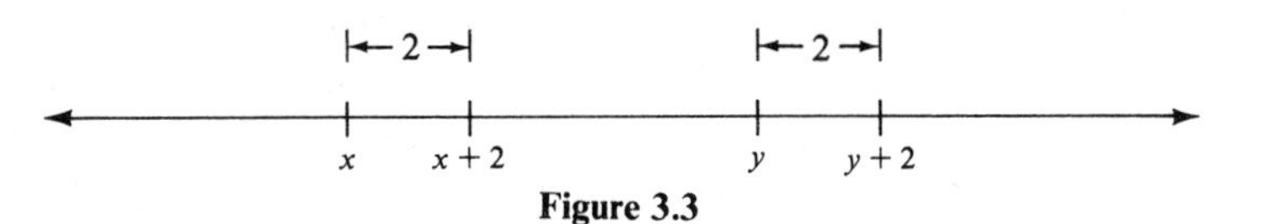

Figure 3.3

If you add some number c, there are two possibilities:

$c > 0$ ($c > 0$ is read "c is positive.")
$c < 0$ ($c < 0$ is read "c is negative.")

If $c > 0$, then $x + c$ is still to the left of $y + c$, as shown in Figure 3.4.

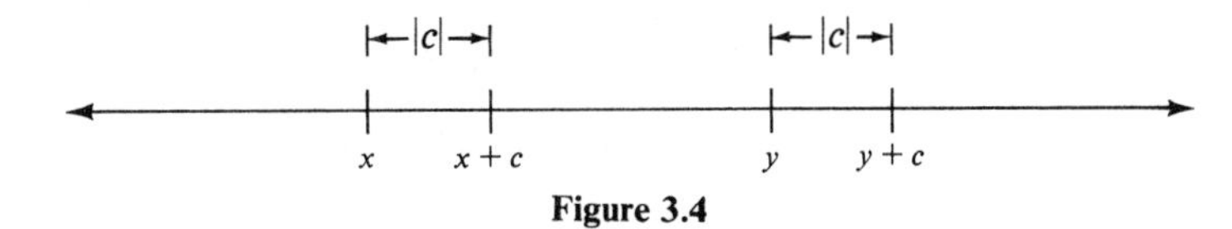

Figure 3.4

If $c < 0$, then $x + c$ is still to the left of $y + c$, as shown in Figure 3.5.

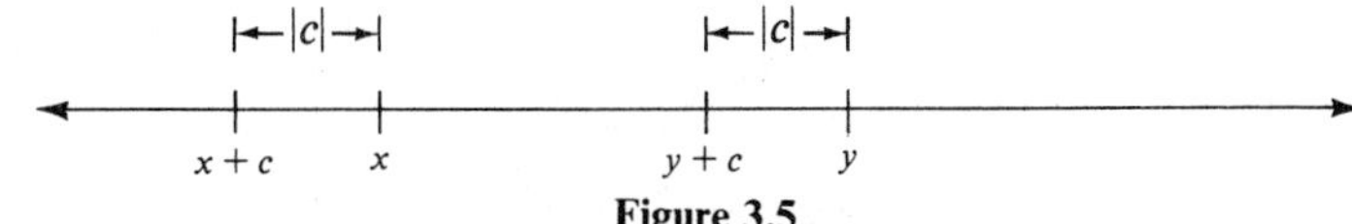

Figure 3.5

In both cases, $x < y$, which justifies the following property.

Addition Property of Inequality

If $x < y$, then

$$x + c < y + c$$

Also, if $x \leq y$, then $x + c \leq y + c$
if $x > y$, then $x + c > y + c$
if $x \geq y$, then $x + c \geq y + c$

Because this ***Addition Property of Inequality*** is essentially the same as the addition property of equality, you might expect that there is also a multiplication property of inequality. We would hope that we could multiply both sides of an inequality by some number c without upsetting the inequality. Consider some examples. Let $x = 5$ and $y = 10$ so that $5 < 10$.

Let $v = 2$: $5 \cdot 2 < 10 \cdot 2$
$10 < 20$ *True*

Let $c = 0$: $5 \cdot 0 < 10 \cdot 0$
$0 < 0$ *False*

Let $c = -2$: $5(-2) < 10(-2)$
$-10 < -20$ *False*

You can see that you cannot multiply both sides of an inequality by a constant and be sure that the result is still true. However, if you restrict c to a positive value, then you can multiply both sides of an inequality by c. On the other hand, if c is a negative number, then the order of the inequality should be reversed. This is summarized by the ***Multiplication Property of Inequality.***

Multiplication Property of Inequality

Positive multiplication ($c > 0$)

If $x < y$, then

$cx < cy$ ← **order unchanged**

Also for $c > 0$,

if $x \leq y$, then $cx \leq cy$
if $x > y$, then $cx > cy$
if $x \geq y$, then $cx \geq cy$

Negative multiplication ($c < 0$)

If $x < y$, then

$cx > xy$ ← **order reversed**

Also for $c < 0$,

if $x \leq y$, then $cx \geq cy$
if $x > y$, then $cx < cy$
if $x \geq y$, then $cx \leq cy$

The same properties hold for positive and negative division. We can summarize with the following statement.

Solution of First-Degree Inequalities

The procedure for solving first-degree inequalities is the same as the procedure for solving first-degree equations except that, if you multiply or divide by a negative, you reverse the order of the inequality.

In summary, given $x < y, x \le y, x > y$, or $x \ge y$, the inequality symbols are the *same* if we

1. add the same number to both sides;
2. subtract the same number from both sides;
3. multiply both sides by a positive number;
4. divide both sides by a positive number.

} This works the way equations do.

The inequality symbols are *reversed* if we

1. multiply both sides by a negative number;
2. divide both sides by a negative number;
3. reverse the x and the y.

} This is where inequalities differ from equations.

Solve and graph the following inequalities.

Example 4. $5x - 3 \ge 7$

Solution

$$5x - 3 + \mathbf{3} \ge 7 + \mathbf{3}$$
$$5x \ge 10$$
$$\frac{5x}{\mathbf{5}} \ge \frac{10}{\mathbf{5}}$$
$$x \ge 2$$

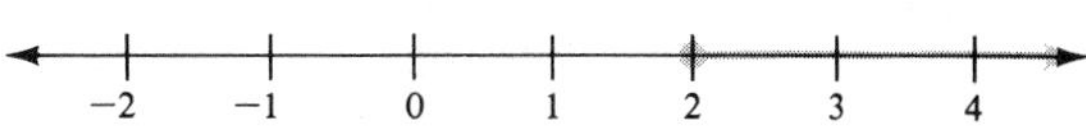

■

Example 5. $-x \ge 2$

Solution Multiply both sides by -1 and remember to reverse the order of inequality:

$$x \le -2$$

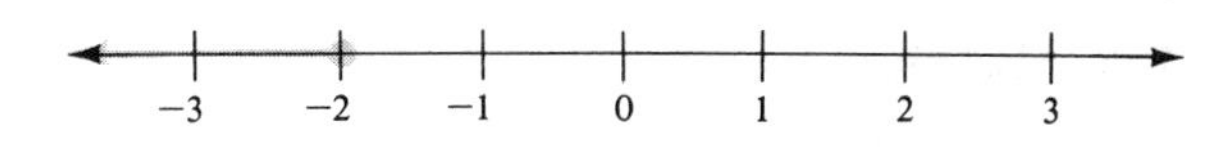

■

Example 6. $2(3t + 4) < 5t + 2$

Solution

$$6t + 8 < 5t + 2$$
$$6t + 8 - \mathbf{2} < 5t + 2 - \mathbf{2}$$
$$6t + 6 < 5t$$
$$6t + 6 - 6t < 5t - 6t$$
$$6 < -t$$
$$6(\mathbf{-1}) > (-t)(\mathbf{-1}) \quad \textit{Reverse the order.}$$
$$-6 > t$$

or

$$t < -6$$

−8 −7 −6 −5 −4 −3 −2

■

Problem Set 3.7

Graph the solution sets in Problems 1–9. See Examples 1–3.

1. $x < 5$
2. $x \geq 6$
3. $x \geq -3$
4. $x \leq -2$
5. $4 \geq x$
6. $-1 < x$
7. $-5 \leq x$
8. $1 > x$
9. $-3 > x$

Solve and graph the solutions to the inequalities in Problems 10–30. See Examples 4–6.

10. $x + 7 \geq 3$
11. $x - 2 \leq 5$
12. $x - 2 \geq -4$
13. $10 < 5 + y$
14. $-4 < 2 + y$
15. $-3 < 5 + y$
16. $2 > -s$
17. $-t \leq -3$
18. $-m > -5$
19. $5 \leq 4 - y$
20. $3 > 2 - x$
21. $5 \leq 1 - w$
22. $2x + 6 \leq 8$
23. $3y - 6 \geq 9$
24. $4s + 3 > s + 9$
25. $s + 2 < 2s + 7$
26. $3a + 4 \leq a + 2$
27. $4b - 3 > 2b - 13$
28. $4a - 7 > 3(a + 1)$
29. $6(2v - 1) \leq 3(v + 4)$
30. $2(4 - 3x) > 4(3 - x)$
31. Suppose seven times a number is added to 35 and the result is positive. What are the possible numbers?
32. Suppose five times a number is subtracted from 15 and the result is negative. What are the possible numbers?
33. An architect is designing a facing for a building that uses both equilateral triangles and squares. If the side of the triangle is 2 ft longer than the side of the square, and the perimeter of the square is greater than the perimeter of the triangle, what are the possible lengths for the side of the square?
34. Current postal regulations state that no package may be sent if its combined length, width, and height exceed 72 in. What are the possible dimensions of a box to be mailed with equal height and width if the length is four times the height?

3.8 Review

In this chapter we introduced a main idea in the study of algebra. We have presented the techniques necessary to solve first-degree equations. If you master the techniques developed in this chapter, you will not only be able to apply algebra to a wide variety of topics, but will also have a firm foundation for the remainder of the course.

Self-Test

[3.1] **1.** Determine whether the given value is a root of the equation.
a. $-x = -3$; $x = -3$
b. $6x - 3 = 27$; $x = 5$
c. $15 - x = 18$; $x = 3$
d. $2(x + 3) + 4(2 - 3x) = 64$; $x = 5$

Solve the equations or inequalities for Problems 2–6. Also graph the solutions of inequalities.

[3.2] **2.** **a.** $x + 7 = -10$
b. $3 + y = 3$
c. $-5 = 14 - z$
d. $6 + 3m - 10 - 2m = 2 + 3 \cdot 4$

[3.3] **3.** **a.** $-18x = 90$ **b.** $-y = 12$

c. $10 = 12 - z$ **d.** $\frac{t}{-12} = -4$

[3.3] **4.** **a.** $2x + 5 = 1$ **b.** $6y - 1 = 11$

c. $3 = 5 - 2x$ **d.** $\frac{u}{3} - 5 = 0$

[3.4] **5.** **a.** $5x - 2 = 3x + 4$ **b.** $(-1)(y - 2) = 4 - 2y$

c. $2(a + 2) - 2(a + 4) = 3$ **d.** $3(g - 3) + 2(5 - g) = 2g - 7$

[3.7] **6.** **a.** $2p + 3 \geq 7$ **b.** $-3y \geq 0$

c. $5x - (2x + 6) < 0$ **d.** $-5 > 7 - 2y$

[3.5] **7.** **a.** Solve $A = bh$ for b.

b. Solve $3x - y + 2 = 0$ for y.

c. Solve $6x + 2y + 12 = 0$ for y.

d. Solve $A = b + (n + 2)p$ for n.

[3.6] **8.** The sum of twice a number and 6 is equal to the number. What is the number?

[3.6] **9.** The sum of three consecutive odd integers is 8 more than twice the first integer. What is the middle integer?

[3.6] **10.** The scale on a floor plan is drawn 1 in. to 4 ft. If the perimeter of a room on the plan is 24 in. and the length of the room is twice the width, what are the dimensions of the room?

Review Problems

Solve the equations and inequalities in Problems 1–81.

1. $x - 5 = 8$ **2.** $y - 8 = -4$ **3.** $z - 2 = 14$

4. $t - 9 = 6$ **5.** $s + 9 = 10$ **6.** $x + 5 = 19$

7. $y + 8 = 2$ **8.** $z + 5 = 5$ **9.** $6 + x = 15$

10. $19 + y = 5$ **11.** $-m = 4$ **12.** $-n = 6$

13. $12 - p = 6$ **14.** $5 - x = 6$ **15.** $6 = w - 3$

16. $5 = u + 7$ **17.** $6x = 18$ **18.** $2y = 14$

19. $5t = 20$ **20.** $-3t = 6$ **21.** $3m = -15$

22. $0 = 5k$ **23.** $-6 = -2h$ **24.** $40 = -4x$

25. $\frac{m}{6} = 4$ **26.** $\frac{n}{-2} = 5$ **27.** $\frac{p}{16} = 5$

28. $\frac{y}{3} = -42$ **29.** $2x + 5 = 9$ **30.** $3x - 4 = 11$

31. $5x - 2 = 23$ **32.** $4x + 2 = 18$ **33.** $2 - 5y = 12$

34. $5 - 3y = 23$ **35.** $6 + 2y = -10$ **36.** $4 + 6y = -20$

37. $5 = 7 - 2z$ **38.** $12 = 3 - 9z$ **39.** $18 = 5z + 18$

40. $12 = 12 - 3z$ **41.** $\frac{w}{2} + 2 = 6$ **42.** $6 - \frac{w}{3} = 1$

43. $5 - \frac{w}{4} = 3$ **44.** $\frac{w}{5} - 2 = 0$ **45.** $-12 = -4 - 8p$

46. $-14 = -2 - 6p$ **47.** $-62 = -6 - 7p$ **48.** $-43 = -3 - 5p$

49. $6(x - 2) = 6$ **50.** $4(x - 1) = 12$ **51.** $3(x + 5) = 24$

52. $5y + 2(y + 1) = 16$ **53.** $5y + 3(y - 6) = 6$ **54.** $9y + 4(y - 5) = -7$

55. $3z - (z - 1) = 15$ **56.** $4z - 2(z + 1) = 0$ **57.** $5z - 3(z - 2) = 14$

58. $12 - (w + 3) = 9$ **59.** $16 - (2w + 6) = 0$ **60.** $10 - (3w - 5) = 0$

61. $3x + 5 = 2x + 1$
62. $5x + 3 = 4x - 1$
63. $4x - 5 = 3x - 3$
64. $2x - 3 = 5x + 6$
65. $3x + 1 = 8x - 4$
66. $4x - 3 = 8x + 5$
67. $5 - 3x = 7 - 2x$
68. $6 - 5x = 5 - 4x$
69. $8 - x = 12 - 2x$
70. $5x - 1 < 4$
71. $3x - 2 \geq 7$
72. $4x - 5 \leq 15$
73. $-5y > 0$
74. $-2y \leq 0$
75. $-8y \geq 0$
76. $6 < -3p$
77. $4 \leq -2p$
78. $21 > -7p$
79. $-7 > 5 - 2z$
80. $-8 \leq -3z$
81. $-3 < 7 - 5z$

Solve each equation in Problems 82–100 for y.

82. $2x + y = 6$
83. $3x + y = 5$
84. $4x + y = 8$
85. $5x - y = -3$
86. $7x - y = -8$
87. $6x - y = 3$
88. $x + 2y = 6$
89. $3x + 2y = 10$
90. $4x + 5y = 20$
91. $3x - 2y = 18$
92. $5x - 4y = 20$
93. $2x - 3y = 21$
94. $4x + 5y - 15 = 0$
95. $3x - 2y + 8 = 0$
96. $5x - 4y + 12 = 0$
97. $x - 3 = 3(y - 2)$
98. $x - 6 = 2(y + 1)$
99. $3(x - 5) = 2(y + 1)$
100. $2(x - 1) = 3(y + 1)$

Polynomials

How often have I said to you that
when you have eliminated the impossible
whatever remains, however improbable,
must be the truth?

Sherlock Holmes
in *Sign of the Four*
by Sir Arthur Conan Doyle

4.1 Addition and Subtraction of Polynomials

We have already worked with a variety of expressions in algebra—that is, expressions that contained numbers, variables, operations, and symbols of inclusion, such as parentheses, brackets, and braces. But they have all been just algebraic expressions. Now we want to classify these expressions more carefully so that they can be referred to easily. The first classification we discussed is similar terms. A ***term*** is an algebraic expression that contains numbers and variables and the operation of multiplication only. Two similar terms are combined into a single term by addition, but only multiplication is indicated in the term. For example, each of the following is a term: $3x$, xy, $2x^2$, $5xy^2z$, y, 4, 3 and could be rewritten as: $3 \cdot x$, $x \cdot y$, $2 \cdot x \cdot x$, $5 \cdot x \cdot y \cdot y \cdot z$, $1 \cdot y$, $4 \cdot 1$, $3 \cdot 1$. Notice that each term is a number, a variable, or a ***product*** of a number of factors, both constants and variables. The number of variable factors is called the ***degree*** of term, and the numerical factors are multiplied to find the ***numerical coefficient*** of the term.

Examples *Find the degree and numerical coefficient of each of the given terms.*

	Term	Degree	Numerical Coefficient
1.	$3x$	1	3
2.	$5x^2$	2	5
3.	$6x^2yz$	4	6
4.	x^3	3	1
5.	y	1	1
6.	4	0	4

■

We have often taken several terms and combined them into a single expression by adding or subtracting:

$$3x + 2x, \qquad 3x + 2y, \qquad 3x + 2y - z$$

We gave no particular name to these new expressions. They were just terms, numbers, or multinumbers—or polynumbers or polynomials. The word ***polynomial,*** then, simply means many numbers or many terms. But it is actually one term or the sum or difference of terms. Because every subtraction can be written as a sum, we can simply say a sum of terms.

Polynomial

> A *polynomial* is a term or the sum of terms.

Because the majority of algebraic expressions with which we deal are polynomials, it would be convenient to classify these expressions further in order to talk about them more easily. First, polynomial means one or more terms. We can further classify the polynomial by the number of terms it has.

Name	*Number*	*Examples*
Monomial	One term	$x,\ 3y,\ 4,\ 7x^2,\ 5x^2y^3z$
Binomial	Two terms	$x + 3y,\ 7x^2 - 4,\ 5x^2y^2 + 2z^3$
Trinomial	Three terms	$x^2 + 2x + 1,\ 3x + y - 2z,\ 5x^2 - 7y^3 + xy$

Quadnomial, quintnomial (or pentanomial), or hexanomial *could* be used to describe polynomials with four, five, or six terms, respectively. This classification is based on addition.

We also classify a polynomial according to the degree of its terms.

Degree of a Polynomial

The *degree of a polynomial* is the largest degree of any of its terms.

Examples *Classify each of the following by type and degree.*

	Solutions
7. $32x^2$	**7.** monomial, degree 2
8. $4x^2 + 3x - 2$	**8.** trinomial, degree 2
9. $x^5 - ab$	**9.** binomial, degree 5
10. 139	**10.** monomial, degree 0
11. $\frac{2}{x}$	**11.** not a polynomial (involves division by a variable)
12. $\frac{x}{2}$ or $\frac{1}{2}x$	**12.** monomial, degree 1 $\left(\text{coefficient } \frac{1}{2}\right)$
13. $.5x - .2y - .01$	**13.** trinomial, degree 1
14. xyz	**14.** monomial, degree 3 ■

To combine similar terms, we added and subtracted polynomials. Let us review the procedures with some examples.

Examples *Simplify by combining similar terms.*

15. $(3x + 2y) + (5x - 6y) + 4x$
$= 3x + 5x + 4x + 2y - 6y$
$= 12x - 4y$

Mentally associate similar terms when doing addition.

16. $(a + 2b) + (3a - 4b) = a + 3a + 2b - 4b$
$= 4a - 2b$

17. $(3x^2 - 1) + (2x + 4) + (x^2 + x + 1) = 3x^2 - 1 + 2x + 4 + x^2 + x + 1$
$= 4x^2 + 3x + 4$ ■

You must take a little more care when subtracting polynomials. The opposite of a number is -1 times that number: the opposite of b is $-b = (-1)b$. This concept allows you to change a subtraction to an addition as shown by Example 18.

Example **18.** $(2x + 5) - (x + 7)$ *This step is usually a mental step.*

$= (2x + 5) + (-1)(x + 7)$

$= 2x + 5 - x - 7$ *Notice that subtracting a polynomial is the same*

$= x - 2$ *as subtracting each term.* ■

By relying on addition and multiplication, you gain the advantage of the commutative and associative properties of those operations. If you are adding, terms may be put in any order and grouped in the most convenient way. One of the most common errors in beginning algebra is incorrectly subtracting polynomials, so study each of the following examples very carefully.

Examples *Simplify by combining similar terms.*

19. $(5x - 3) - (3x - 9) = 5x - 3 - 3x + 9$
$= 2x + 6$

20. $(2x + 5) - (3x + 8) = 2x + 5 - 3x - 8$
$= -x - 3$

21. $(2x^2 - 5x + 3) - (x^2 + 8x - 6) = 2x^2 - 5x + 3 - x^2 - 8x + 6$
$= x^2 - 13x + 9$

22. $(3x^2 - 1) - (2x + 4) - (x^2 + x + 1) = 3x^2 - 1 - 2x - 4 - x^2 - x - 1$
$= 2x^2 - 3x - 6$ ■

Notice that it is customary to write a polynomial in the order of terms with decreasing degree.

Earlier in the course the distributive property was used to change products to sums. For example,

$$2(x + y) = 2x + 2y$$
$$-6(3x - 4y) = -18x + 24y$$

If these polynomials are combined as in

$$2(x + y) - 6(3x - 4y)$$

you must first change the products to sums, and then combine similar terms. However, you must be very careful with the signs. First *think*

$$2(x + y) + (-6)(3x - 4y)$$

and write

$$2(x + y) - 6(3x - 4y) = 2x + 2y - 18x + 24y$$
$$= -16x + 26y$$

If terms are the same degree then it is customary to list them in alphabetical order according to their variables.

Examples *Simplify.*

23. $2(a + b) + 5(a - b) = 2a + 2b + 5a - 5b$
$= 7a - 3b$

24. $3(s-2t)-4(2s-3t) = 3s-6t-8s+12t$
$= -5s+6t$

25. $4(2m-3n)-4(m-6n) = 8m-12n-4m+24n$
$= 4m+12n$ ■

Problem Set 4.1

For Problems 1–3, give the degree and numerical coefficient of each term. See Examples 1–6.

1. **a.** $3x^4$ **b.** $5x^2$ **c.** $8x^5$ **d.** $123x$ **e.** 46
2. **a.** $2x^6$ **b.** $12x$ **c.** $4x^3$ **d.** -23 **e.** x
3. **a.** $8x$ **b.** $7x^3$ **c.** $-4x$ **d.** 10 **e.** $-5x^3$

For Problems 4–6, identify the polynomial as a monomial, binomial, or trinomial, if possible, and find the degree of the polynomial. See Examples 7–14.

4. **a.** x^2-4 **b.** $12{,}345$ **c.** $23hk^2-29$
5. **a.** y^3+z **b.** $a-b-c$ **c.** xyz
6. **a.** $abcd$ **b.** $x+y+z$ **c.** $13hk+17hk^2$

Simplify by combining similar terms in Problems 7–16. See Examples 15–17.

7. $(2a+3b)+(5a-6b)$
8. $(h-2k)+(3h-5k)$
9. $(m-n)+(2m-3n)$
10. $(p+q)+(3p-5q)$
11. $(2s-t)+(s-3t)+5t$
12. $(3x+2y)+(x-3y)+5y$
13. $(x-4y)+(3x+5y)+2x$
14. $(4x-y)+(x-3y)+6y$
15. $(2x^2-1)+(3x-4)+(x^2-3x+3)$
16. $(3y^2+4)+(2y-5)+(y^2-5y+7)$

Simplify by combining similar terms in Problems 17–28. See Examples 18–22.

17. $(r-2s)-(3r-5s)$
18. $(2s-t)-(s-3t)$
19. $(3x+2)-(2x+1)$
20. $(5x+3)-(3x-5)$
21. $(y-3)-(2y+4)$
22. $(y-4)-(3y-5)$
23. $(3z-5)-(2z+3)$
24. $(5z+3)-(3z+5)$
25. $(3x^2-2x+4)-(x^2+2x+1)$
26. $(2x^2+3x-6)-(3x^2-x+3)$
27. $(w^2+4w-1)-(3w^2-4w-1)$
28. $(5p^2-3p+4)-(7p^2+3p-4)$

Simplify the expressions in Problems 29–54. See Examples 23–25.

29. $(x+y)-2(x-y)$
30. $3(y+2)-2(3+y)$
31. $5(b-4)-4(4-b)$
32. $7(c-1)-2(c-6)$
33. $2(h+2)-3(h+3)$
34. $6(k+5)-2(3-k)$
35. $6(3-x)-5(x+8)$
36. $2(m-2n)-4(3m-n)$
37. $4(3x-1)-2(2x+1)$

38. $5(3y + 2) - 3(2y - 3)$
39. $2(3x + 2) - (2x + 1) + 2(x - 3)$
40. $3(x + 2) - (2x + 1) - 3(x - 3)$
41. $7(5 - m) - (7 - 5m) - 6(m - 2)$
42. $2(3 - n) - (1 - 3n) - 5(n - 2)$
43. $3(s - 2) - 2(3s - 4) + (s + 3)$
44. $2(4 - 2t) - (t - 3) - 3(4 - t)$
45. $2(x - 5) - 3(6 - 2x) - (10 - 2x)$
46. $3(x + 3) - 4(x - 5) - (5 - 7x)$
47. $3x - 2 - (2x - 1) + x - 3$
48. $3x - 2 - 2(x - 1) - (x - 3)$
49. $3x - 2 - (2y - 1) - 3y$
50. $3v - 2 - 2(w - 1) - (v - 3w)$
51. $(2x^2 - 1) - (3x + 4) - (x^2 + 2x - 1)$
52. $2x^2 - 1 - (3x + 4 - x^2) + (2x - 1)$
53. $(2x^2 - 1) - 3x + 4 - (x^2 - 2x - 1)$
54. $2x^2 - 1 - (3x + 4 - x^2) - (2x - 1)$

4.2 Multiplication of Monomials and Laws of Exponents

Recall the definition of exponents given in Chapter 1:

Definition of Exponent

b^1 means b, and for counting numbers greater than 1,

$$b^n \text{ means } \underbrace{b \cdot b \cdot b \cdot \cdots b}_{n \text{ factors}}$$

b is called the ***base,*** and n is called the ***exponent.*** The exponent is the number of times the base is used as a ***factor.***

In Section 1.3 we stated a ***First Law of Exponents.***

First Law of Exponents

$b^m \cdot b^n = b^{m+n}$ where m and n are positive integers. To multiply two numbers with like bases, add the exponents.

We can now show that this is true in general.

$$\begin{aligned} b^m \cdot b^n &= \underbrace{(b \cdot b \cdot b \cdot \cdots b)}_{m \text{ factors}}\underbrace{(b \cdot b \cdot b \cdot \cdots b)}_{n \text{ factors}} \\ &= \underbrace{b \cdot b \cdot b \cdot \cdots b \cdot b \cdot b \cdot b \cdot \cdots b}_{m+n \text{ factors}} \\ &= b^{m+n} \end{aligned}$$

Examples *Write each product in exponent form.*

1. $7^5 \cdot 7^4 = 7^{5+4}$
$= 7^9$

2. $(-1)(-1)^2 = (-1)^{1+2}$
$= (-1)^3$
or $(-1)(-1)(-1) = (-1)(+1)$
$= -1$

3. $(216)(36)(1296) = 6^3 \cdot 6^2 \cdot 6^4$
$= 6^{3+2+4}$
$= 6^9$

4. $t^2 \cdot t^5 \cdot t^3 = t^{2+5+3}$
$= t^{10}$

5. $(2rs^3)(3r^2s) = 2 \cdot 3rr^2s^3s$
$= 6r^{1+2}s^{3+1}$
$= 6r^3s^4$

6. $(hk^2)(h^2k^4 + hk^3 + 1) = (hk^2)(h^2k^4) + (hk^2)(hk^3) + (hk^2)(1)$
$= h^3k^6 + h^2k^5 + hk^2$

7. $2xy^2(3x - x^2y) + 3x^2y(y + xy^2) = 2xy^23x - 2xy^2x^2y + 3x^2yy + 3x^2yxy^2$
$= 6x^2y^2 - 2x^3y^3 + 3x^2y^2 + 3x^3y^3$
$= 6x^2y^2 + 3x^2y^2 - 2x^3y^3 + 3x^3y^3$
$= 9x^2y^2 + x^3y^3$ ■

The terms ab^2, $(ab)^2$, a^2b, a^2b^2 all seem to be very similar, but let us express each without its exponents.

$$abb, \quad \underbrace{abab, \quad aab, \quad aabb}$$
(abab and aabb: similar terms)

Only two of the terms are equivalent. The exponent applies only to the variable to which it is affixed, unless parentheses are used. Hence, $ab^2 \neq (ab)^2$, which suggests a ***Second Law of Exponents.***

$$(ab)^n = \underbrace{ab \cdot ab \cdot ab \cdots ab}_{n \text{ factors of } ab}$$
$$= \underbrace{a \cdot a \cdot a \cdots a}_{n \text{ factors}} \cdot \underbrace{b \cdot b \cdot b \cdots b}_{n \text{ factors}}$$
$$= a^nb^n$$

Second Law of Exponents

> $(ab)^n = a^nb^n$ where n is a positive integer. To raise a product to a power, raise each factor to that power.

Examples *Simplify.*

8. $(2x)^3 = 2^3x^3$
$= 8x^3$

Note: Compare this with $2x^3$ which is simplified.

9. $(-x)^3 = [(-1)x]^3$ *This step would be done mentally.*
$= (-1)^3x^3$
$= -x^3$

10. $(-x)^4 = [(-1)x]^4$
$= (-1)^4x^4$
$= x^4$ *Note: Compare this with $-x^4$ which is simplified.*

11. $(2mn)^2(2mn^2) = 4m^2n^2(2mn^2)$
$= 4 \cdot 2m^2mn^2n^2$
$= 8m^3n^4$ ■

Suppose that the quantity being raised to a power is itself raised to some power.

$$(3^2)^3 = 3^2 \cdot 3^2 \cdot 3^2 = 3 \cdot 3 \cdot 3 \cdot 3 \cdot 3 \cdot 3 = 3^6$$
$$(2^4)^2 = 2^4 \cdot 2^4 = 2 \cdot 2 \cdot 2 \cdot 2 \cdot 2 \cdot 2 \cdot 2 \cdot 2 = 2^8$$
$$(5^3)^2 = 5^3 \cdot 5^3 = 5 \cdot 5 \cdot 5 \cdot 5 \cdot 5 \cdot 5 = 5^6$$
$$(b^m)^n = \underbrace{b^m \cdot b^m \cdot b^m \cdots b^m}_{n \text{ factors}}$$
$$= \underbrace{\underbrace{b \cdot b \cdot b \cdots b}\cdot\underbrace{b \cdot b \cdot b \cdots b}\cdots\underbrace{b \cdot b \cdot b \cdots b}}_{\substack{n \text{ groups of } m \text{ factors} \\ m \cdot n \text{ factors}}}$$
$$= \underbrace{b \cdot b \cdot b \cdots b}_{mn \text{ factors}}$$
$$= b^{mn}$$

Thus, we have the ***Third Law of Exponents.***

Third Law of Exponents

> $(b^m)^n = b^{mn}$ where m and n are positive integers. To raise a power to a power, multiply the exponents.

Examples *Simplify.*

12. $(p^2)^3 = p^6$

13. $(m^4)^2 = m^8$

14. $(x^2y^3)^4 = (x^2)^4(y^3)^4$ *Second Law of Exponents; this is often a mental step.*
$= x^8y^{12}$ *Third Law of Exponents*

15. $w^5(u^3w)^2 = w^5(u^3)^2w^2$ *Second Law of Exponents*
$= w^5u^6w^2$ *Third Law of Exponents*
$= w^7u^6$ *First Law of Exponents*

16.
$$\begin{aligned}(-3mn)^3 + (-2m^2n)^2 &= (-3)^3m^3n^3 + (-2)^2(m^2)^2n^2 \\ &= -27m^3n^3 + 4m^4n^2\end{aligned}$$ ■

Note the differences among the following terms.

$$-ab^2 = -abb = -(ab^2) = -a(b^2)$$

but

$$-(ab)^2 = -abab = -a^2b^2 = -(a^2b^2)$$

and

$$(-ab)^2 = (-ab)(-ab) = abab = a^2b^2$$

Thus

$$-ab^2 = -(ab^2) = (-a)b^2$$

but

$$-(ab)^2 \neq (-ab)^2$$

Expressions may appear to be very similar, but the change of a set of parentheses may make them completely different. Just as the exponent in ab^2 applies only to the variable b, the negative signs in $-a^2$ and $(-a)^2$ apply to quite different quantities.

Example 17. Compare $(-2)^3$ and -2^3.

$$\begin{aligned}(-2)^3 &= (-2)(-2)(-2) \quad \text{and} \quad & -2^3 &= -(2\cdot 2\cdot 2) \\ &= -8 & &= -8\end{aligned}$$ ■

Example 18. Compare $(-2)^2$ and -2^2.

$$\begin{aligned}(-2)^2 &= (-2)(-2) \quad \text{and} \quad & -2^2 &= -(2\cdot 2) \\ &= 4 & &= -4\end{aligned}$$ ■

Example 19.
$$\begin{aligned}(-2)^3(2)^3 &= (-2)(-2)(-2)(2^3) \\ &= (-1)(2)(-1)(2)(-1)(2)(2^3) \\ &= (-1)(-1)(-1)(2)(2)(2)(2^3) \\ &= -2^6\end{aligned}$$ ■

Problem Set 4.2

For Problems 1–15 express each number in exponent form, and simplify the expression if possible. See Examples 1–5.

1. $2^3\cdot 4\cdot 2^5$ **2.** $3^2\cdot 27\cdot 3^4$ **3.** $1000\cdot 10^2\cdot 100$
4. $10^3\cdot 10{,}000\cdot 10^2$ **5.** $(3^2)^3\cdot 3^5$ **6.** $2^5\cdot(2^3)^4$
7. $25\cdot 125\cdot 625$ **8.** $49\cdot 343\cdot 2401$ **9.** $x^2x^3x^5$
10. $y^3y^2y^6$ **11.** $(xy^2)(x^2y^3)$ **12.** $(x^3y)(x^2y)$
13. $(2ab)(3ab^3)$ **14.** $(5hk)(7h^2k)$ **15.** $m^2n^2(3m^3n^2)$

Simplify each expression in Problems 16–33 and combine similar terms if possible. See Examples 6 and 7.

16. $2a(a + b)$
17. $3b(a - b)$
18. $4a(b - 2a)$
19. $e(e + 2) + 3(e - 1)$
20. $f(f - 1) + 5(f + 2)$
21. $h(2 - h) - 3(h - 2)$
22. $4(2 - k) - k(k + 5)$
23. $x(x - 2) - 3(x + 1)$
24. $y(y + 1) - 2(y - 3)$
25. $5(m - 3) - m(2 - m)$
26. $3(n + 5) - n(7 - n)$
27. $h(h - 1) - (h - 1)$
28. $k(k - 2) - 2(k - 2)$
29. $m^2(m - n) - n(n - m^2)$
30. $n(m + n) - n^2(m - 3)$
31. $3r^2s(rs^2 - 2r^2s^2 + r^2s)$
32. $2s^3t(s^2t + 3s^2t^2 - st^3)$
33. $4t^2u^3(tu^2 - 2t^2u^2 - u)$

Simplify each expression in Problems 34–45. See Examples 8–11.

34. $(pq)^3(5pq^3)$
35. $u^2(uv^3)^2$
36. $(mn)^2(2m^3n^2)$
37. $(xy)^2(5x^3y^4)$
38. $(3x^2y)(2x^3y^2)$
39. $(3xy)^2(2xy)^3$
40. $3h(2h^2k)^3$
41. $2m^2(3mn^3)^2$
42. $x^3(x^2y)^3$
43. $y^4[x(xy)^2]^2$
44. $[xy^2(xy)^2]^2$
45. $u^2[u^2(uv^2)^3]$

In Problems 46–60, use the laws of exponents to simplify the given expressions. See Examples 12–19.

46. $(-1)^4(-2)^3(-3)^2$
47. $(-1)^2(-2)^3(-3)^4$
48. $[(-1)(-2)^2]^3$
49. $[(-1)(-2)^3]^2$
50. $[(-1)(-2)^2]^3$
51. $-[(-1)(-2)^2]^3$
52. $-x - (-x)^2$
53. $(-x)^2 - (-x) - 1$
54. $(-x)^4 - (-x)^3$
55. $(-x)^2 - (-x)^3 - 1$
56. $(-x)^3 - (-x)^2 + 1$
57. $[(-x)^2(-y)^2]^2$
58. $[(-1)^3(xy)^4]^2$
59. $[(-1)^5(x^2y)^3]^3$
60. $[x^5(-xy)^2]^3$

4.3 Multiplication of Binomials

Thus far, whenever we have multiplied polynomials, one of the polynomials has usually been a monomial. To extend beyond single-term multipliers, consider the product of two binomials. As it happens, this particular example is very common and important to your later work. Consider $(3x + 2)(2x + 1)$, and apply the distributive property.

$$\begin{aligned} \mathbf{A} \cdot (\mathbf{B} + \mathbf{C}) &= \mathbf{A} \cdot \mathbf{B} + \mathbf{A} \cdot \mathbf{C} \\ (\mathbf{3x + 2})(2x + 1) &= (\mathbf{3x + 2})(2x) + (\mathbf{3x + 2})(1) \\ &= 6x^2 + 4x + 3x + 2 \\ &= 6x^2 + 7x + 2 \end{aligned}$$

This process is mathematically correct, but it is also lengthy. Because finding the product of binomials is a very common algebraic process, we need to find a "shortcut" that will allow you to write down the product *directly* while doing all the arithmetic efficiently in your head:

$$(3x + 2)(2x + 1) = 6x^2 + 7x + 2$$

This shortcut process is carried out in three steps. To describe this process, we name the terms of the binomial and product as follows:

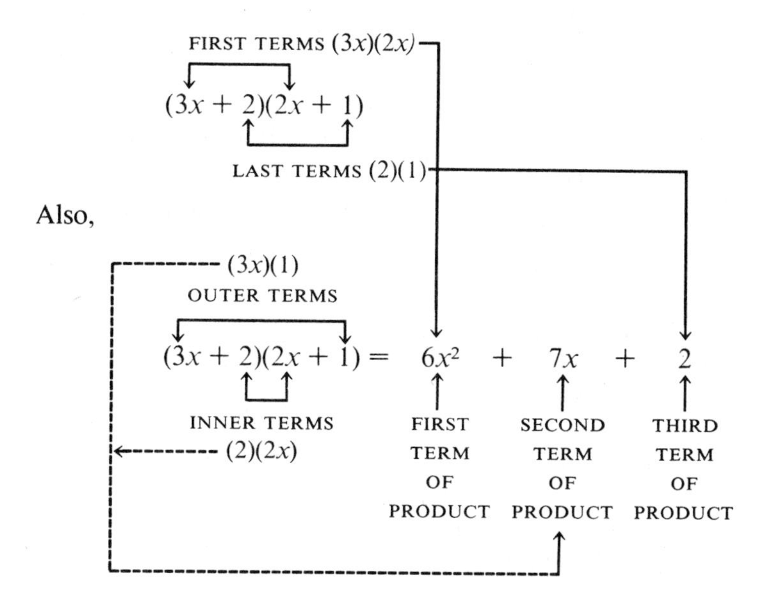

To Multiply Binomials (shortcut method: FOIL)

1. The first term of the product is the product of FIRST terms
2. The second term of the product is (product of OUTER terms + product of INNER terms)
3. The third term of the product is the product of LAST terms

We describe this process as FOIL because:

Step 1	F	First terms
Step 2	O + I	Outer + Inner terms
Step 3	L	Last terms
	or	
	FOIL	

Examples *Simplify.*

1. $(2x - 5)(3x + 6)$

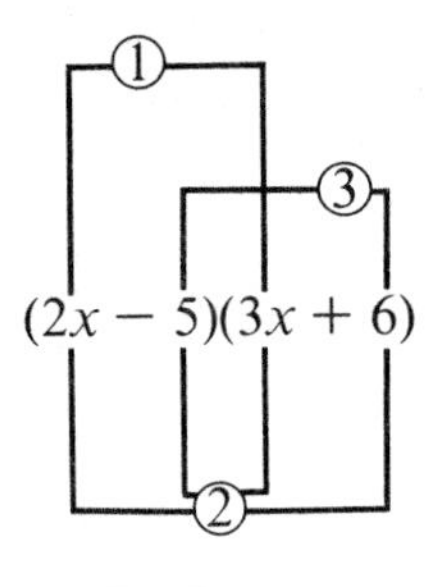

① The first term is the product $(2x)(3x)$.
② The second term is the sum $(-15x + 12x)$.
③ The third term is the product $(-5)(6)$.

$= 6x^2 - 3x - 30$

2. $(x - 7)(x - 5)$

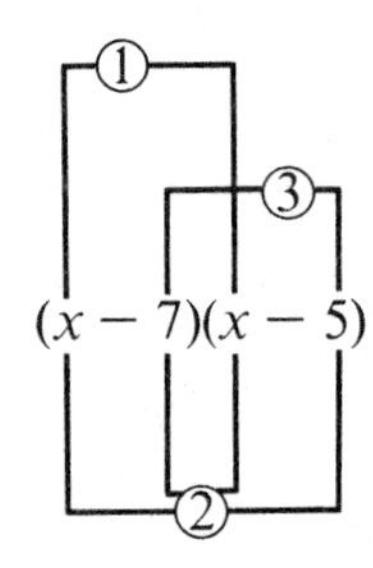

① $(x)(x)$
② $(-5x) + (-7x)$
③ $(-7)(-5)$

$= x^2 - 12x + 35$

3. $(7x - 3)(2x + 3)$
$= 14x^2 + 15x - 9$

4. $(x + a)(x + b)$
$= x^2 + (a + b)x + ab$

5. $(x - 5)(x + 5)$
$= x^2 - 25$

6. $(x + a)(x - a)$
$= x^2 - a^2$

■

It is vital that you find the following products of binomials *mentally* and not rely on some alternative method that will form a habit. Such habits will be difficult to alter later.

Examples 5 and 6 illustrate a special type of binomial product that deserves special attention. It is called the ***difference of squares,*** and it will be helpful if you can identify this type by inspection.

Difference of Squares

$$(x + a)(x - a) = x^2 - a^2$$

$$\begin{pmatrix}\text{SUM OF}\\\text{TERMS}\end{pmatrix}\begin{pmatrix}\text{DIFFERENCE}\\\text{OF TERMS}\end{pmatrix} = \begin{pmatrix}\text{FIRST}\\\text{TERM}\end{pmatrix}^2 - \begin{pmatrix}\text{SECOND}\\\text{TERM}\end{pmatrix}^2$$

Examples

7. $(x + 2)(x - 2) = x^2 - 2^2$
$= x^2 - 4$

8. $(3x + 5)(3x - 5) = (3x)^2 - (5)^2$
$= 9x^2 - 25$

9. $(xy + 3z)(xy - 3z) = (xy)^2 - (3z)^2$
$= x^2y^2 - 9z^2$ ■

Sometimes a binomial product is presented as a binomial squared. These products are called **perfect squares** and are also multiplied using the FOIL shortcut.

Examples *Multiply the perfect squares using FOIL.*

10. $(2x + 3)^2 = (2x + 3)(2x + 3)$
$= 4x^2 + 12x + 9$

11. $x(6x - 5)^2 = x(6x - 5)(6x - 5)$
$= x(36x^2 - 60x + 25)$
$= 36x^3 - 60x^2 + 25x$

12. $(x + a)^2 = (x + a)(x + a)$
$= x^2 + 2ax + a^2$ ■

Example 12 illustrates a general pattern that is sometimes used instead of FOIL.

The Perfect Square

$$(x + a)^2 = (x + a)(x + a) = x^2 + 2ax + a^2$$

$$\left(\begin{matrix}\text{FIRST}\\\text{TERM}\end{matrix} + \begin{matrix}\text{SECOND}\\\text{TERM}\end{matrix}\right)^2 = \left(\begin{matrix}\text{FIRST}\\\text{TERM}\end{matrix}\right)^2 + 2\left(\begin{matrix}\text{FIRST}\\\text{TERM}\end{matrix}\right)\left(\begin{matrix}\text{SECOND}\\\text{TERM}\end{matrix}\right) + \left(\begin{matrix}\text{SECOND}\\\text{TERM}\end{matrix}\right)^2$$

Examples *Multiply using the perfect square pattern.*

13. $(x + 3)^2 = x^2 + 2(3)(x) + 3^2$
$= x^2 + 6x + 9$

14. $(3x + 7)^2 = (3x)^2 + 2(3x)(7) + (7)^2$
$= 9x^2 + 42x + 49$

15. $(9x - 2)^2 = (9x)^2 + 2(9x)(-2) + (-2)^2$
$= 81x^2 - 36x + 4$ ■

Problem Set 4.3

Multiply each of Problems 1–39 mentally. See Examples 1–6.

1. $(x + 1)(x + 2)$
2. $(x + 3)(x + 5)$
3. $(x + 5)(x + 7)$
4. $(x + 8)(x + 9)$
5. $(x - 4)(x - 5)$
6. $(x - 7)(x - 2)$
7. $(x + 7)(x - 7)$
8. $(x - 8)(x + 8)$
9. $(x - 12)(x + 7)$
10. $(x + 10)(x - 6)$
11. $(x + 8)(x - 3)$
12. $(x - 5)(x + 7)$
13. $(x + 6)(x - 9)$
14. $(x - 9)(x + 5)$
15. $(x - 3)(x - 8)$
16. $(x - 5)(x - 6)$
17. $(x + 2)(2x - 1)$
18. $(x - 1)(2x + 1)$
19. $(x - 3)(3x - 1)$
20. $(x + 2)(3x - 1)$
21. $(x + 1)(2x - 3)$

22. $(x + 2)(3x - 2)$
23. $(2x + 1)(x + 3)$
24. $(x + 2)(3x + 1)$
25. $(3x + 2)(2x - 1)$
26. $(3x - 2)(2x + 1)$
27. $(4x - 5)(3x + 4)$
28. $(5x - 6)(2x + 3)$
29. $(7x + 10)(4x - 5)$
30. $(9x + 11)(3x - 4)$
31. $(3x - 2)(5x + 4)$
32. $(4x - 3)(7x + 6)$
33. $(6x - 7)(2x + 3)$
34. $(5x - 3)(6x + 5)$
35. $(5a - 4)(5a + 4)$
36. $(7b + 3)(7b - 3)$
37. $(4c - 1)^2$
38. $(5d + 2)^2$
39. $(3e - 8)(2e + 5)$

Problems 40–48 involve differences of squares or perfect squares. Find each product mentally. See Examples 7–15.

40. $(x + 5)^2$
41. $(x + 8)^2$
42. $(y - 6)^2$
43. $(ab + 10)(ab - 10)$
44. $(mn + 12)(mn - 12)$
45. $(2rs + t)^2$
46. $(5uv - w)^2$
47. $(3h^2 - 7k)(3h^2 + 7k)$
48. $(2k - 5j^2)(2k + 5j^2)$

Find the products in Problems 49–72. Work the problems mentally whenever possible. See Examples 1–15.

49. $(2x^2 + 3y^3)^2$
50. $(3x^3 - 4y^2)^2$
51. $(ab + 2)(ab - 5)$
52. $(a - 2b)(a + 5b)$
53. $(c - 8d)(c + 5d)$
54. $(cd - 8)(cd + 5)$
55. $(f^2 + 7)(f^2 + 8)$
56. $(e^2 - 5)(e^2 + 5)$
57. $h(3h - 4)(5h + 6)$
58. $k(7k - 4)(3k + 2)$
59. $(2p - 3q)(3p - 2q)$
60. $(3r - 4s)(4r - 5s)$
61. $(5r + 2t)(7r - 3t)$
62. $(8s - 3t)(5s + 2t)$
63. $(2u + 7v)^2$
64. $(3v - 8w)^2$
65. $(7x + 5y)(9x - 11y)$
66. $(5x - 13y)(7x + 19)$
67. $(5h - 9k)^2$
68. $(7m + 11n)^2$
69. $(3x^2 - 13)(5x^2 + 17)$
70. $(13y^2 - 6)(11y^2 + 5)$
71. $(x^2 - y^2)(x^2 + y^2)$
72. $(x^3 + y^3)(x^3 - y^3)$

4.4 Multiplication of Polynomials

We began this chapter by introducing words to describe polynomials. It is easier to talk about polynomials if we classify them accurately. We also began to perform operations with polynomials. In this section we continue that discussion to concentrate on the product of a binomial and a polynomial with more than two terms. The process depends upon the distributive property.

Simplify the given expressions.

$$\mathbf{A \cdot (B + C + D) = A \cdot B + A \cdot C + A \cdot D}$$

Example 1.
$$\begin{aligned}\overbrace{(x + 2)}(x^2 + 3x + 4) &= \overbrace{(x + 2)}x^2 + \overbrace{(x + 2)}3x + \overbrace{(x + 2)}4\\ &= x^3 + 2x^2 + 3x^2 + 6x + 4x + 8\\ &= x^3 + 5x^2 + 10x = 8\end{aligned}$$

$$\mathbf{A \cdot (B + C + D) = A \cdot B + A \cdot C + A \cdot D}$$

Example 2.
$$\begin{aligned}\overbrace{(x + y)}(x^2 + 5xy + y^2) &= \overbrace{(x + y)}x^2 + \overbrace{(x + y)}5xy + \overbrace{(x + y)}y^2\\ &= x^3 + x^2y + 5x^2y + 5xy^2 + xy^2 + y^3\\ &= x^3 + 6x^2y + 6xy^2 + y^3\end{aligned}$$ ■

Example 3.
$$\begin{aligned}(2x - 3)(x + 1)(3x + 1) &= (2x^2 - x - 3)(3x + 1)\\ &= 2x^2(3x + 1) - x(3x + 1) - 3(3x + 1)\\ &= 6x^3 + 2x^2 - 3x^2 - x - 9x - 3\\ &= 6x^3 - x^2 - 10x - 3\end{aligned}$$ ■

These examples look long and involved, but they are not overly complex. If you multiply, no matter how large and involved the factors are, you must multiply one term at a time. Just as (2)(3)(5) can be done only as (6)(5), (10)(3), or (2)(15), any multiplication process involves only two factors. You do not find the product of three factors in a single step. So, $(x + 1)\cdot(x^2 + 2x + 3)$ is found by several small multiplications, not one giant step.

$$\begin{aligned}(\mathbf{x} + \mathbf{1})(x^2 + 2x + 3) &= (\mathbf{x} + \mathbf{1})x^2 + (\mathbf{x} + \mathbf{1})2x + (\mathbf{x} + \mathbf{1})3 \\ &= (\mathbf{x})x^2 + (\mathbf{1})x^2 + (\mathbf{x})2x + (\mathbf{1})2x + (\mathbf{x})3 + (\mathbf{1})3\end{aligned}$$

We actually have six products of two terms instead of any special binomial-trinomial scheme. Be careful with the parts—especially the signs—and the finished product will emerge correctly. Study the previous examples again before beginning the problem set.

Problem Set 4.4

Find the products in Problems 1–30 and combine like terms, where possible, to simplify the expressions. See Examples 1–3.

1. $(x + 1)(x^2 - 3x - 2)$
2. $(y - 1)(y^2 + 2y + 3)$
3. $(a + 3)(a^2 + a + 1)$
4. $(b - 2)(b^2 + b + 2)$
5. $(3h^2 - 5h + 9)(2h + 5)$
6. $(2k^2 - 7k + 5)(3k - 4)$
7. $(2w + 3)(2w^2 - 3w + 4)$
8. $(3x + 4)(3x^2 - 4x + 2)$
9. $(5y - 4)(y^2 - 1)$
10. $(7z - 3)(z^2 - 4)$
11. $(x + 1)(x + 2)(x - 3)$
12. $(x - 1)(x - 3)(x + 4)$
13. $(2x + y)(x^2 - xy + y^2)$
14. $(a + 2b)(a^2 + ab - b^2)$
15. $(x^2y - z)(x^4y^2 - x^2yz + z^2)$
16. $(st^3 + r^2)(s^2t^6 - st^3r^2 + r^4)$
17. $(2m - n)^2(3m + 5)$
18. $(p + 3q)^2(4p - 1)$
19. $(3u - 4)^3$
20. $(2v + 5)^3$
21. $(a + 5)(a - 2) - a(a + 1)$
22. $(b - 7)(b + 2) - b(b - 3)$
23. $(x + 2)(x^2 + 1) - x^2(x + 2)$
24. $(y - 1)(y^2 + 2) - y^2(y - 1)$
25. $(a + 1)(a + 5) + (a - 5)(a - 4)$
26. $(b + 2)(b - 3) + (b + 4)(b - 7)$
27. $(r^2 - 3)(r + 1) + r^2(r + 1)$
28. $s^2(s - 1) + (s^2 - 5)(s - 1)$
29. $(x - 1)(x^2 + x + 1) - (x^2 - 3x - 1)(x + 3)$
30. $(y^2 + 2y + 4)(y - 2) + (y + 5)(y^2 - 5y + 2)$

4.5 Dividing Polynomials

So far in this chapter we have been concerned with the operations of addition, subtraction, and multiplication of polynomials. We conclude this chapter with a discussion of division of polynomials. The process of long division of polynomials is essentially the same process you learned for long division of numbers. To divide a polynomial by a monomial, divide each term of the polynomial by the monomial.

Example **1.** Find the quotient: $(x^3 + 2x^2 - 5x) \div x$

Solution

$$\begin{array}{rll} & x^2 + 2x \; - 5 & \\ x\,\big) & \overline{x^3 + 2x^2 - 5x} & \textit{Divide x into the first term.} \\ & \underline{x^3 \qquad\quad} & \textit{Multiply } x^2 \cdot x \\ & 2x^2 \qquad & \textit{Subtract and bring down the next term, } 2x^2. \\ & \underline{2x^2 \qquad} & \\ & -5x & \\ & \underline{-5x} & \textit{Repeat the process until the last term is found.} \\ & 0 & \end{array}$$

■

If a term divides evenly into a polynomial, then it is a ***factor*** of that polynomial. This terminology comes from the check. Remember in arithmetic,

$$\begin{array}{r} 2 \\ 5\,\big)\overline{10} \end{array}$$

The check was *multiplication:* $5 \times 2 = 10$. The numbers 5 and 2 are *factors* of 10. Notice from Example 1

$$x(x^2 + 2x - 5) = x^3 + 2x^2 - 5x$$

so that x and $(x^2 + 2x - 5)$ are factors of $x^3 + 2x^2 - 5x$.

The process of dividing a polynomial by a binomial is similar. Just remember the distributive property on the multiplication step.

Example **2.** Find the quotient: $(x^2 + 6x + 8) \div (x + 2)$.

Solution

$$x + 2\,\big)\overline{x^2 + 6x + 8}$$

Just as the digits in a number are ordered by place value, we arrange the terms by degree in descending order.

$$\begin{array}{rl} & x \\ x + 2\,\big) & \overline{x^2 + 6x + 8} \end{array}$$

Divide the first term of the divisor into the first term of the dividend. That is, x goes into x^2 an x number of times.

$$\begin{array}{rl} & x \\ x + 2\,\big) & \overline{x^2 + 6x + 8} \\ & \underline{x^2 + 2x} \\ & 4x + 8 \end{array}$$

Multiply x times the divisor, x + 2, and subtract. The x^2 term is eliminated, so we bring down the constant, 8, and prepare to repeat the process.

$$\begin{array}{rl} & x + 4 \\ x + 2\,\big) & \overline{x^2 + 6x + 8} \\ & \underline{x^2 + 2x} \\ & 4x + 8 \end{array}$$

Divide the first terms again: x goes into 4x four times.

$$\begin{array}{rl} & x + 4 \\ x + 2\,\big) & \overline{x^2 + 6x + 8} \\ & \underline{x^2 + 2x} \\ & 4x + 8 \\ & \underline{4x + 8} \\ & 0 \end{array}$$

Multiply 4 times the divisor, x + 2, and subtract, leaving a remainder of 0. Thus, x + 2 is a factor of $x^2 + 6x + 8$, and x + 4 is the other factor.

■

We may, therefore, divide to determine if a given binomial is a factor of another polynomial. In the next example, we take a polynomial of larger degree. Notice that we arrange the terms of the polynomial in descending order and insert zero coefficients for the missing terms.

Example **3.** Is $2x + 3$ a factor of $4x^4 - 3x^2 + 5x - 6$?

Solution We answer this question by doing long division.

$$\begin{array}{r} 2x^3 \phantom{{}- 3x^2 + 5x - 6} \\ 2x + 3\overline{)4x^4 + 0x^3 - 3x^2 + 5x - 6} \\ \underline{4x^4 + 6x^3} \phantom{{}- 3x^2 + 5x - 6} \\ -6x^3 - 3x^2 \phantom{{}+ 5x - 6} \end{array}$$

Find the first term of the quotient by dividing first terms, multiply and subtract, and then bring down the next term. Notice the added third-degree term with zero coefficient.

$$\begin{array}{r} 2x^3 - 3x^2 \phantom{{}+ 5x - 6} \\ 2x + 3\overline{)4x^4 + 0x^3 - 3x^2 + 5x - 6} \\ \underline{4x^4 + 6x^3} \phantom{{}- 3x^2 + 5x - 6} \\ -6x^3 - 3x^2 \phantom{{}+ 5x - 6} \\ \underline{-6x^3 - 9x^2} \phantom{{}+ 5x - 6} \\ 6x^2 + 5x \phantom{{}- 6} \end{array}$$

Repeat the process for the next term. Remember that you are subtracting: $-9x^2$ from $-3x^2$ is $+6x^2$.

$$\begin{array}{r} 2x^3 - 3x^2 + 3x - 2 \\ 2x + 3\overline{)4x^4 + 0x^3 - 3x^2 + 5x - 6} \\ \underline{4x^4 + 6x^3} \phantom{{}- 3x^2 + 5x - 6} \\ -6x^3 - 3x^2 \phantom{{}+ 5x - 6} \\ \underline{-6x^3 - 9x^2} \phantom{{}+ 5x - 6} \\ 6x^2 + 5x \phantom{{}- 6} \\ \underline{6x^2 + 9x} \phantom{{}- 6} \\ -4x - 6 \\ \underline{-4x - 6} \\ 0 \end{array}$$

Continue the process. Watch your signs!

Yes, $2x + 3$ is a factor, because it divides evenly, and
$4x^4 - 3x^2 + 5x - 6 = (2x + 3)(2x^3 - 3x^2 + 3x - 2)$. ■

Study this process in Examples 4–6. You may check the results by multiplying out the factors.

Example **4.** $(15x^3 - 62x^2 + 32) \div (5x - 4) = ?$

Solution

$$\begin{array}{r} 3x^2 - 10x - 8 \\ 5x - 4\overline{)15x^3 - 62x^2 + 0x + 32} \\ \underline{15x^3 - 12x^2} \phantom{{}+ 0x + 32} \\ -50x^2 + 0x \phantom{{}+ 32} \\ \underline{-50x^2 + 40x} \phantom{{}+ 32} \\ -40x + 32 \\ \underline{-40x + 32} \\ 0 \end{array}$$

The quotient is $3x^2 - 10x - 8$

$(15x^3 - 62x^2 + 32) \div (5x - 4) = 3x^2 - 10x - 8$ ■

Example 5. $(8x^3 + 27) \div (2x + 3) = ?$

Solution

$$\begin{array}{r} 4x^2 - 6x + 9 \\ 2x + 3 \overline{)8x^3 + 0x^2 + 0x + 27} \\ \underline{8x^3 + 12x^2} \qquad\qquad \\ -12x^2 + 0x \qquad \\ \underline{-12x^2 - 18x} \qquad \\ 18x + 27 \\ \underline{18x + 27} \\ 0 \end{array}$$

Notice the terms with the zero coefficients.

The quotient is $4x^2 - 6x + 9$. ■

Example 6. Is $5x - 3$ a factor of $15x^4 + 11x^3 - 17x^2 - 7x - 6$?

Solution

$$\begin{array}{r} 3x^3 + 4x^2 - x - 2 \\ 5x - 3 \overline{)15x^4 + 11x^3 - 17x^2 - 7x - 6} \\ \underline{15x^4 - 9x^3} \qquad\qquad\qquad\qquad \\ 20x^3 - 17x^2 \qquad\qquad \\ \underline{20x^3 - 12x^2} \qquad\qquad \\ -5x^2 - 7x \qquad \\ \underline{-5x^2 + 3x} \qquad \\ -10x - 6 \\ \underline{-10x + 6} \\ -12 \end{array}$$

No. $5x - 3$ is not a factor, because the remainder is not 0. ■

Unfortunately not every division comes out evenly and we must have a way to write the result when this occurs. For instance, 4 does not divide evenly into 17. To express the result, we show that the remainder is also divided by 4.

$$\begin{array}{r} 4 \\ 4\overline{)17} \\ \underline{16} \\ 1 \text{ R} \end{array}$$

$$17 \div 4 = 4 + \text{remainder } 1$$

or if you show the remainder as also divided by 4:

$$17 \div 4 = 4 + \frac{1}{4}$$

Using this same idea for Example 6, we can write the quotient as follows:

$$3x^3 + 4x^2 - x - 2 + \frac{-12}{5x - 3}$$

Because the remainder is not 0, $5x - 3$ is not a factor of the dividend. Moreover, the answer is not a polynomial, because it includes division by $5x - 3$.

Example 7. Divide and write the remainder in fractional form.

$$(6x^4 - 8x^3 - 5x^2 + 3x - 9) \div (3x - 7)$$

Solution

$$
\begin{array}{r}
2x^3 + \ 2x^2 + \ 3x + \ 8 \\
3x - 7\overline{)6x^4 - \ 8x^3 - \ 5x^2 + \ 3x - \ 9} \\
\underline{6x^4 - 14x^3} \qquad\qquad\qquad\quad \\
6x^3 - \ 5x^2 \qquad\qquad \\
\underline{6x^3 - 14x^2} \qquad\qquad \\
9x^2 + \ 3x \qquad \\
\underline{9x^2 - 21x} \qquad \\
24x - \ 9 \\
\underline{24x - 56} \\
47
\end{array}
$$

$$\frac{6x^4 - 8x^3 - 5x^2 + 3x - 9}{3x - 7} = 2x^3 + 2x^2 + 3x + 8 + \frac{47}{3x - 7}$$ ■

Problem Set 4.5

Find the quotients in Problems 1–8. See Example 1.

1. $(5x^3 - 2x^2 + 3x) \div x$

2. $(9x^4 + 3x^3 - 7x^2 + x) \div x$

3. $(15x^3 - 10x^2 - 5x) \div 5$

4. $(9x^4 - 3x^3 + 6x^2) \div 3$

5. $(10x^3 - 6x^2 + 4x) \div 2x$

6. $(12x^4 + 6x^3 - 3x^2 - 18x) \div 3x$

7. $(12x^4 - 6x^2) \div 2x^2$

8. $(15x^6 - 5x^4) \div 5x^2$

Divide Problems 9–18 using the long-division process. See Examples 2, 4, and 5.

9. $(2x^2 + 2x - 12) \div (x + 3)$
10. $(3x^2 - 4x - 15) \div (x - 3)$
11. $(8x^2 + 2x - 21) \div (2x - 3)$
12. $(24x^2 - x - 10) \div (3x - 2)$
13. $(9a^4 - 16a^2 - 3a - 20) \div (3a - 5)$
14. $(9b^4 + 11b^2 + 7b - 2) \div (3b + 2)$
15. $(u^3 + 27) \div (u + 3)$
16. $(u^3 - 27) \div (u - 3)$
17. $(v^3 - 8) \div (v - 2)$
18. $(v^3 + 8) \div (v + 2)$

In Problems 19–24, tell whether the first polynomial is a factor of the second. See Examples 3 and 6.

19. $2x - 3;\ 6x^3 + x^2 - 19x + 6$

20. $3x - 1;\ 6x^3 + x^2 - 19x + 6$

21. $x + 2;\ 6x^3 + x^2 - 19x + 6$

22. $3y - 2;\ 6y^3 + 11y^2 - 19y - 6$

23. $2y - 1;\ 6y^3 + 11y^2 - 19y + 6$

24. $y + 3;\ 6y^3 + 11y^2 - 19y + 6$

In Problems 25–30, divide and write the remainder, if any, in fractional form. See Example 7.

25. $(2z^3 - z^2 - 17z + 15) \div (z + 3)$
26. $(3x^3 + x^2 - 2x + 14) \div (x + 2)$
27. $x^4 + 3x^3 + x^2 + 2x + 2$ divided by $x + 1$
28. $6x^3 - 23x^2 - 5x + 6$ divided by $2x + 1$
29. $y - 1$ divided into $y^4 - 4y^3 + 3y^2 - 3y + 2$
30. $3x - 1$ divided into $12x^3 - 4x^2 - 3x$
31. If $(5x - 4)$ and $(x + 1)$ are factors of $35x^3 + 32x^2 - 23x - 20$, what is the other binomial factor?

32. If $(4y - 3)$ and $(y - 1)$ are factors of $24y^3 - 22y^2 - 17y + 15$, what is the remaining binomial factor?
33. If $2x - 5$ is a factor of $8x^3 - 16x^2 - 26x + h$, then what is the value of h?
34. If $3y - 4$ is a factor of $9y^3 - 9y^2 - 34y + k$, then what is the value of k?

4.6 Review

In this chapter, you refined the vocabulary of algebraic expressions and concentrated on the polynomial. You should now feel comfortable adding, subtracting, multiplying, and dividing polynomials. The next chapter will lean heavily on your understanding of the multiplication process for polynomials.

Self-Test

[4.1] **1.** Identify the polynomial as a monomial, binomial, or trinomial, if possible, and find the degree of the expression.
a. x^5y^2 **b.** $x^3 + x$ **c.** $3x^4 + 2x^2 + 1$
d. 72 **e.** $xyz + z^2$

[4.1] **2.** Simplify by combining similar terms.
a. $3h + 2k - 4h + 5h - 6k$ **b.** $(3x - 4) - (2x - 1)$
c. $(m + 2n) + (3m - n) - (2m + 3n)$ **d.** $(a + 3) + (2b - 5) - (3a + 4b)$
e. $(2x - 3) - (2 + 3x + 1 + x)$

[4.2] **3.** Express each number in exponent form and simplify.
a. $8 \cdot 2^5 \cdot 4$ **b.** $b(b^2)(3b^4)$ **c.** $(2^3)^2 \cdot 2^2$
d. $(-1)^2(-1)^3(-1)^5$ **e.** $-2^4 \cdot 2^2$

Simplify each expression in Problems 4–10.

[4.2] **4.** **a.** $5a(a - b)$ **b.** $h(h - 3) - 5(h + 1)$
c. $m^3(m^2 - n) - mn(m^2 + 1)$ **d.** $2x^2y(xy + 2x^2y - xy^2)$
e. $4(3 - x) - 2x(1 - 2x) + 6$

[4.3] **5.** **a.** $(x + 3)(x + 4)$ **b.** $(x - 1)(x + 5)$
c. $(x - 2)(x - 7)$ **d.** $(2x - 3)(x - 5)$
e. $(3x - 2)(5x + 3)$

[4.3] **6.** **a.** $(x + 7)(x + 7)$ **b.** $(x + 7)(x - 7)$
c. $(2x - 3)(2x + 3)$ **d.** $(2x - 3)^2$
e. $(x - 2y)^2$

[4.4] **7.** **a.** $(x + 1)(x^2 + x + 1)$ **b.** $(x - 1)(x^2 + x + 1)$
c. $(2y - 5)(3y^2 - y + 2)$

[4.4] **8.** **a.** $(2z - 3)(4z^2 - 12z + 9)$ **b.** $(3x - 1)(9x^2 - 6x + 1)$
c. $(x - 2)(x^3 + 2x^2 - x + 3)$

[4.5] **9.** **a.** $(x^3 - 7x^2 + 4x + 12) \div (x + 1)$ **b.** $(6x^3 + x^2 - 7x - 3) \div (2x + 1)$

[4.5] **10.** **a.** $(y^3 + 64) \div (y + 4)$ **b.** $(y^3 - 64) \div (y - 4)$

Review Problems

Simplify each expression in Problems 1–100.

1. $3x + 6 - 2x - 10$
2. $5y + 1 - 3y - 6$
3. $5p - 3 + 2p + 8$
4. $6q + 5 - 10q - 8$

5. $3x - 2y + 5 - x + 3y - 4$
6. $3t + 5u - 2t - 8u + 6$
7. $6x + 4y - 3x + 2y - 5x$
8. $4t - 6s + 3t + 10s$
9. $(5x + 2) + (7x - 5)$
10. $(3x - 7) + (2x + 3)$
11. $(3t + 8) + (4t - 7)$
12. $(2u - 5) + (6u + 9)$
13. $(4y + 3) - (3y - 2)$
14. $(3y + 2) - (5y - 6)$
15. $(a + b) - (3a - 2b)$
16. $(6a - 5b) - (3b + 2a)$
17. $(a + 2b - 5c) + (6a - 2b - 3c)$
18. $(4c - 2d + 3) + (3x + 5d - 8)$
19. $(s + 3t + 6) + (4s - 5t - 8)$
20. $(m - 3n + 5) + (5m + 5n - 6)$
21. $(3x - 5) - (2x^2 + 5x + 3)$
22. $(5x - 3) - (3x^2 - 6x + 5)$
23. $(6x^2 + 5x - 5) - (3x - 8)$
24. $(2x^2 + x + 6) - (5x^2 - 6)$
25. $2x(x + y)$
26. $3y(x + 2y)$
27. $3y(y + z)$
28. $5z(x - 2z)$
29. $6x(x^2 - 2x + 3)$
30. $3y(y^2 + 2y + 1)$
31. $3a(a + 5) - 2(a + 5)$
32. $2m(m + 6) + 5(m + 6)$
33. $x(x + 6) - 3(x - 5)$
34. $y(y - 5) + 2(3y + 1)$
35. $x^2(x^3 - 1) - 2x(x^3 - 1)$
36. $x^3(x^2 - x) + 3x(x^2 + x)$
37. $x^3(x + 2) + x(x + 2) + 2(x + 2)$
38. $y(y - 3) + y(y + 5) - y^2(y - 3)$
39. $3x^2y(x^2y + xy^2 + x)$
40. $2xy^2(xy^2 + x^2y + xy)$
41. $2x^5y(6xy^3 + 2xy + 3)$
42. $6xy^3(2x^2 + 3x + 2)$
43. $k(h - k) - 2k(h + 3k)$
44. $3s(s - t) + 2t(s - t)$
45. $s(s - 5) + 3s(s + 9)$
46. $2m(m - 2n) + 6(m - 2n)$
47. $(a + 2)(a + 5)$
48. $(b + 3)(b + 5)$
49. $(c - 3)(c - 1)$
50. $(d - 2)(d - 7)$
51. $(m - 5)(m + 2)$
52. $(p + 3)(p - 2)$
53. $(2x - 1)(3x + 5)$
54. $(3x + 2)(6x - 5)$
55. $(2x + 1)(3x - 10)$
56. $(5x + 6)(x - 2)$
57. $(3x - 4)(2x + 5)$
58. $(2x - 3)(3x + 4)$
59. $(12x + 1)(x - 10)$
60. $(y - 4)(y + 4)$
61. $(t - 8)(t + 8)$
62. $(x + 6)(x - 6)$
63. $(y - 5)(y + 5)$
64. $(2x + 1)(2x - 1)$
65. $(6z - 1)(6z + 1)$
66. $(5x + 3)(5x - 3)$
67. $(3t + 5)(3t - 5)$
68. $(x + 4)^2$
69. $(x + 3)^2$
70. $(y - 3)^2$
71. $(y - 5)^2$
72. $(2x - 3)^2$
73. $(3x - 2)^2$
74. $(4x + 1)^2$
75. $(x + y)^2$
76. $(x - y)^2$
77. $(2x + 3)^2$
78. $(5x - 1)^2$
79. $(x - 3y)^2$
80. $(x + 2y)^2$
81. $(x + 1)(x - 2) + 6$
82. $(y - 3)(y + 2) + 6$
83. $(3x + 2)(x - 5) + 10$
84. $(2x - 5)(x + 1) + 3$
85. $(2x + 7)(3x - 2) + 8$
86. $(3x - 1)(2x + 11) + 9$
87. $(x^2 + x)(x + 3)$
88. $(y^3 - 1)(y^2 + 2)$
89. $(k + 1)(k^2 + 2k + 1)$
90. $(t - 1)(t^2 + t + 1)$
91. $(z - 3)(z^2 - 6z + 9)$
92. $(z + 2)(z^2 + 4z + 4)$
93. $(p - 1)(p^2 + 3p + 2)$
94. $(q - 1)(q^2 + 3q + 5)$
95. $(x + 1)(x^3 + 3x^2 - 2x + 4)$
96. $(x - 1)(x^3 - 2x^2 + 3x - 5)$
97. $(z^3 + 8) \div (z + 2)$
98. $(z^2 - 16) \div (z - 4)$
99. $(z^2 - 25) \div (x + 5)$
100. $(z^3 - 8) \div (z - 2)$

5

Factoring

Whatever one man is
capable of conceiving,
other men will be able to achieve.

Jules Verne

Genius may conceive,
but patient labor
must consummate.

Horace Mann

5.1 Common Monomial Factors

At the beginning of the course we defined the numbers that are multiplied to obtain a product as *factors.* We use the distributive property to change a basic product to a basic sum and vice versa. In one case, such as $3x(4 + 5x) = 12x + 15x^2$, we multiply to obtain the product. In another case, we are given the product and must determine the factors used to obtain the product—for example, $6x^2 + 9x = \mathbf{3x} \cdot 2x + \mathbf{3x} \cdot 3 = 3x(2x + 3)$. This latter method is called ***factoring*** the expression.

In Section 1.3 a *prime* number was defined as a counting number that has exactly two distinct factors. For example, 11 is prime and there is only one way to factor a prime: $1 \cdot 11$. Because 1 does not have two *distinct* factors, it is not a prime, but every other counting number has at least two distinct factors (itself and 1). If it has more than two distinct factors, it is called *composite.* For example, $12 = 2 \cdot 2 \cdot 3 = 2^2 \cdot 3$ and $90 = 2 \cdot 3^2 \cdot 5$ are composite numbers. In this section we will extend the notion of factoring a number to factoring an algebraic expression. Just as we found a unique prime factorization of numbers in Section 1.3, we will try to find a unique prime factorization of algebraic expressions in this chapter.

Examples *Determine the prime factorization, if possible.*

1. $220 = 22 \cdot 10$
$= 2 \cdot 11 \cdot 2 \cdot 5$
$= 2^2 \cdot 5 \cdot 11$

2. $231 = 3 \cdot 77$
$= 3 \cdot 7 \cdot 11$

3. 241 is prime.

■

Just as constants have factors, variable expressions may have factors. A term is a product of constants and/or variables. Terms, therefore, have constant or variable factors or both, and several terms may have such factors in common. Terms are polynomials that are called monomials. If several terms share a factor, that factor is called a ***common monomial factor.***

The binomial $3x^2 + 6x$ has three common monomial factors: 3, x, and $3x$. We can change this sum to a product in several ways.

$$3x^2 + 6x = 3(x^2 + 2x)$$
$$3x^2 + 6x = x(3x + 6)$$
$$3x^2 + 6x = 3x(x + 2)$$

The last of these possibilities has the greatest common monomial factor as a factor of the basic product and is said to be ***completely factored.***

We now extend the idea of factoring to algebraic expressions. First factor each term and then find all the factors common to each term.

Examples *Completely factor, if possible.*

4. $25a + 40b = \mathbf{5} \cdot 5 \cdot a + \mathbf{5} \cdot 2 \cdot 2 \cdot 2 \cdot b$
$= 5(5a + 8b)$

5. $2x + 2 = \mathbf{2} \cdot x + \mathbf{2} \cdot 1$ *Note that 1 is a factor here.*
$= 2(x + 1)$

6. $6hk + 2k = \mathbf{2} \cdot 3 \cdot h \cdot \mathbf{k} + \mathbf{2} \cdot \mathbf{k}$
$= \mathbf{2k}(3h) + \mathbf{2k}(1)$
$= \mathbf{2k}(3h + 1)$

7. $x^2y - xy^2 = \mathbf{x} \cdot x \cdot \mathbf{y} - \mathbf{x} \cdot \mathbf{y} \cdot y$
$= \mathbf{xy}(x - y)$

8. $\frac{1}{2}bh + 2b = \mathbf{b}\left(\frac{1}{2}h + 2\right)$

9. $7rs^2 + 28rs - 35r^2 = \mathbf{7} \cdot \mathbf{r} \cdot s \cdot s + 2 \cdot 2 \cdot \mathbf{7} \cdot \mathbf{r} \cdot s - 5 \cdot \mathbf{7} \cdot \mathbf{r} \cdot r$
$= \mathbf{7r}(s^2 + 4s - 5r)$

10. $\frac{4}{3}\pi r^3 - \frac{\pi}{2}r^2 + 2\pi r = \boldsymbol{\pi r}\left(\frac{4}{3}r^2 - \frac{1}{2}r + 2\right)$

11. $\frac{1}{2}bh + 2(b + h) = \frac{1}{2}bh + 2b + 2h$

There is no factor common to all 3 factors. ■

Factoring to Find an Area Formula (Optional)

A trapezoid is a four-sided plane figure with one pair of opposite sides parallel. Consider the trapezoid with height h and bases b and B shown in Figure 5.1a. The figure can be divided into two triangles by a diagonal, as shown in Figure 5.1b.

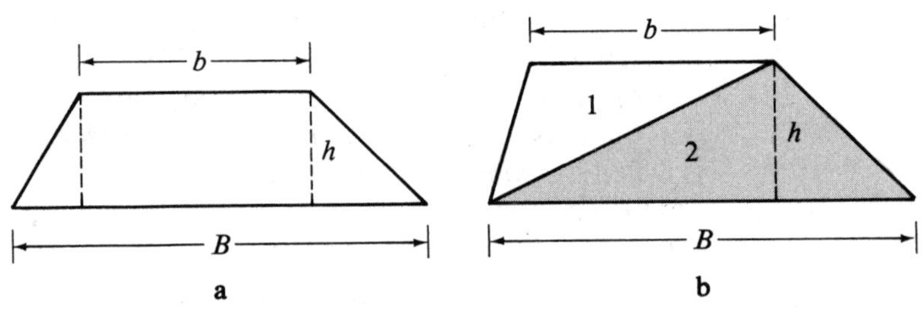

Figure 5.1
Area of a Trapezoid

Each of the two triangles still has height h, and the bases of the trapezoid are the bases of the two triangles. Calculate the area of the trapezoid from the area of the triangles.

(AREA OF TRAPEZOID) = (AREA OF TRIANGLE 1) + (AREA OF TRIANGLE 2)

$$A \quad = \quad \frac{1}{2}bh \quad + \quad \frac{1}{2}Bh$$

$$A = \frac{1}{2}h(b + B) \qquad \textit{Common factor } \frac{1}{2}h$$

Notice that if we remove the common monomial factor $\frac{1}{2}h$, the formula is more easily described. The area of a trapezoid is one-half the height times the sum of the bases, or the height times the average height of the bases.

Example 12. Write an expression for the shaded area in Figure 5.2 and write in factored form, if possible.

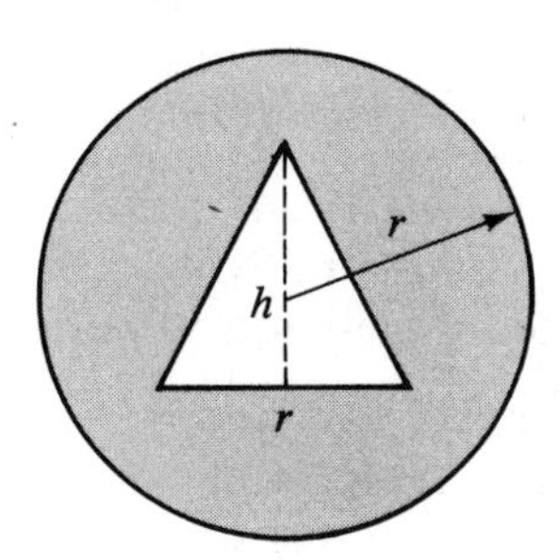

Figure 5.2

Solution It is a circle with a triangular section removed.

(TOTAL AREA) = (AREA OF CIRCLE) − (AREA OF TRIANGLE)

$$A = \pi r^2 - \frac{1}{2}rh$$

$$= r\left(\pi r - \frac{1}{2}h\right)$$

The area may be written as a difference or as a product by identifying the common factor r and factoring the expression. ■

Problem Set 5.1

In Problems 1–10, write the prime factorization in exponent form. See Examples 1–3.

1. 32
2. 27
3. 40
4. 54
5. 36
6. 75
7. 9000
8. 49,000
9. 392
10. 363

In Problems 11–25 determine the greatest common monomial factor of the terms of the expression and factor completely. See Examples 4–8.

11. $3a^2 + 15b^2$
12. $5c^2 + 10d^2$
13. $4s^2 + 12t^2$
14. $27e^2 - 25e$
15. $14f - 15f^2$
16. $5g^3 - 6g^2$
17. $3x^2y + 5xy^2$
18. $2vw^3 + 7v^2w$
19. $4m^2n + 5mn^2$
20. $abc^2 - ab^2$
21. $mn^3 - m^2n$
22. $2\pi r + \pi r^2$
23. $\frac{4}{3}\pi r^3 - \pi r^2$
24. $\frac{1}{2}bh + b^2$
25. $2\pi r - \frac{2}{3}\pi r^2$

In Problems 26–39 rewrite each expression in completely factored form. See Examples 9–11.

26. $12x^2 + 6x - 9$
27. $8y^2 - 18y + 6$
28. $2x^3 + 3x^2 + 5x$
29. $3y^4 - 7y^3 - 4y^2$
30. $6a^2 + 18b^2 + 9c^2$
31. $8x^2 + 12y^2 + 6z^2$
32. $2m^2n^3 - 6m^2n^2 + 4m^3n^2$
33. $3rs^3 + 9r^2s - 6rs^2$

34. $14u^3v^2 - 28u^2v + 35uv$

35. $15wx^3 - 25w^2x^2 + 35w^3x$

36. $\frac{4}{3}\pi r^4 + \frac{1}{3}\pi r^3 + \pi r^2$

37. $\frac{1}{3}\pi r^3 + \pi r^2 + 2\pi r$

38. $2\pi r - \frac{2}{3}\pi r^2 + 3\pi$

39. $\pi r^2 + \frac{4}{3}\pi r^3 - 2\pi r$

Area Problems (Optional)

In Problems 40–47 write an expression for the shaded area shown in each of the illustrations, and write the expression in factored form if possible. See Example 12. Recall these formulas from Section 1.4:

Area of a circle	Area of a rectangle	Area of a square
$A = \pi r^2$	$A = lw$	$A = s^2$

40.

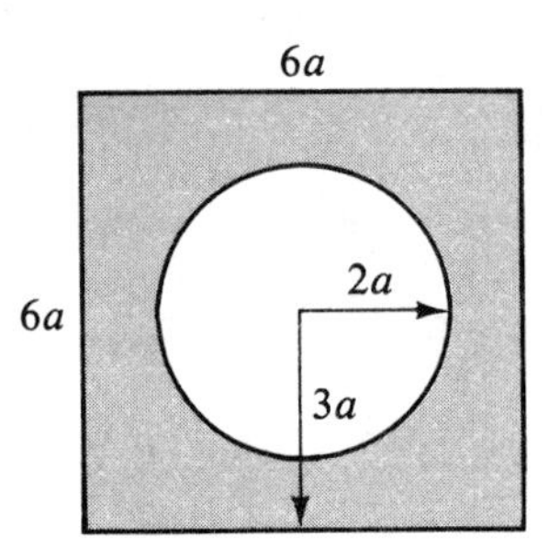

41.

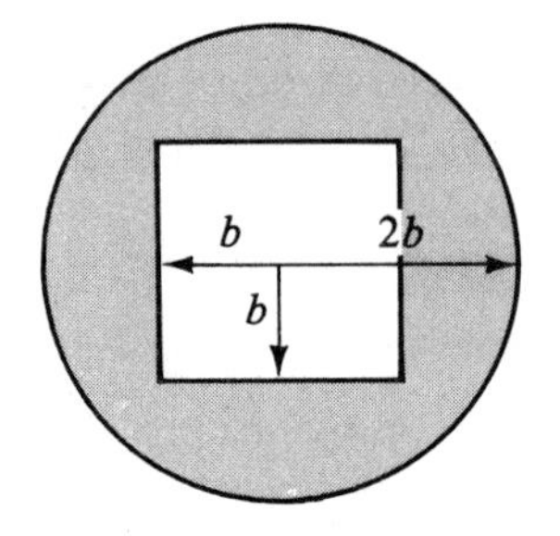

42.

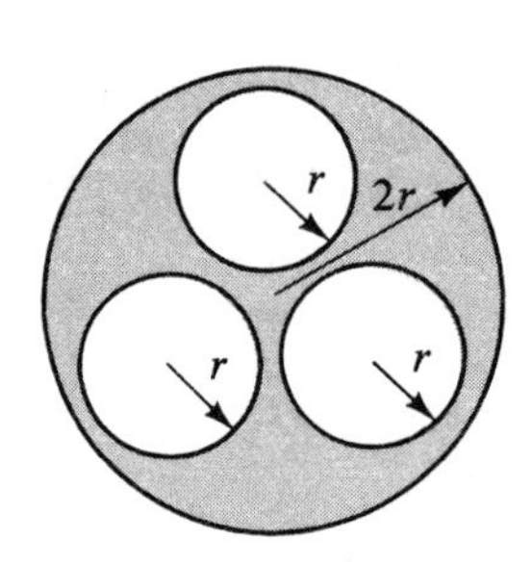

43.

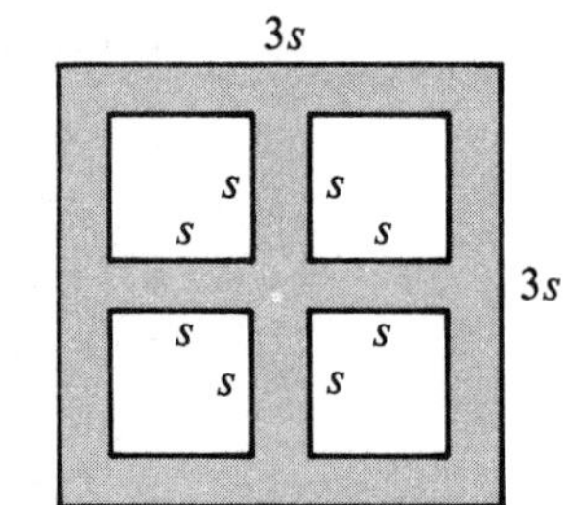

44.

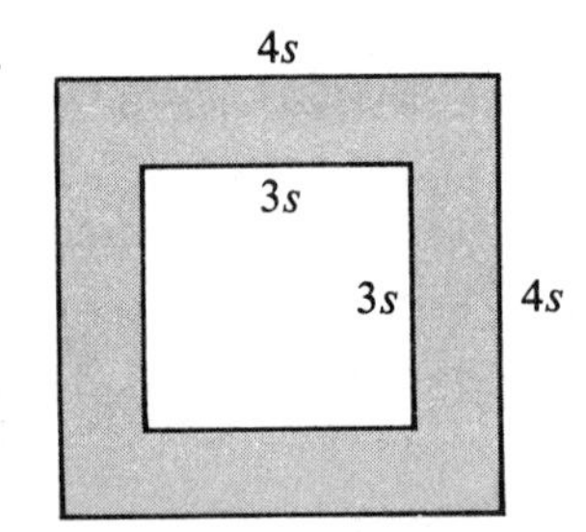

45.

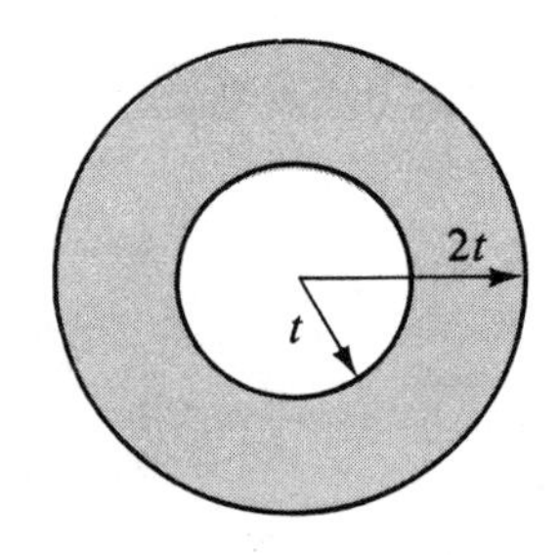

46.

47.

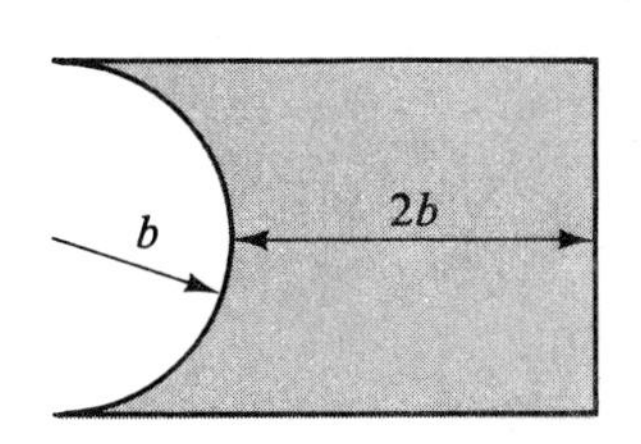

5.2 Factoring Trinomials

In this section we will concentrate on a particular product and its factors, the second-degree trinomial in a single variable with a leading coefficient of 1. We will also consider only integers as coefficients. The process we will use relates to FOIL, which was discussed in Section 4.3.

Consider this example:

$$(x + 2)(x - 5) = x^2 - 3x - 10.$$

First examine the second-degree, or leading, term and the last term or constant.

$$x^2 \qquad -10$$

$$(x + 2)(x - 5)$$

These terms are the products of the variable terms and the constants of the binomial factors. Recall now the origin of the first-degree term.

$$(x + 2)(x - 5)$$

$$2x$$

$$-5x$$

The so-called middle term of the product is the sum of the products of the variable and constant terms in the binomial factors. Thus,

$$(x + 2)(x - 5) = x^2 - 3x - 10$$

Now consider a product and reverse the procedure to determine the factors.

$$x^2 - 4x - 21$$

First, find two factors whose product is x^2. They determine the variable terms of the factors and hence the form of the factors.

$$(x \quad)(x \quad)$$

Second, factor the constant term.

$$-21 = (-1)(21) = (1)(-21) = (-3)(7) = (3)(-7)$$

These factors yield all the possible pairs of factors, which may be listed.

$$(x - 1)(x + 21)$$
$$(x + 1)(x - 21)$$
$$(x - 3)(x + 7)$$
$$(x + 3)(x - 7)$$

Third, check each to see which of the possibilities gives the correct middle term.

$$(x + 3)(x - 7) = x^2 - 4x - 21$$

We were able to factor the trinomial by reversing our knowledge of multiplication. Not all trinomials are this simple, but the same procedure can be followed.

Procedure for Factoring a Trinomial

1. Find the factors of the second-degree term, and set up the binomials.
2. Find the factors of the constant term, and consider all possible binomials.
3. Determine the factors that yield the correct middle term. If no pair of factors produces the correct full product, then the trinomial is nonfactorable using integers.

We will call this type of factoring FOIL.

In Examples 1–3, factor each trinomial.

Example 1. $x^2 + 9x + 8$

Solution *First,* factor x^2.

$$(x \quad)(x \quad)$$

Second, factor 8.

$$(x \quad 1)(x \quad 8)$$
$$(x \quad 2)(x \quad 4)$$

The signs must be positive, because the middle term is positive.

$$(x + 1)(x + 8)$$
$$(x + 2)(x + 4)$$

Third, select the correct combination.

$$x^2 + 9x + 8 = \mathbf{(x + 1)(x + 8)}$$ ■

Example 2. $x^2 - 5x - 24$

Solution *First,*

$$(x \quad)(x \quad)$$

The second step should be a *mental* enumeration of the possibilities. You certainly are not expected to write them out as shown in these examples.

Second,

$$\begin{array}{l} (x \quad 1)(x \quad 24) \\ (x \quad 2)(x \quad 12) \\ (x \quad 3)(x \quad 8) \\ (x \quad 4)(x \quad 6) \end{array}$$

The signs must be different, because the constant is negative.

$$\begin{array}{ll} (x+1)(x-24) & (x-1)(x+24) \\ (x+2)(x-12) & (x-2)(x+12) \\ (x+3)(x-8) & (x-3)(x+8) \\ (x+4)(x-6) & (x-4)(x+6) \end{array}$$

Third,

$$x^2 - 5x - 24 = \mathbf{(x + 3)(x - 8)}$$

■

Example 3. $x^2 + 5x + 8$

Solution *First,*

$$(x \quad\quad)(x \quad\quad)$$

Second,

$$\begin{array}{l} (x \quad 1)(x \quad 8) \\ (x \quad 2)(x \quad 4) \end{array}$$

The signs must be the same, because the constant is positive:

$$\begin{array}{ll} (x+1)(x+8) & (x-1)(x-8) \\ (x+2)(x+4) & (x-2)(x-4) \end{array}$$

None of these possibilities gives the correct middle term, so this trinomial is ***not factorable.*** ■

Factoring and Areas (Optional)

A trinomial can represent something such as area, and its factors are simply the length and width. Suppose the area of a rectangle is $x^2 + 5x + 6$. What are the possible dimensions of the rectangle? The dimensions are the length and the width, and the area is the product of the length and the width. The factors of the area are, therefore, the dimensions.

$$x^2 + 5x + 6 = (x + 3)(x + 2)$$

By factoring we see that the dimensions of the rectangle are $x + 3$ by $x + 2$.

$x + 3$

$x + 2$

More importantly, we can see that there is an application of factoring trinomials whenever the quantity is the product of two or more other quantities. Just glancing at a list of formulas, we would find many product relationships.

In Examples 4–6, illustrate the factored equations as areas.

Example 4. Illustrate the equation $x^2 + 4x + 3 = (x + 3)(x + 1)$ geometrically.

Solution *Step 1* Draw a rectangle with sides of dimensions indicated by the equation in factored form.

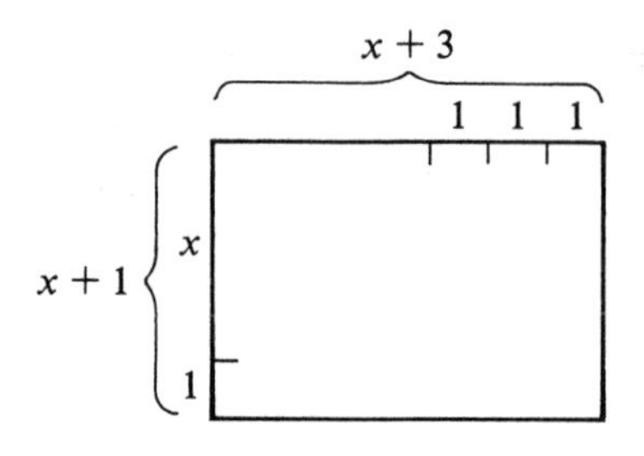

Step 2 Divide the rectangle into the component parts.

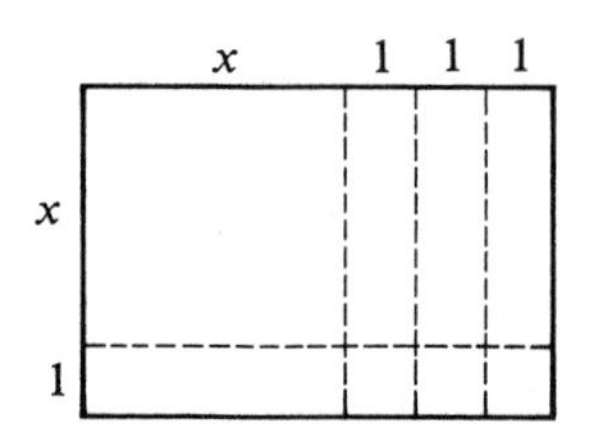

Step 3 Label the area of each part: $x^2 + x + x + x + x + 1 + 1 + 1$

These pieces have area $x^2 + 4x + 1$.

Thus, $(x + 3)(x + 1) = x^2 + 4x + 1$. ■

Example 5. Factor $x^2 + 3x + 2$ using areas.

Solution *Step 1* Draw an area diagram for each term.

Step 2 Place all the preceding pieces into the form of a rectangle. Place them one at a time as shown by the following sequence of steps:

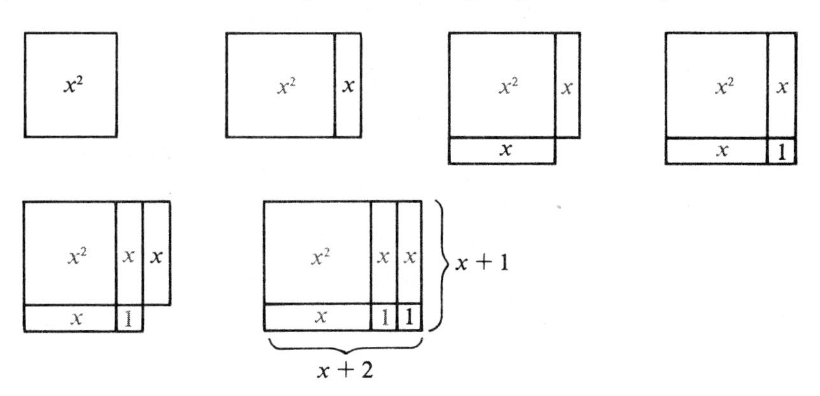

Step 3 $x^2 + 3x + 2 = (x + 1)(x + 2)$ ■

Example 6. Example 3 showed that $x^2 + 5x + 8$ is not factorable. Illustrate this using areas.

Solution *Step 1*

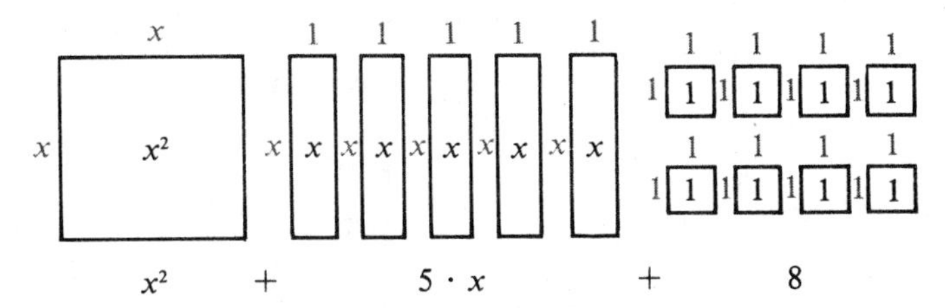

Step 2 There are no other ways of arranging the pieces. (Three pieces on the bottom and two on the side is considered the same as the one shown.)

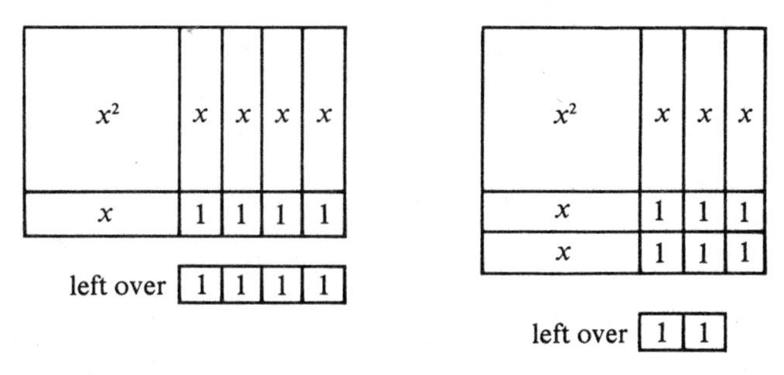

Step 3 It is not possible to arrange the pieces into a rectangular form, so the trinomial is not factorable. ■

Problem Set 5.2

Factor each of the expressions in Problems 1–30, if possible. See Examples 1–3.

1. $x^2 + 3x + 2$ **2.** $x^2 - 3x + 2$ **3.** $x^2 + x - 2$
4. $x^2 + 4x + 3$ **5.** $x^2 - 4x + 3$ **6.** $x^2 - 2x - 3$

7. $y^2 - 5y + 6$
8. $y^2 + y - 6$
9. $y^2 - y - 6$
10. $y^2 - 7y + 12$
11. $y^2 + 7y + 12$
12. $y^2 - 4y - 12$
13. $z^2 - 2z - 8$
14. $z^2 + 7z - 8$
15. $z^2 - 7z - 8$
16. $z^2 - 8z - 9$
17. $z^2 - 10z + 9$
18. $z^2 - 6z + 9$
19. $a^2 + 5a - 14$
20. $a^2 + 2a - 15$
21. $a^2 - a - 30$
22. $b^2 - 11b + 24$
23. $b^2 + 3b - 18$
24. $b^2 - 19b + 34$
25. $c^2 - 15c + 26$
26. $c^2 - c - 12$
27. $c^2 - 5c + 24$
28. $d^2 + 9d - 14$
29. $d^2 - 2d - 143$
30. $d^2 - 4d - 165$
31. The area of a rectangle is $r^2 - 2r - 24$. What are the dimensions of the figure?
32. What are the dimensions of a rectangle whose area is given by $a^2 - 4a - 32$?
33. The total number of trees in an orchard is $t^2 - 9t + 14$. How many rows are there and how many trees in each row?
34. How many rows of how many desks are there if there are $d^2 - 10d + 24$ desks in the room?

Area Problems (Optional)

Illustrate the given equation in Problems 35–38 using the figure supplied for each. See Example 4.

35. $(x + 2)(x + 3) = x^2 + 5x + 6$

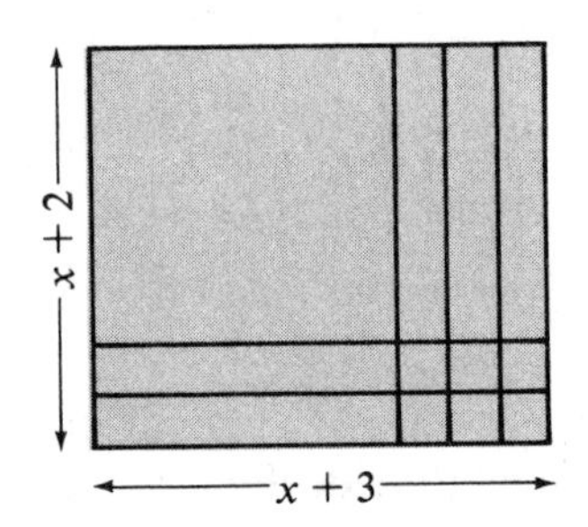

36. $(x + 3)(x + 4) = x^2 + 7x + 12$

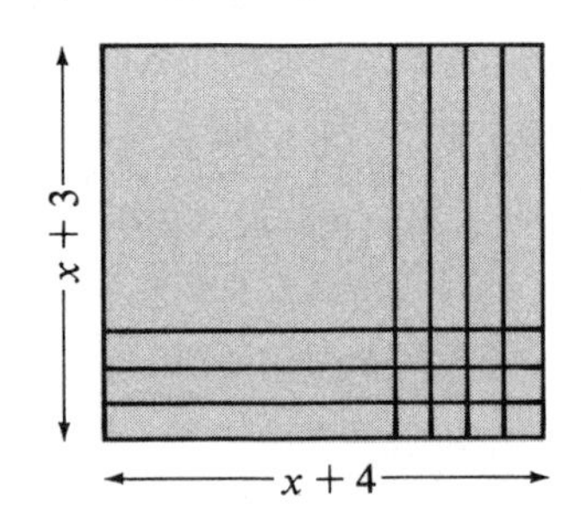

37. $y^2 + y - 2 = (y - 1)(y + 2)$

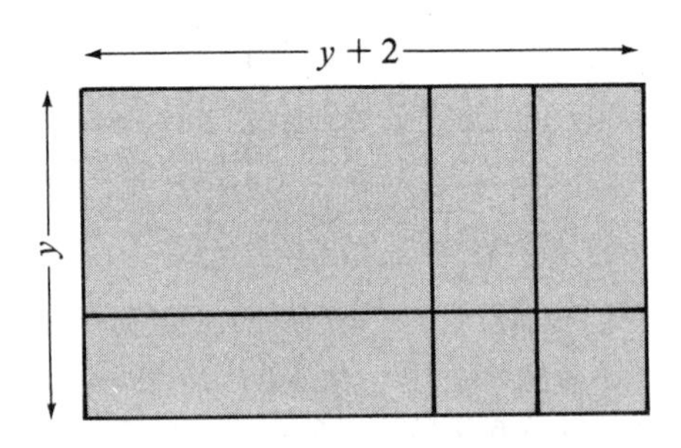

38. $z^2 - z - 2 = (z + 1)(z - 2)$

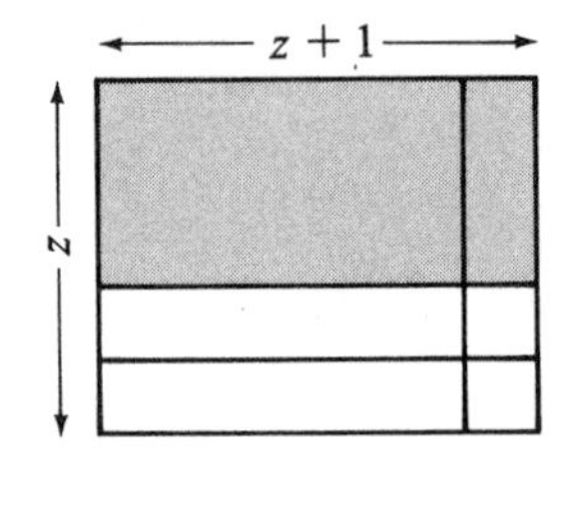

Factor the equations given in Problems 39–41 using areas. See Example 5.

39. $x^2 + 4x + 3$
40. $x^2 + 5x + 4$
41. $x^2 + 6x + 8$

Illustrate that the trinomials given in Problems 42–44 are not factorable by using areas. See Example 6.

42. $x^2 + 2x + 2$
43. $x^2 + 3x + 4$
44. $x^2 + 4x + 2$

5.3 More Factoring

The process of using FOIL extends to trinomials that do not have a leading coefficient of 1, as well as to trinomials with more than one variable.

In Examples 1–5, factor each trinomial.

Example 1. $3x^2 + 7x + 2$

Solution *First,* factor $3x^2$.

$(x \quad)(3x \quad)$

Second, factor 2.

$(x \quad 1)(3x \quad 2)$
$(x \quad 2)(3x \quad 1)$

The signs must be positive, because the middle term is positive.

$(x + 1)(3x + 2) = 3x^2 + 5x + 2$
$(x + 2)(3x + 1) = 3x^2 + 7x + 2$

Third, select the correct combination.

$\mathbf{(x + 2)(3x + 1)}$ ■

Example 2. $6x^2 - 13x + 5$

Solution *First,*

$(x \quad)(6x \quad)$
$(2x \quad)(3x \quad)$

Second,

$(x - 1)(6x - 5)$
$(x - 5)(6x - 1)$
$(2x - 1)(3x - 5)$
$(2x - 5)(3x - 1)$

Note that the signs are negative, because the constant term is positive and the middle term negative.

Third,

$\mathbf{(2x - 1)(3x - 5)} = 6x^2 - 13x + 5$ ■

Example 3. $3x^2 + 7x - 20$

Solution *First,*

$(x \quad)(3x \quad)$

Second,

$(x \quad 1)(3x \quad 20)$ or $(x \quad 20)(3x \quad 1)$
$(x \quad 2)(3x \quad 10)$ $\quad (x \quad 10)(3x \quad 2)$
$(x \quad 4)(3x \quad 5)$ $\quad (x \quad 5)(3x \quad 4)$

The negative constant term (−20) means that its factors must have opposite signs. We omit the signs in the second step until they are correctly determined in the trial-and-error stage.

Third,

$$(x + 4)(3x - 5) = 3x^2 + 7x - 20$$ ■

Example 4. $2x^2 - 11x - 15$

Solution *First,*

$(x \quad)(2x \quad)$

Second,

$(x - 1)(2x + 15)$ $(x + 1)(2x - 15)$
$(x - 3)(2x + 5)$ $(x + 3)(2x - 5)$
$(x - 5)(2x + 3)$ $(x + 5)(2x - 3)$
$(x - 15)(2x + 1)$ $(x + 15)(2x - 1)$

Note: $(x - 3)(2x - 5) = 2x^2 - 11x + 15$, which differs only in sign. Be careful about binomials that are almost factors.

Third,
None of the possibilities yields the proper product, thus $2x^2 - 11x - 15$ is **not factorable.** ■

Example 5. $2x^2 + xy - y^2$

Solution *First,*

$(2x \quad)(x \quad)$

Second,

$(2x \quad y)(x \quad y)$

The signs must be opposites because the third term is negative:

$(2x - y)(x + y)$
$(2x + y)(x - y)$

Third,
The correct factorization is $\mathbf{(2x - y)(x + y)}$. ■

In Section 5.1 we worked with polynomials whose terms have a common factor—a constant, a variable, or both. Now we combine this type of factoring with factoring trinomials.

Example 6. Completely factor $6x^3 - 21x^2 - 12x$.

Solution

$6x^3 - 21x^2 - 12x = 3x(2x^2 - 7x - 4)$ *Common factor first.*

$= 3x(x \quad)(2x \quad)$
$(x \quad 1)(2x \quad 4)$
$(x \quad 2)(2x \quad 2)$
$(x \quad 4)(2x \quad 1)$

FOIL (Don't write this down—this is a mental step.)

$= \mathbf{3x(x - 4)(2x + 1)}$ *Completely factored form* ■

Problem Set 5.3

Factor each of the expressions in Problems 1–30, and factor each completely, if possible. See Examples 1–6.

1. $2y^3 - 7y + 3$
2. $3y^2 - 10y + 3$
3. $2u^2 + 9u + 4$
4. $2a^2 + a - 6$
5. $5v^2 - 3v - 2$
6. $7b^2 + 4b - 3$
7. $3w^2 + w - 2$
8. $3c^2 + 7c + 2$
9. $7x^2 + x - 6$
10. $3y^2 + 8y + 4$
11. $6z^2 - 27z - 15$
12. $8t^2 + 12t + 4$
13. $5u^3 + 7u^2 - 6u$
14. $7v^4 - 11v^3 - 6v^2$
15. $3w^4 + 3w^3 - 36w^2$
16. $2x^3y + 8x^2y - 24xy$
17. $xy^2 - 17xy + 72x$
18. $yz^2 + 18yz + 72y$
19. $6z^2 + 5z - 4$
20. $6z^2 - 7z - 3$
21. $15t^2 + t - 6$
22. $15u^2 + 13u - 6$
23. $8v^2w + 16vw + 6w$
24. $12vw^2 + 12vw - 9v$
25. $12x^2 + 2 - 11x$
26. $5y - 2 + 12y^2$
27. $9h^2 + 9h - 10$
28. $5mn^2 + 15mn + 10m$
29. $(3xy)^2 + 3x^2y - 2x^2$
30. $(2xy)^3 + 2x^2y^3 - 3xy^3$
31. The area of a rectangle is $6r^2 - r - 2$. What are the dimensions of the figure?
32. What are the dimensions of a rectangle whose area is given by $10r^2 + 7r - 12$?
33. What is the time needed to travel a distance of $8s^2 - 10s + 3$, traveled at a rate of $2s - 1$?
34. If a distance of $9s^2 - 6s - 8$ is traveled in a time of $3s + 2$, what is the rate?

5.4 Factoring Special Types

In this section we will look at two very special cases of factoring. These two examples are important because there are several applications that require an understanding of one or the other of these cases.

For the first type, consider binomial factors that produce a binomial instead of a trinomial.

$$(a + b)(a - b) = a^2 - b^2 \qquad (z + 3)(z - 3) = z^2 - 9$$
$$(x + 1)(x - 1) = x^2 - 1 \qquad (2x + 1)(2x - 1) = 4x^2 - 1$$
$$(y + 2)(y - 2) = y^2 - 4 \qquad (4x + 5)(4x - 5) = 16x^2 - 25$$

The product in each example is the difference of two terms, and each of the terms is the square of one of the terms in the binomials. This distinctive case is called the ***difference of squares,*** and its factors are a sum and difference.

Difference of Squares

$$x^2 - a^2 = (x + a)(x - a)$$

$$\begin{pmatrix}\text{FIRST}\\\text{TERM}\end{pmatrix}^2 - \begin{pmatrix}\text{SECOND}\\\text{TERM}\end{pmatrix}^2 = \begin{pmatrix}\text{SUM}\\\text{OF TERMS}\end{pmatrix}\begin{pmatrix}\text{DIFFERENCE}\\\text{OF TERMS}\end{pmatrix}$$

The binomial factors $x + a$ and $x - a$ are called ***conjugates.*** Thus, $x + 6$ and

$x - 6$, $2t - 5$ and $2t + 5$, and $2v + 3w$ and $2v - 3w$ are pairs of conjugates. The conjugate of $3r + 8$ is $3r - 8$.

Examples *Factor each of the following, examples if possible.*

1. $x^2 - 36 = (x)^2 - (6)^2$
   $= (x + 6)(x - 6)$

   The first step in each of these examples is usually not shown but is done mentally.

2. $9r^2 - 64 = (3r)^2 - (8)^2$
   $= (3r + 8)(3r - 8)$

3. $8t^2 - 50 = 2(4t^2 - 25)$
   $= 2[(2t)^2 - (5)^2]$
   $= 2(2t + 5)(2t - 5)$

   Don't forget to common factor first.

4. $4x^2 + 81$ *not a difference*

5. $9y^2 - 125$ *125 not a perfect square*

6. $4v^2 - 9w^2 = (2v)^2 - (3w)^2$
   $= (2v + 3w)(2v - 3w)$ ■

There is a very easy geometric interpretation to factoring the difference of squares. Consider a square with side x and area x^2. If a square with area a^2 is removed, the remaining area is $x^2 - a^2$.

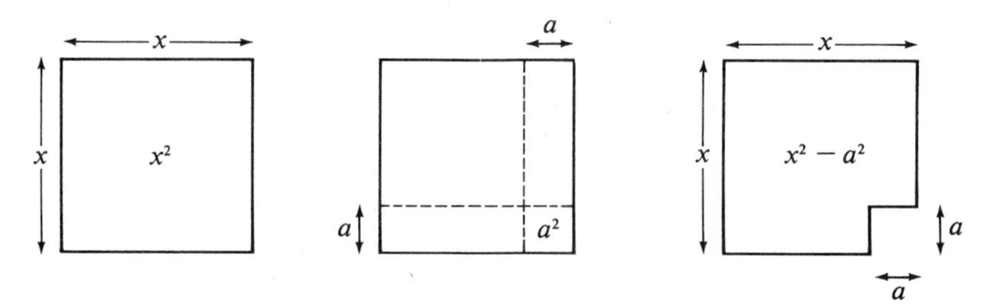

The area is unchanged when the shaded piece in Figure 5.3 is moved to form a rectangle with dimensions $x + a$ by $x - a$. This demonstrates that $x^2 - a^2 = (x - a)(x + a)$.

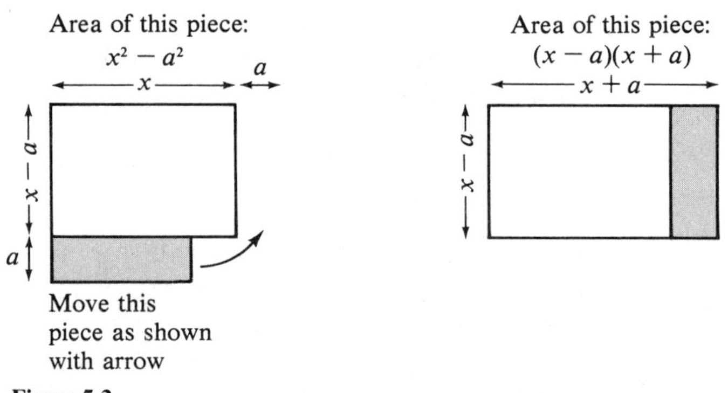

Figure 5.3
A Geometric Interpretation of $x^2 - a^2 = (x - a)(x + a)$

The second special type involves the product called a ***perfect square.***

$$(x+1)^2 = x^2 + 2(x)(1) + 1^2 \qquad (x-3)^2 = x^2 - 6x + 9$$
$$= x^2 + 2x + 1 \qquad (x+4)^2 = x^2 + 8x + 16$$
$$(x+2)^2 = x^2 + 4x + 4 \qquad (x-5)^2 = x^2 - 10x + 25$$

We can see that the product does take a special form. The first and last terms are perfect squares, and the middle term is twice the product of the numbers being squared.

Perfect Square

$$x^2 + 2ax + a^2 = (x+a)^2 = (x+a)(x+a)$$

$$\begin{pmatrix}\text{FIRST}\\\text{TERM}\end{pmatrix}^2 + 2\begin{pmatrix}\text{FIRST}\\\text{TERM}\end{pmatrix}\begin{pmatrix}\text{SECOND}\\\text{TERM}\end{pmatrix} + \begin{pmatrix}\text{SECOND}\\\text{TERM}\end{pmatrix}^2 = \begin{pmatrix}\text{FIRST}\\\text{TERM}\end{pmatrix} + \begin{pmatrix}\text{SECOND}\\\text{TERM}\end{pmatrix}^2$$

Examples *Factor each expression in Examples 7–10 to see if the trinomial is a perfect square.*

7. $x^2 + 12x + 36 = x^2 + 2(6x) + (6)^2$
$= (x+6)^2$

8. $4x^2 - 4x + 1 = (2x)^2 - 2(2x) + (1)^2$
$= (2x-1)^2$

9. $z^2 + 10z + 100 = (z)^2 + \quad + (10)^2$
The middle term is not twice $10 \cdot z$.
This trinomial is not a perfect square.

10. $18x^2 - 60x + 50 = 2(9x^2 - 30x + 25)$
$= 2[(3x)^2 - 2(15x) + (5)^2]$
$= 2(3x-5)^2$ ■

As before, there is a geometric interpretation for a perfect square. Consider the following figure. The total area of this square is the sum of the areas of the four regions created by the divisions.

$$\begin{pmatrix}\text{AREA OF}\\\text{REGION I}\end{pmatrix} + \begin{pmatrix}\text{AREA OF}\\\text{REGION II}\end{pmatrix} + \begin{pmatrix}\text{AREA OF}\\\text{REGION III}\end{pmatrix} + \begin{pmatrix}\text{AREA OF}\\\text{REGION IV}\end{pmatrix} = \begin{pmatrix}\text{TOTAL AREA}\\\text{OF SQUARE}\end{pmatrix}$$

Now look at Figure 5.4 on the next page.

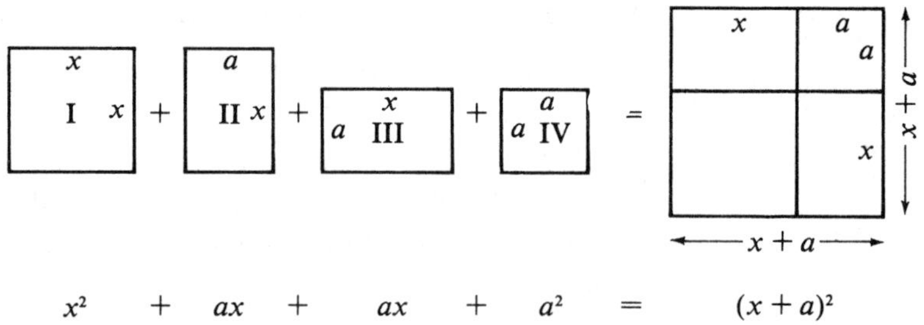

$$x^2 + ax + ax + a^2 = (x + a)^2$$

Figure 5.4
A Geometric Interpretation of $x^2 + 2ax + a^2 = (x + a)^2$

We multiply to find a product, and sometimes we are given the product and must find its factors. In this section, we have considered two very special types of products with very distinctive factors. Throughout this chapter, we have used an understanding of the form of the product to find the factored form.

Procedure in Factoring

1. Is there a ***common monomial factor?*** If so, factor the greatest common monomial from the polynomial, and consider the polynomial factor.
2. If the factor is a binomial, is it the ***difference of squares?*** You can factor such a form.
3. If the factor is a trinomial, is it a ***perfect square?*** You can factor this form.
4. If the factor is a trinomial and is not a perfect square, begin to look for the ***binomial factors.***
   a. Factor the leading term, and set up the possible binomial factors.
   b. Factor the last term, and list all of the possibilities.
   c. Select the pair of binomial factors that produce the correct middle term.
5. If all the possibilities fail, then the polynomial is ***nonfactorable*** with integer coefficients. Such a nonfactorable polynomial is called *prime* over the integers. In this book, when we say *nonfactorable* we mean nonfactorable over the integers.

To see how the checklist is used, examine the following examples. Notice, especially, the last three examples where the expression must be simplified *before* it may be factored.

Examples *Write each of the following Examples 11–15 in completely factored form.*

11. $2x^3 - 12x^2 + 18x = 2x(x^2 - 6x + 9)$ *Common factor first.*
$= 2x(x - 3)^2$ *Perfect square.*

12. $6x^3y + 18x^2y^2 + 12xy^3 = 6xy(x^2 + 3xy + 2y^2)$
$= 6xy(x + y)(x + 2y)$

13. $(3z + 8)(2z - 3) - 7z = (6z^2 + 7z - 24) - 7z$
$= 6z^2 - 24$
$= 6(z^2 - 4) = 6(z + 2)(z - 2)$

14. $$\begin{aligned}(h^3 - 2) - (2h - 1)(h + 2) &= h^3 - 2 - (2h^2 + 3h - 2)\\ &= h^3 - 2 - 2h^2 - 3h + 2\\ &= h^3 - 2h^2 - 3h\\ &= h(h^2 - 2h - 3)\\ &= h(h + 1)(h - 3)\end{aligned}$$

15. $$\begin{aligned}(2a - 1)(a + 3) - (a + 1)(a + 7) &= (2a^2 + 5a - 3) - (a^2 + 8a + 7)\\ &= 2a^2 + 5a - 3 - a^2 - 8a - 7\\ &= a^2 - 3a - 10\\ &= (a + 2)(a - 5)\end{aligned}$$

■

Problem Set 5.4

Factor each of the polynomials in Problems 1–15, if possible. If any is nonfactorable, say so. See Examples 1–6.

1. $x^2 - 64$ **2.** $y^2 - 169$ **3.** $x^2 - 9$
4. $p^2 - 4$ **5.** $z^2 - 121$ **6.** $4t^2 - 1$
7. $4w^2 - 9$ **8.** $9m^2 + 4$ **9.** $25n^2 + 4$
10. $16y^2 - 25$ **11.** $9z^2 - 4$ **12.** $9x^2 - 16$
13. $x^4 - 1$ **14.** $y^6 - 1$ **15.** $z^8 - 1$

Factor each of the polynomials in Problems 16–24, if possible. See Examples 7–10.

16. $x^2 + 2x + 1$ **17.** $y^2 - 2y + 1$ **18.** $x^2 - 10x + 25$
19. $y^2 + 14y + 49$ **20.** $4z^2 + 12z + 9$ **21.** $9z^2 - 12z + 4$
22. $4y^2 - 28y + 49$ **23.** $4x^2 - 12x + 9$ **24.** $25t^2 - 10t + 1$

Factor each of the polynomials in Problems 25–45, if possible. See Examples 11–15.

25. $2x^2 - 32x + 128$ **26.** $3x^2 - 147$ **27.** $2x^2 - 7x - 4$
28. $4x^2 + 7x - 2$ **29.** $4y^2 - 4$ **30.** $9y^2 + 9$
31. $2a^3 + 5a^2 + 3a$ **32.** $3a^3 - a^2 - 2a$ **33.** $6b^2 + 15b + 9$
34. $6m^3 - 13m^2 + 2m$ **35.** $4m^3 + 9m^2 + 2m$ **36.** $10x^2 + 3x - 1$
37. $21x^2 + 4x - 1$ **38.** $9y^2 + 15y + 25$ **39.** $4y^2 - 14y + 49$
40. $4z^3 - 12z^2 + 9z$ **41.** $2z^2 - 12z + 18$ **42.** $5y^3 - 30y^2 + 45y$
43. $3x^4 + 30x^3 + 75x^2$ **44.** $h^4k^2 - 4h^2k^2$ **45.** $2x^3y - 2xy^3$

5.5 Solving Quadratic Equations by Factoring

Factoring has far-reaching applications in mathematics. Chapter 6, which discusses rational expressions, relies very heavily on factoring. In this section we use factoring to solve certain types of equations. To solve these equations, we will also need a property of products. If the product of two numbers—call them a and b—equals 0, then what do you know about the numbers? That is, if $a \cdot b = 0$, what can you say about a and b? Even a quick glance at a multiplication table reveals that 0 appears as an answer in only one row and only one column, the 0 row and the 0 column. To get 0 as a product means that one (or perhaps both) of the factors is 0.

Property of Products

> If $a \cdot b = 0$, then $a = 0$ or $b = 0$, or $a = b = 0$.
>
> If the product of two numbers is zero, then at least one of the factors must be zero.

Using this simple but important result, we can easily arrange the nonzero terms on one side of an equation. To solve such an equation, we must be able to find the factors of the expression and determine when they are zero.

Solve each of the following equations in Examples 1–5.

Example 1. $x^2 = x$

Solution

$x^2 - x = 0$
$x(x - 1) = 0$
$x = 0 \quad \text{or} \quad x - 1 = 0$
$x \qquad\qquad = 1$
$\mathbf{x = 0 \quad or \quad x = 1}$

Arrange nonzero terms on one side of the equation. Find the factors. Set the factors equal to zero. Solve for the variable. ■

Example 2. $x^2 = 9$

Solution

$x^2 - 9 = 0$ — *Equal to zero.*
$(x + 3)(x - 3) = 0$ — *Factor.*
$x + 3 = 0 \quad \text{or} \quad x - 3 = 0$ — *Set equal to zero.*
$\mathbf{x = -0 \quad or \quad x = 3}$ — *Solve.* ■

Example 3. $y^2 + 3y - 10 = 0$

Solution

$(y - 2)(y + 5) = 0$
$y - 2 = 0 \quad \text{or} \quad y + 5 = 0$
$\mathbf{y = 2 \quad or \quad y = -5}$ ■

Example 4. $6x^2 = 4 - 5x$

Solution

$6x^2 + 5x - 4 = 0$ — *First rearrange the terms to obtain a zero on one side.*
$(2x - 1)(3x + 4) = 0$ — *Factor.*
$2x - 1 = 0 \quad \text{or} \quad 3x + 4 = 0$
$2x = 1 \quad \text{or} \quad 3x = -4$

$$\mathbf{x = \frac{1}{2} \quad or \quad x = -\frac{4}{3}}$$ ■

Example 5. $z(z - 8) = 4(z - 9)$ — *The factors $z(z - 8)$ and $4(z - 9)$ are not equal to zero, so these factors do not help.*

Solution

$z^2 - 8z = 4z - 36$
$z^2 - 12z + 36 = 0$ — *Arrange the terms so they are equal to zero.*
$(z - 6)(z - 6) = 0$ — *Factor.*
$z - 6 = 0 \quad \text{or} \quad z - 6 = 0$ — *Set the factors equal to zero.*
$\mathbf{z = 6 \quad or \quad z = 6}$ — *Solve and state the solution.* ■

It seems that each factor produces a root. The previous examples show second-degree equations. Second-degree polynomials are called ***quadratics,*** and second-degree polynomial equations are called ***quadratic equations.*** The quadratics in the examples have two first-degree factors; therefore, the equations have *two* roots or solutions. We became accustomed to one solution when solving first-degree equations. As in Example 5, the solutions of a quadratic may not be unique, and in this case the two solutions are equal.

Let us now review this factoring procedure. We arrange the equation as a polynomial equal to zero and factor the polynomial. The factors are set equal to zero, and these equations are solved. What is important about knowing this procedure is that we are able to solve a greater number of problems. In particular, a number of its applications involve products that could produce quadratic or higher-degree polynomial equations. For instance, to compute a rectangular area, we multiply the dimensions.

Example 6. The area of a room is 108 sq ft. If the room is 3 ft longer than it is wide, find the dimensions of the room.

Solution

(LENGTH OF ROOM)·(WIDTH OF ROOM) = (AREA OF ROOM)

Let w = WIDTH OF ROOM; then $w + 3$ = LENGTH OF ROOM

$$(w + 3)w = 108$$
$$w^2 + 3w = 108$$
$$w^2 + 3w - 108 = 0$$
$$(w - 9)(w + 12) = 0$$
$$w - 9 = 0 \quad \text{or} \quad w + 12 = 0$$
$$w = 9 \quad \text{or} \quad w = -12$$

$w + 3 = 12$ *(Width cannot be negative.)*

The dimensions of the rectangular room are 9 ft by 12 ft. ■

Problem Set 5.5

Solve each of the quadratic equations in Problems 1–30 by factoring. See Examples 1–5.

1. $x^2 + 2x - 15 = 0$
2. $x^2 + x - 6 = 0$
3. $2x^2 - 3x - 9 = 0$
4. $3y^2 + y = 4$
5. $y^2 = 108 + 3y$
6. $y^2 + 9 = 6y$
7. $z^2 - 5z + 6 = 0$
8. $z^2 - 8z + 15 = 0$
9. $3z^2 + 10z + 7 = 0$
10. $x^2 - 10x = 0$
11. $x^2 - 14x = 0$
12. $2x^2 - x = 0$
13. $5x + 66 = x^2$
14. $9x + 10 = -2x^2$
15. $6x^2 = 7x + 5$
16. $y^2 - 25 = 0$
17. $y^2 = 16$
18. $y^2 = 169$
19. $41y = 14y^2 + 15$
20. $7y + 12 = 10y^2$
21. $9y^2 + 10 = 21y$
22. $4z = z^2 + 3$
23. $21z^2 + 4z = 1$
24. $15z^2 + 4z = 4$
25. $18r^2 = 6r + 60$
26. $15r^2 = 27r + 6$

27. $3r - r^2 = 10r + 10$
28. $2(2x - 3) = 5x(1 - 3x)$
29. $4x(x - 9) = 9(1 - 4x)$
30. $4(9x - 1) = 9x(4 - x)$

In Problems 31–36 design an equation and solve it to answer the stated question. See Example 6.

31. A rectangle is 4 ft longer than it is wide. The area is 45 sq ft. What are the dimensions of the rectangle?
32. A rectangle has an area of 48 sq yd. The width is 2 yd shorter than the length. What are the dimensions of the rectangle?
33. Find the dimensions of a rectangle whose area is 24 sq in. if the sum of the length and width is 11 in.
34. Find the dimensions of a rectangle whose area is 18 sq cm if the difference between length and width is 3 cm.
35. A rectangular field measures 40 by 60 yd. A strip is added to the field on all sides that produces a rectangular field with twice the area. How wide is the strip?
36. A rectangular lawn is 40 by 60 ft. A strip is tilled around the lawn and planted with flowers. If the lawn area is reduced to 1500 sq ft, how wide is the flower bed?

5.6 Review

Factoring determines the factors of a given product. The process depends on a clear understanding of the multiplication of polynomials, particularly binomials. Using area, you can illustrate the process geometrically. Certain polynomial equations can now be solved by factoring, enabling us to tackle more substantive problems. Factoring will continue to play a key role throughout the remainder of the course and deserves whatever effort is necessary for you to achieve proficiency.

Self-Test

[5.1] **1.** Write the prime factorization in exponent form.
a. 64 **b.** 1000 **c.** 196 **d.** $x^2 - 12xy + 36y^2$

[5.1] **2.** Determine the greatest common monomial factor of the terms of each expression and factor completely.
a. $5xy + 5$ **b.** $\frac{2}{3}hk + 2h$ **c.** $5x^2 + 15x + 20$ **d.** $3y^4 - 5y^3 + 7y^2$

[5.2] **3.** Factor completely, if possible.
a. $x^2 - 6x + 5$ **b.** $x^2 - 4x - 5$ **c.** $x^2 + 4x + 5$ **d.** $x^2 - 5x + 4$

[5.2] **4.** Factor completely, if possible.
a. $x^2 - 5x - 6$ **b.** $x^2 + 5x - 6$ **c.** $x^2 + 5x + 6$ **d.** $x^2 - 5x + 6$

In Problems 5–8, factor each expression completely, if possible.

[5.3] **5.** **a.** $2x^2 - 11x + 15$ **b.** $3x^2 + 5xy - 2y^2$ **c.** $18x^2 + 21x - 60$
[5.3] **6.** **a.** $2x^2 + 7x - 4$ **b.** $x^2 - 10xy - 24y^2$ **c.** $12x^2 + 7x + 1$
[5.4] **7.** **a.** $x^2 - 49$ **b.** $25w^2 + 36$ **c.** $x^2 - 22x + 121$ **d.** $x^2 - 12xy + 36y^2$
[5.4] **8.** **a.** $3x^2 - 27$ **b.** $8x^2 + 24x + 16$ **c.** $8x^2 + 24x + 18$ **d.** $3x^4 + x^3 - 14x^2$

[5.5] **9.** Solve each equation.

a. $a^2 + 10a + 25 = 0$ **b.** $b^2 - 169 = 0$

c. $2c^2 = 3c + 35$ **d.** $d = 28 - 15d^2$

[5.5] **10.** The area of a rectangular room is 176 sq ft. If the room is 5 ft longer than it is wide, what are the dimensions?

Area Problems

[5.1] **11.** Write an expression for the shaded area shown and write the expression in factored form.

[5.2] **12.** Factor $x^2 + 5x + 6$ using areas.

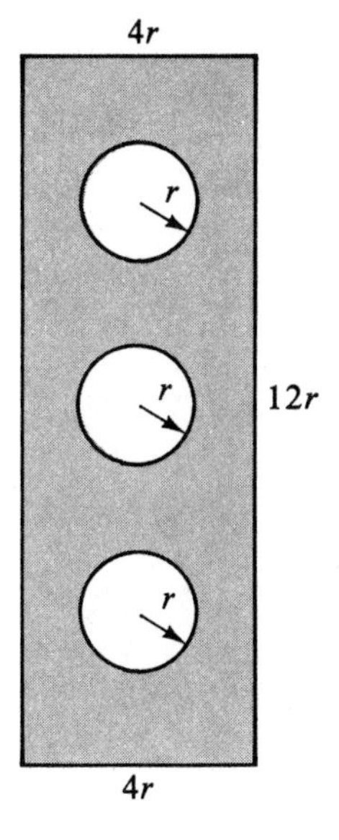

Review Problems

Factor each expression in Problems 1–75.

1. $3x + 18$
2. $24x - 6$
3. $20x + 4$
4. $6x^2 + 5x$
5. $3x^2 - 2x$
6. $9x^2 - 6x$
7. $3xy^2 + 5x^2y$
8. $9s^3t^2 - 3s^2t^3$
9. $5x^3y^3 - 5xy$
10. $x^2 + 8x + 15$
11. $x^2 + 7x + 10$
12. $x^2 + 7x + 12$
13. $x^2 - 7x + 6$
14. $x^2 - 6x + 8$
15. $x^2 - 8x + 15$
16. $x^2 + x - 20$
17. $x^2 + 3x - 10$
18. $x^2 + 2x - 15$
19. $6x^2 + 11x + 4$
20. $6x^2 + 13x + 6$
21. $5x^2 + 17x + 6$
22. $5x^2 - 17x + 6$
23. $12x^2 + 5x - 2$
24. $9x^2 + 3x - 20$
25. $6x^2 + 11x - 2$
26. $8x^2 - 7x - 1$
27. $24x^2 + 2x - 5$
28. $x^3 + 3x^2 + 2x$
29. $x^3 + x^2 - 2x$
30. $x^3 - 2x^2 - 3x$
31. $x^4 - 4x^3 + 3x^2$
32. $x^4 - x^3 - 6x^2$
33. $x^4 + 2x^3 - 8x^2$
34. $3x^2 + 3x - 6$
35. $4x^2 + 4x - 24$
36. $5x^2 + 20x - 25$
37. $2x^2 + 2x - 24$
38. $3x^2 - 9x - 12$
39. $3x^2 - 9x - 30$
40. $3x^2 + 6x - 45$
41. $2x^2 + 4x - 30$
42. $5x^2 + 15x - 20$
43. $x^2 - 36$
44. $x^2 - 100$
45. $x^2 - 1$
46. $x^2 - 121$
47. $x^2 - 144$
48. $x^2 - 49$
49. $5x^2 - 5$
50. $3x^2 - 12$
51. $2x^2 - 18$
52. $2 - 2x^2$
53. $16 - 4x^2$
54. $36 - 4x^2$

55. $4x^2 - 9$
56. $9x^2 - 16$
57. $25x^2 - 49$
58. $x^2 - 6x + 9$
59. $x^2 - 8x + 16$
60. $x^2 + 10x + 25$
61. $4x^2 + 4x + 1$
62. $9x^2 - 12x + 4$
63. $16x^2 - 40x + 25$
64. $x^3 + 4x^2 + 4x$
65. $x^4 - 6x^3 + 9x^2$
66. $x^5 + 2x^4 + x^3$
67. $3x^2 + 24x + 48$
68. $4x^2 - 16x + 16$
69. $5x^2 - 10x + 5$
70. $4x^3 + 4x^2 + x$
71. $3x^3 + 5x^2 + 2x$
72. $2x^3 - 7x^2 + 3x$
73. $6x^2 - 21x - 12$
74. $12x^2 - 14x - 6$
75. $12x^3 - 14x^2 - 6x$

Solve each equation in Problems 76–100.

76. $x^2 + 7x + 10 = 0$
77. $x^2 + 8x + 15 = 0$
78. $x^2 - 4x + 3 = 0$
79. $x^2 - 9x + 14 = 0$
80. $x^2 + 3x - 10 = 0$
81. $x^2 + x - 2 = 0$
82. $x^2 = 36$
83. $x^2 = 25$
84. $4x^2 = 1$
85. $36x^2 = 1$
86. $25x^2 - 9 = 0$
87. $9x^2 - 25 = 0$
88. $x^2 + 4x = -4$
89. $x^2 + 9 = 6x$
90. $4x^2 + 9 = 12x$
91. $9x^2 + 4 = 12x$
92. $3x^2 + 3x = 6$
93. $5x^2 + 10x = 15$
94. $4x^2 = 4x + 24$
95. $4x^2 + 4x = 24$
96. $4x^2 = 12x + 40$
97. $3x^2 = 12x + 36$
98. $6x^2 = 12x - 6$
99. $9x^2 = 36x - 36$
100. $72x^2 = 8$

6

Rational Expressions

I think it would suit me nicely,
If I knew one-tenth
of the little I know,
But knew that tenth precisely.

Who Did Which? or Who Indeed?
by Ogden Nash

6.1 Rational Numbers

The central reason for introducing integers was to provide a set of numbers that includes all possible answers for subtraction. In the same way, we need to consider a set of numbers that includes all possible answers for the operation of division. So, to the integers we append the set of fractions. With this larger set we can add, subtract, multiply, and divide any two numbers (except division by zero) and get answers in the set.

We call this enlarged set *the set of* ***rational numbers,*** denoted by Q. Q is the set consisting of integers and fractions; that is, numbers that can be written as $\frac{p}{q}$ where p is an integer and q is a nonzero integer.

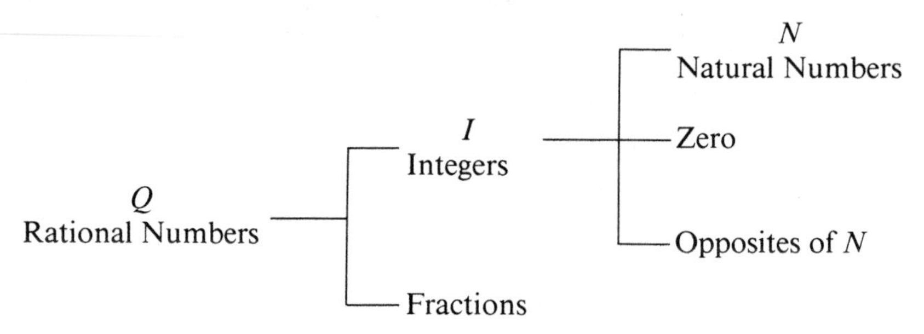

Consider any number in N—say 3. Then 3 is also contained in I and Q. Consider some number in Q—say $\frac{1}{2}$. Since $\frac{1}{2}$ is in Q but not in I, we say $\frac{1}{2}$ is a rational number that is not an integer.

Equality

$$\frac{a}{b} = \frac{c}{d} \text{ if and only if } ad = bc,\ b \neq 0,\ d \neq 0.$$

Equality of fractions helps us to determine when and if two fractions are equal.

Examples *Decide if the given fractions are equal.*

Use the definition of equality to write:

1. $\frac{2}{3} \stackrel{?}{=} \frac{37}{51}$ $\quad 2(51) \stackrel{?}{=} 3(37)$

   $102 \neq 111$

   fractions are not equal

2. $\frac{4}{5} \stackrel{?}{=} \frac{28}{35}$ $\quad 4(35) \stackrel{?}{=} 5(28)$

   $140 = 140$

   fractions are equal

3. $\frac{23}{27} \stackrel{?}{=} \frac{92}{107}$ $\quad 23(107) \stackrel{?}{=} 27(92)$

   $2461 \neq 2484$

   fractions are not equal ■

Zero offers some special problems with fractions. Consider a fraction with numerator zero. Let k be any nonzero, integer; then

$$\frac{0}{k} = 0 \text{ since } 0 = k \cdot 0.$$

However, let k be any integer and consider a zero denominator. If there is a number n that checks, then

$$\frac{k}{0} = n$$

but $k = 0 \cdot n = 0$ means that regardless of the value of n, $k = 0$. Thus $\frac{k}{0}$ is meaningless, because no number checks.

Consider the fraction with both numerator and denominator zero. If there is a value n for such a fraction, then

$$\frac{0}{0} = n$$

Which of the following is n?

A.	1	A number divided by itself is one.
B.	0	Zero divided by anything is zero.
C.	undefined	$0 \div 0$ is meaningless.
D.	7.5	You have to have five choices for multiple-choice questions
E.	none of these	E is always "none of these"!

Who put 7.5 in there? Well, check and see how silly it is.

$$\frac{0}{0} = 7.5? \qquad 0 = (0)(7.5) = 0 \text{ checks!}$$

Does $0 \div 0 = 7.5$? Try any number n.

$$\frac{0}{0} = n? \qquad 0 = (0)(n) = 0, \text{ so any number } n \text{ checks.}$$

Since every number checks, $\frac{0}{0}$ is meaningless because it has no unique value. All fractions with a zero denominator are undefined.

Dealing with fractions often involves reducing them to their lowest terms. To reduce a fraction, we find a ***common factor*** in the numerator and the denominator.

Reducing Fractions

$$\frac{ac}{bc} = \frac{a}{b}$$ To reduce a fraction, divide the numerator and the denominator by a common factor.

To reduce to ***lowest terms,*** find the greatest common factor; in other words, be sure there are no common factors remaining except 1.

Examples *Reduce each fraction in Examples 4–7 to lowest terms.*

4. $\dfrac{6}{8} = \dfrac{3 \cdot \not{2}}{4 \cdot \not{2}} = \dfrac{3}{4}$

5. $\dfrac{35}{28} = \dfrac{5 \cdot \not{7}}{4 \cdot \not{7}} = \dfrac{5}{4}$

6. $\dfrac{84}{90} = \dfrac{\not{2} \cdot 2 \cdot \not{3} \cdot 7}{\not{2} \cdot 3 \cdot \not{3} \cdot 5} = \dfrac{2 \cdot 7}{3 \cdot 5} = \dfrac{14}{15}$

7. $\dfrac{715}{825} = \dfrac{\not{5} \cdot \not{11} \cdot 13}{3 \cdot \not{5} \cdot 5 \cdot \not{11}} = \dfrac{13}{3 \cdot 5} = \dfrac{13}{15}$ ■

Notice how, in the Examples, diagonal lines are sometimes used to indicate a factor common to the numerator and denominator is "cancelled" when reducing the fraction. Do you recall how to compute with fractions?

Operations with Fractions

MULTIPLICATION

$\dfrac{a}{b} \cdot \dfrac{c}{d} = \dfrac{ac}{bd}$ To multiply two fractions, multiply the numerators and multiply the denominators.

DIVISION

$\dfrac{a}{b} \div \dfrac{c}{d} = \dfrac{a}{b} \cdot \dfrac{d}{c}$ To divide two fractions, multiply by the reciprocal of the divisor.

where $\dfrac{m}{n}$ is the reciprocal of $\dfrac{n}{m}$, and $\dfrac{1}{n}$ is the reciprocal of n.

ADDITION/SUBTRACTION

$\dfrac{a}{b} + \dfrac{c}{d} = \dfrac{ad + bc}{bd}$ To add (or subtract) two fractions, write each with a common denominator and add (or subtract) the numerators.

You may be familiar with the fundamental operations with fractions, but the following examples will help refresh your memory.

Examples *In Examples 8–17, perform the indicated operations and leave the answer in lowest terms.*

8. $\dfrac{3}{4} \cdot \dfrac{1}{5} = \dfrac{3}{20}$

9. $\dfrac{2}{3} \cdot \dfrac{5}{3} = \dfrac{10}{9}$

10. $$\frac{4}{9}\cdot\frac{15}{8} = \frac{4\cdot 15}{9\cdot 8} = \frac{\not{2}\cdot\not{2}\cdot\not{3}\cdot 5}{\not{3}\cdot 3\cdot\not{2}\cdot\not{2}\cdot 2} = \frac{5}{2\cdot 3} = \frac{5}{6}$$

11. $$\frac{33}{34}\cdot\frac{51}{55} = \frac{33\cdot 51}{34\cdot 55} = \frac{3\cdot\not{11}\cdot 3\cdot\not{17}}{2\cdot\not{17}\cdot 5\cdot\not{11}} = \frac{3\cdot 3}{2\cdot 5} = \frac{9}{10}$$

12. $$\frac{3}{4}\div\frac{1}{5} = \frac{3}{4}\cdot\frac{5}{1} = \frac{15}{4}$$

13. $$\frac{2}{3}\div\frac{5}{3} = \frac{2}{3}\cdot\frac{3}{5} = \frac{2}{5}$$

14. $$\frac{1}{\frac{2}{3}} = 1\div\frac{2}{3} = 1\cdot\frac{3}{2} = \frac{3}{2}$$

15. $$\frac{2}{3}+\frac{5}{3} = \frac{2+5}{3} = \frac{7}{3}$$

16. $$\frac{3}{4}-\frac{1}{5} = \frac{15}{20}-\frac{4}{20} = \frac{15-4}{20} = \frac{11}{20}$$

17. $$\frac{3}{5}\left(1-\frac{5}{6}\right) = \frac{3}{5}\left(\frac{6}{6}-\frac{5}{6}\right) = \frac{3}{5}\left(\frac{1}{6}\right) = \frac{3}{30} = \frac{1}{10}$$ ■

Because the signs of the numerator and denominator of a fraction may be either positive or negative, there are several equivalent forms of any fraction. Suppose p and q are positive integers, then

$$\frac{p}{q} = \frac{-p}{-q} = -\frac{-p}{q} = -\frac{p}{-q}$$

which you can verify with the property of equality. Likewise, the fraction itself can be negative, then

$$-\frac{p}{q} = \frac{-p}{q} = \frac{p}{-q} = -\frac{-p}{-q}$$

Because of this variety of forms, we call $\frac{p}{q}$ and $\frac{-p}{q}$ the ***standard forms of a fraction*** and reduce all other forms to one of these.

Standard Forms of a Fraction

1. $\dfrac{p}{q} = \dfrac{-p}{-q} = -\dfrac{-p}{q} = -\dfrac{p}{-q}$

2. $\dfrac{-p}{q} = \dfrac{p}{-q} = -\dfrac{p}{q} = -\dfrac{-p}{-q}$

The first form on each line is the **standard form.**

Examples *Write each nonstandard form in standard form.*

	Solutions
18. $\dfrac{-5}{-7}$	**18.** $\dfrac{5}{7}$
19. $\dfrac{5}{-7}$	**19.** $\dfrac{-5}{7}$
20. $-\dfrac{-5}{7}$	**20.** $\dfrac{5}{7}$
21. $-\dfrac{5}{-7}$	**21.** $\dfrac{5}{7}$
22. $-\dfrac{5}{7}$	**22.** $\dfrac{-5}{7}$
23. $-\dfrac{-5}{-7}$	**23.** $\dfrac{-5}{7}$ ■

Problem Set 6.1

Determine if the fractions in Problems 1–10 are equal or not equal. See Examples 1–3.

1. $\dfrac{3}{4} \stackrel{?}{=} \dfrac{4}{5}$ **2.** $\dfrac{6}{7} \stackrel{?}{=} \dfrac{7}{8}$ **3.** $\dfrac{4}{6} \stackrel{?}{=} \dfrac{12}{18}$ **4.** $\dfrac{6}{8} \stackrel{?}{=} \dfrac{15}{20}$

5. $\dfrac{14}{22} \stackrel{?}{=} \dfrac{21}{33}$ **6.** $\dfrac{20}{44} \stackrel{?}{=} \dfrac{25}{55}$ **7.** $\dfrac{36}{81} \stackrel{?}{=} \dfrac{24}{54}$ **8.** $\dfrac{36}{84} \stackrel{?}{=} \dfrac{24}{56}$

9. $\dfrac{22}{26} \stackrel{?}{=} \dfrac{33}{39}$ **10.** $\dfrac{45}{51} \stackrel{?}{=} \dfrac{30}{34}$

Reduce the fractions in Problems 11–30 to lowest terms. See Examples 4–7.

11. $\dfrac{2}{4}$ **12.** $\dfrac{3}{9}$ **13.** $\dfrac{3}{6}$ **14.** $\dfrac{4}{10}$

15. $\dfrac{6}{9}$ **16.** $\dfrac{4}{6}$ **17.** $\dfrac{9}{12}$ **18.** $\dfrac{8}{10}$

19. $\dfrac{8}{12}$ **20.** $\dfrac{14}{24}$ **21.** $\dfrac{12}{28}$ **22.** $\dfrac{15}{25}$

23. $\frac{24}{26}$ **24.** $\frac{24}{33}$ **25.** $\frac{34}{51}$ **26.** $\frac{39}{65}$

27. $\frac{63}{84}$ **28.** $\frac{124}{128}$ **29.** $\frac{ak}{bk}$ **30.** $\frac{ak}{ah}$

Perform the indicated operations in Problems 31–60 and reduce to lowest terms. See Examples 8–17.

31. $\frac{2}{5}\cdot\frac{1}{7}$ **32.** $\frac{7}{8}\cdot\frac{3}{2}$ **33.** $\frac{3}{4}\cdot\frac{8}{9}$

34. $\frac{11}{18}\cdot\frac{3}{2}$ **35.** $\frac{4}{5}\div\frac{3}{7}$ **36.** $\frac{7}{9}\div\frac{2}{5}$

37. $\frac{3}{10}\div\frac{9}{25}$ **38.** $\frac{4}{9}\div\frac{10}{21}$ **39.** $\frac{1}{\frac{1}{4}}$

40. $\frac{\frac{1}{3}}{4}$ **41.** $\frac{\frac{2}{3}}{\frac{6}{7}}$ **42.** $\frac{\frac{3}{5}}{\frac{1}{9}}$

43. $\frac{2}{7}+\frac{3}{7}$ **44.** $\frac{1}{5}+\frac{2}{5}$ **45.** $\frac{1}{2}+\frac{3}{4}$

46. $\frac{2}{3}+\frac{1}{6}$ **47.** $\frac{5}{6}-\frac{1}{4}$ **48.** $\frac{4}{9}-\frac{1}{6}$

49. $2-\frac{1}{6}-\frac{3}{8}$ **50.** $\frac{2}{3}+\frac{1}{5}-1$ **51.** $\frac{3}{4}-\frac{2}{9}+\frac{1}{6}$

52. $\frac{1}{4}-\frac{3}{14}+\frac{5}{7}$ **53.** $\frac{11}{12}-\frac{5}{9}-\frac{3}{8}$ **54.** $\frac{17}{18}-\frac{3}{4}-\frac{5}{27}$

55. $\left(\frac{2}{9}\right)\left(\frac{3}{4}\right)+\frac{5}{6}$ **56.** $\left(\frac{5}{14}\right)\left(\frac{7}{10}\right)+\frac{3}{4}$ **57.** $\frac{1}{5}\left(2-\frac{1}{3}\right)$

58. $\frac{1}{8}\left(3+\frac{1}{5}\right)$ **59.** $\frac{7}{9}\left(\frac{1}{7}-\frac{2}{5}\right)$ **60.** $\frac{5}{11}\left(\frac{2}{7}-\frac{3}{5}\right)$

Write the fractions in Problems 61–70 in standard form. See Examples 18–23.

61. $-\frac{1}{2}$ **62.** $\frac{2}{-3}$ **63.** $\frac{-3}{-5}$

64. $\frac{5}{-8}$ **65.** $-\frac{-8}{13}$ **66.** $-\frac{13}{21}$

67. $-\frac{-21}{-34}$ **68.** $-\frac{34}{-55}$ **69.** $\frac{55}{-89}$

70. $\frac{89}{-144}$

6.2 Rational Expressions

Working with polynomials in Chapter 4, we saw that adding, subtracting, and multiplying two polynomials always produced another polynomial. However, dividing two polynomials may or may not produce a polynomial quotient. For example,

$$\frac{1}{x}, \frac{x+y}{x-y}, \frac{x+1}{x+4}, \frac{x^2-2x-3}{x^2+x-12}$$

are polynomials divided by polynomials that do not "divide evenly." Such expressions are examined in this section. However, when dividing by a polynomial, we still must make sure that we do not divide by zero. Therefore, we speak of a *nonzero polynomial* as a polynomial that has excluded from the domain all values for which it is zero.

Definition of Rational Expression

A ***rational expression*** is a polynomial divided by a nonzero polynomial.

The following examples compare integers with polynomials.

$5 + (-3) = 2$
integer + integer = integer

$(5x) + (4-3x) = 2x+4$
polynomial + polynomial = polynomial

$5 - (-3) = 8$
integer − integer = integer

$(5x) - (4-3x) = 8x-4$
polynomial − polynomial = polynomial

$5 \cdot (-3) = -15$
integer · integer = integer

$(5x) \cdot (4-3x) = 20x - 15x^2$
polynomial · polynomial = polynomial

$5 \div (-3) = -\frac{5}{3}$
integer ÷ integer = ↑
This is not an integer. We call it a rational number.

$(5x) \div (4-3x) = \frac{5x}{4-3x}$
polynomial ÷ polynomial = ↑
This is not a polynomial. We call it a rational expression.

The properties of rational numbers were stated for numbers of the form $\frac{a}{b}$ where a and b represent integers. Those same properties may now be used with rational expressions where a and b represent polynomials. An important procedure is the simplification of a rational expression. The procedure is of such basic importance that it is called the ***Fundamental Principle of Rational Expressions.***

Fundamental Principle of Rational Expressions

If a is any polynomial and b and x are nonzero polynomials, then

$$\frac{ax}{bx} = \frac{a}{b}$$

This says that given any rational expression, the form can be changed by multiplying both the numerator and denominator by any nonzero polynomial x. It also allows us to reduce rational expressions to lowest terms. We say that a rational expression is ***completely reduced*** if the largest factor common to both numerator and denominator is 1. Thus, the key to reducing rational expressions is finding the factors of both numerator and denominator.

Examples *Simplify the following expressions.*

1. $$\frac{6xy}{10x^2} = \frac{2\cdot 3xy}{2\cdot 5xx} = \frac{(2x)3y}{(2x)5x} = \frac{3y}{5x}$$

2. $$\frac{3xy}{7z}$$ *This is in simplified form, because the only factor common to both numerator and denominator is 1.*

3. $$\frac{21m^3n^2}{28n^2} = \frac{3\cdot \not{7}m^3n^2}{4\cdot \not{7}n^2} = \frac{3m^3}{4}$$

4. $$\frac{30a^2bc}{45abc^3} = \frac{2\cdot 3\cdot 5a^2bc}{3\cdot 3\cdot 5abc^3} = \frac{(3\cdot 5abc)2a}{(3\cdot 5abc)3c^2} = \frac{2a}{3c^2}$$ ∎

Technically, we should say that $a \neq 0$, $b \neq 0$, and $c \neq 0$ in Example 4. However, in this book we will assume that all values for the variables that cause division by zero are excluded.

Recall when dealing with positive and negative numerators that it was necessary to introduce a standard form for fractions. If the numerator and denominator are polynomials, the standard form can be somewhat obscure. Several forms may be acceptable. For example, recall that $a - b = -(b - a)$, so that

$$\frac{-a}{b-a} = \frac{a}{-(b-a)} = \frac{a}{a-b}$$

Although the first and second forms are acceptable, the third is usually preferred.

Examples *Write the following in standard form.*

5. $$-\frac{-x}{y} = \frac{x}{y}$$

6. $$-\frac{x+y}{2x+y} = \frac{-(x+y)}{2x+y} = \frac{-x-y}{2x+y}$$

7. $$\frac{x-2y}{-(x-y)} = \frac{-(x-2y)}{x-y} = \frac{-x+2y}{x-y}$$

Other acceptable forms for Example 7 are

$$\frac{2y-x}{x-y} \quad \text{or} \quad \frac{x-2y}{y-x}$$ ∎

A knowledge of the standard forms of a fraction can be helpful when simplifying rational expressions.

Examples *Simplify the given expressions.*

8. $$\frac{2(x+y)}{-5} = \frac{-2(x+y)}{5}$$

9. $$\frac{x+y}{-x-y} = \frac{\cancel{x+y}}{(-1)(\cancel{x+y})} = \frac{1}{-1} = -1$$

10. $$\frac{(x+y)(x-y)}{(y-x)(x+y)} = \frac{(\cancel{x+y})(-1)(\cancel{y-x})}{(\cancel{y-x})(\cancel{x+y})} = -1$$

11. $$\frac{x^2-4}{x^2+5x+6} = \frac{(x-2)(\cancel{x+2})}{(x+3)(\cancel{x+2})} = \frac{x-2}{x+3}$$

12. $$\frac{x^2-4x-5}{-2x^2-2x} = \frac{(x-5)(\cancel{x+1})}{-2x(\cancel{x+1})} = \frac{x-5}{-2x} = \frac{5-x}{2x}$$ ■

Example 10 illustrates a special case that occurs frequently enough to deserve additional attention. Notice that $x - y$ and $y - x$ are opposites and as a result their quotient is -1. That is,

$$\frac{x-y}{y-x} = \frac{x-y}{-(x-y)} = \frac{1(x-y)}{-1(x-y)} = -1$$

A quantity divided by its nonzero opposite is equal to -1. Alternately, it could be said that a number divides into its opposite -1 times, and this is frequently shown by the diagonal slashes used in Example 10. Further examples should clarify.

Examples *Simplify the given expressions.*

13. $$\frac{6ab^2c}{-6bc^3} = \frac{\overset{-1}{\cancel{6}}abbc}{\cancel{-6}bccc} = \frac{-ab}{c^2}$$

14. $$\frac{ax-ay}{3y-3x} = \frac{a(\overset{-1}{\cancel{x-y}})}{3(\cancel{y-x})} = \frac{-a}{3}$$

15. $\dfrac{4n - 2n^2}{n^3 - 4n} = \dfrac{2n(2 - n)}{n(n^2 - 4)}$ *Factor out common monomial factors, then factor difference of squares.*

$= \dfrac{2n\overset{-1}{\cancel{(2 - n)}}}{n(n + 2)\cancel{(n - 2)}}$ *Reduce, noting the quotient of opposites.*

$= \dfrac{-2}{n + 2}$ *State result in standard form.* ■

Problem Set 6.2

Reduce each rational expression in Problem 1–20 to lowest terms. See Examples 1–4.

1. $\dfrac{2xy}{4y}$
2. $\dfrac{3ab}{9a^2}$
3. $\dfrac{3r}{6rs}$
4. $\dfrac{4x^2y}{10xy}$
5. $\dfrac{6a^2b^2}{9ab}$
6. $\dfrac{4rs^2}{6r^2s}$
7. $\dfrac{9abc}{12ac}$
8. $\dfrac{8xy}{10xyz}$
9. $\dfrac{8rt}{12r^2st}$
10. $\dfrac{14a^2bc^2}{24ab^2c}$
11. $\dfrac{12rs^2t^2}{28r^2st}$
12. $\dfrac{15x^3yz}{25xy^2}$
13. $\dfrac{24x^4z}{26xz^2}$
14. $\dfrac{24rt^5}{33r^2t^2}$
15. $\dfrac{34a^3b^3}{51ab^2}$
16. $\dfrac{15xy^3z^2}{10x^2yz^3}$
17. $\dfrac{16a^3b^2c}{12ab^3c^2}$
18. $\dfrac{24r^2st^3}{18r^3s^2t}$
19. $\dfrac{39a^2b^5c^4}{65a^2bc^5}$
20. $\dfrac{63x^4y^3z^2}{84x^2y^3z^4}$

Write the fractions given in Problems 21–30 in standard form. Also give values of the variables for which the fraction is not defined. See Examples 5–7.

21. $-\dfrac{y}{3}$
22. $\dfrac{5}{-x}$
23. $-\dfrac{-2a}{-b}$
24. $\dfrac{-5st}{-3w}$
25. $-\dfrac{-x}{y + 1}$
26. $-\dfrac{-m}{m - 1}$
27. $-\dfrac{x}{2 - x}$
28. $-\dfrac{-x}{x - 3}$
29. $\dfrac{a - b}{-(a + b)}$
30. $\dfrac{x + y}{-(x - y)}$

Reduce each rational expression in Problems 31–60 to lowest terms. See Examples 8–15.

31. $\dfrac{x - y}{y - x}$
32. $\dfrac{a - b}{b - a}$
33. $\dfrac{s - t}{t - s}$
34. $\dfrac{3x - 3y}{y - x}$
35. $\dfrac{5x - 5y}{y - x}$
36. $\dfrac{4a - 4b}{b - a}$
37. $\dfrac{(a - b)(2a + b)}{(2a + b)(b - a)}$
38. $\dfrac{(x + y)^2(x - y)^3}{(x + y)(x - y)^4}$
39. $\dfrac{(m - n)^2(m + n)}{m^2 - n^2}$

40. $\dfrac{a^2 - b^2}{a - b}$ 41. $\dfrac{x^2 - y^2}{x + y}$ 42. $\dfrac{m^2 - n^2}{n - m}$

43. $\dfrac{3x^2 - 3x}{9x - 9}$ 44. $\dfrac{5a^2 - 5a}{4(a - 1)}$ 45. $\dfrac{2(m - 1)}{3m - 3m^2}$

46. $\dfrac{a^2 - 3a + 2}{a^2 - a - 2}$ 47. $\dfrac{m^2 - 2m - 3}{m^2 + 4m + 3}$ 48. $\dfrac{x^2 - x - 6}{x^2 + 5x + 6}$

49. $\dfrac{x^2 - 9x + 18}{x^2 - 9}$ 50. $\dfrac{y^2 + 4y + 4}{y^2 - 4}$ 51. $\dfrac{m^2 - 4m + 3}{1 - m^2}$

52. $\dfrac{2n^3 - 8n}{n^2 - 5n + 6}$ 53. $\dfrac{9r - 3r^2}{r^3 - 9r}$ 54. $\dfrac{s^2 - s}{s^2 - 2s + 1}$

55. $\dfrac{x^2 - 4}{4 - x^2}$ 56. $\dfrac{y - y^2}{y^2 - 1}$ 57. $\dfrac{3z^3 - 27z}{9z^3 - 27z^2}$

58. $\dfrac{2m^2 - 14m + 24}{12m^2 - 4m^3}$ 59. $\dfrac{n^3 + 3n^2 - 10n}{4n - n^3}$ 60. $\dfrac{r^2 - 2r - 8}{r^2 + 2r - 8}$

6.3 Multiplication and Division of Rational Expressions

Let us restate the properties of multiplication and division of rational numbers.

MULTIPLICATION	$\dfrac{a}{b}\cdot\dfrac{c}{d} = \dfrac{a\cdot c}{b\cdot d}$	$b, d \neq 0$
DIVISION	$\dfrac{a}{b} \div \dfrac{c}{d} = \dfrac{a}{b}\cdot\dfrac{d}{c}$	$b, c, d \neq 0$

Where a, b, c, and d represented integers, here they will represent polynomials with b, c, and d nonzero polynomials. The following examples illustrate how operations with rational expressions resemble those with rational numbers.

Examples *In Examples 1–6, perform the indicated operations and simplify.*

1. $\dfrac{2a}{b}\cdot\dfrac{b}{3} = \dfrac{2ab}{3b} = \dfrac{2a}{3}$

2. $\dfrac{m}{2n} \div \dfrac{4}{n} = \dfrac{m}{2n}\cdot\dfrac{n}{4} = \dfrac{mn}{8n} = \dfrac{m}{8}$

3. $\dfrac{3ab^2}{10b}\cdot\dfrac{5b^3}{6a^2} = \dfrac{3\cdot 5ab^5}{10\cdot 6a^2b}$ *Multiply.*

$= \dfrac{3\cdot 5abb^4}{2\cdot 5\cdot 2\cdot 3aab}$ *Factor.*

$= \dfrac{(3\cdot 5ab)b^4}{(3\cdot 5ab)4a}$ *Notice common factors (usually a mental step).*

$= \dfrac{b^4}{4a}$ *Reduce.*

4. $$\frac{15m^2}{2n^3} \div \frac{21m}{np^2} = \frac{15m^2}{2n^3}\cdot\frac{np^2}{21m} = \frac{3\cdot 5m^2np^2}{2\cdot 3\cdot 7mn^3} = \frac{(3mn)5mp^2}{(3mn)2\cdot 7n^2} = \frac{5mp^2}{14n^2}$$

5. $$\frac{(x+y)^2}{15}\cdot\frac{12}{x^2-y^2} = \frac{12(x+y)^2}{15(x^2-y^2)} = \frac{3\cdot 4(x+y)(x+y)}{3\cdot 5(x+y)(x-y)} = \frac{4(x+y)}{5(x-y)} = \frac{4x+4y}{5x-5y}$$

6. $$\frac{3n+3}{2n-4} \div \frac{9n+9}{4-2n} = \frac{3n+3}{2n-4}\cdot\frac{4-2n}{9n+9}$$ *Invert and multiply.*

$$= \frac{3(n+1)2(2-n)}{2(n-2)9(n+1)}$$ *Factor.*

$$= \frac{\cancel{2}\cdot\cancel{3}(\cancel{n+1})(\cancelto{-1}{2-n})}{\cancel{2}\cdot\cancel{3}\cdot 3(\cancel{n-2})(\cancel{n+1})}$$ *Reduce, noting that* $\frac{\cancelto{-1}{2-n}}{\cancel{n-2}}$.

$$= \frac{-1}{3}$$ ■

As you can see from the examples, the multiplication or division is the easy part of the problem; the difficult part is reducing the resulting rational expression. We look for the common factors and then simplify the expression. Thus, *the key to working multiplication and division problems with rational expressions is finding the common factors.* To find the common factors, we must first factor both the numerator and the denominator completely.

In Examples 7–10, perform the indicated operations, and simplify the expressions.

Example 7. $$\frac{x^2-9}{x^2-4}\cdot\frac{x-2}{x+3} = \frac{(x-3)(x+3)(x-2)}{(x-2)(x+2)(x+3)} = \frac{(\cancel{x+3})(\cancel{x-2})}{(\cancel{x+3})(\cancel{x-2})}\cdot\frac{(x-3)}{(x+2)} = \frac{x-3}{x+2}$$ ■

Example 8. $$(x^2-25)\cdot\frac{3x-2}{x+5} = \frac{(x-5)(\cancel{x+5})}{1}\cdot\frac{3x-2}{\cancel{x+5}} = \frac{(x-5)(3x-2)}{1} = 3x^2-17x+10$$ ■

Example 9. $$\frac{3x^2-17x+10}{5x+1} \div (3x-2) = \frac{(x-5)(\cancel{3x-2})}{5x+1}\cdot\frac{1}{\cancel{3x-2}} = \frac{x-5}{5x+1}$$ ■

Example 10.

$$\frac{2x+6}{3x+3} \div \frac{x^2-16}{x^2+5x+4} = \frac{(2x+6)(x^2+5x+4)}{(3x+3)(x^2-16)}$$
$$= \frac{2(x+3)\cancel{(x+1)}\cancel{(x+4)}}{3\cancel{(x+1)}(x-4)\cancel{(x+4)}}$$
$$= \frac{2x+6}{3x-12}$$

■

Problem Set 6.3

Find the indicated product or quotient and simplify the expression. See Examples 1–6.

1. $\frac{3a}{b} \cdot \frac{b}{2}$
2. $\frac{m}{3n} \cdot \frac{6n}{5}$
3. $\frac{8x}{3y} \cdot \frac{9}{4x}$
4. $\frac{5m}{7n} \cdot \frac{14n}{10}$
5. $\frac{12x}{5} \cdot \frac{15}{6xy}$
6. $\frac{18ab}{3} \cdot \frac{5b}{30a}$
7. $\frac{2x}{3} \div \frac{4x}{3y}$
8. $\frac{5}{2r} \div \frac{-2}{6rs}$
9. $\frac{3}{7ab} \div \frac{5a}{6b}$
10. $\frac{2xy^2}{15y} \cdot \frac{3y^3}{10x^2}$
11. $\frac{7ab}{6b^2} \div \frac{14a^2}{3b}$
12. $\frac{6r^3}{9rs} \cdot \frac{12rs}{r^2}$
13. $\frac{-3x}{2y} \div \frac{y^2}{6xy}$
14. $\frac{-2b}{5ab^2} \cdot \frac{15a^2}{12b}$
15. $\frac{3m^2}{5n^3} \div \frac{6m}{10n}$
16. $\frac{4xy}{10w} \cdot \frac{5w^2}{8x^3y}$
17. $\frac{3a^2b^3}{7cd^2} \cdot \frac{21cd^3}{3ab}$
18. $\frac{2xy^2}{-5z^3} \div \frac{4x^2y}{25z^2}$
19. $\frac{a+b}{10} \cdot \frac{15}{a^2-b^2}$
20. $\frac{6}{a-b} \div \frac{4}{a^2-b^2}$
21. $\frac{2x-6}{x} \div \frac{3x-9}{y}$
22. $\frac{2a}{5a+10} \cdot \frac{6a+12}{4b}$
23. $\frac{m-n}{n} \cdot \frac{m}{n-m}$
24. $\frac{r}{r-s} \div \frac{s}{s-r}$
25. $\frac{2x-6}{3x+9} \cdot \frac{2x+6}{3x-6}$
26. $\frac{4y+2}{y-4} \div \frac{2y+1}{3y-12}$
27. $\frac{3m-3}{5m+10} \div \frac{2-2m}{3m+6}$
28. $\frac{3-n}{3n+6} \cdot \frac{6n+12}{2n-6}$
29. $\frac{r^2-r}{2r+3} \cdot \frac{4r+6}{r^2-r}$
30. $\frac{3s-2}{s-s^2} \div \frac{6s-4}{3s-3}$

In Problems 31–50 perform the indicated operations, and simplify. See Examples 7–10.

31. $\frac{x^2-4}{x^2-1} \cdot \frac{x-1}{x+2}$
32. $\frac{y-1}{y+3} \cdot \frac{y^2-9}{y^2-1}$
33. $\frac{4-m^2}{m+3} \div \frac{m^2-4}{m^2-9}$
34. $\frac{n^2-4}{n^2-25} \div \frac{2-n}{n-5}$
35. $\frac{r^2-r}{2-r} \cdot \frac{r^2-4}{r^2-1}$
36. $\frac{s^2-36}{s^2-6s} \cdot \frac{6-s}{s^2+6s}$
37. $(x^2-16) \cdot \frac{3x+1}{x+4}$
38. $(x^2-16) \div \frac{x-4}{x+4}$
39. $\frac{x^2+x-2}{x^2+7x+10} \cdot \frac{x+5}{x-1}$
40. $\frac{2x-6}{x^2-2x-3} \cdot \frac{x^2+3x+2}{x+2}$

41. $(2x^2 + 7x + 3)\cdot\dfrac{x-2}{x^2 + x - 6}$

42. $(5x^2 + 4x - 1)\cdot\dfrac{x-5}{x^2 - 4x - 5}$

43. $\dfrac{x^2 - x - 6}{x-3} \div (3x^2 + 7x + 2)$

44. $\dfrac{2x^2 + 3x + 1}{2x+1} \div (6x^2 + 11x + 5)$

45. $\dfrac{2x^2 + 3x + 1}{3x^2 - 19x + 20}\cdot\dfrac{3x^2 - 13x + 12}{x^2 - 2x - 3}$

46. $\dfrac{x^2 + 9x + 20}{3x^2 + 5x - 2}\cdot\dfrac{2x^2 + 5x + 2}{2x^2 + 11x + 5}$

47. $\dfrac{x^2 + 4x + 3}{x^2 - 4x + 4} \div \dfrac{x^2 + 5x + 6}{x^2 - 4}$

48. $\dfrac{x^2 - 9}{x^2 + 2x - 15} \div \dfrac{2x^2 + 5x - 3}{2x^2 - 7x + 3}$

49. $\dfrac{x^2 + 7x + 12}{2 - x} \div \dfrac{4 - 7x - 2x^2}{2x - 1}$

50. $\dfrac{x^2 + 3x + 2}{6 - x - 2x^2} \div \dfrac{x+1}{4x - 6}$

6.4 Addition and Subtraction of Rational Expressions

In this section we apply the addition and subtraction properties for fractions to rational expressions.

> If a and c are any polynomials and b is a nonzero polynomial, then
>
> ADDITION $\quad \dfrac{a}{b} + \dfrac{c}{b} = \dfrac{a+c}{b}$
>
> SUBTRACTION $\quad \dfrac{a}{b} - \dfrac{c}{b} = \dfrac{a-c}{b}$

If you have common denominators, you simply add or subtract numerators. The difficult part is finding the common denominator. Any common denominator will work, but the problem is generally easier if we find the ***least common denominator*** (LCD). Consider some examples before we make a general statement about LCDs.

In Examples 1–4, find the LCD of the given denominators.

Example 1. 24 and 36

Solution *Step 1a* Factor 24 and 36

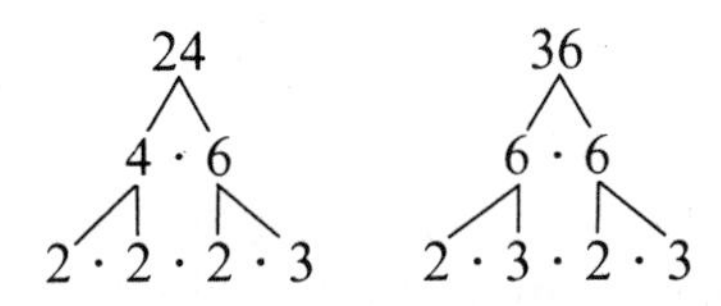

Step 1b Write each of the prime factorizations using exponents.

$$24 = 2^3\cdot 3$$
$$36 = 2^2\cdot 3^2$$

Step 2a Pick out a representative of each base that appears. The one you pick should be the one with the largest exponent. For our example, the two bases are 2 and 3. The terms with the largest exponents (see Step 1b) are 2^3 and 3^2.

Step 2b The product of the repesentatives is the LCD.

$$\begin{aligned} &2^3 \cdot 3^2 \\ &= 8 \cdot 9 \\ &= 72 \end{aligned}$$

■

Example 2. 300, 144, and 108

Solution *Step 1*

$$\begin{aligned} 300 &= 2^2 \cdot 3^1 \cdot \mathbf{5^2} \\ 144 &= \mathbf{2^4} \cdot 3^2 \\ 108 &= 2^2 \cdot \mathbf{3^3} \end{aligned}$$

Step 2

$$\begin{aligned} &\mathbf{2^4 \cdot 3^3 \cdot 5^2} \\ &= 16 \cdot 27 \cdot 25 \\ &= 10{,}800 \end{aligned}$$

300
10 · 30
2 · 5 · 5 · 6
2 · 5 · 5 · 2 · 3

144
12 · 12
3 · 4 · 3 · 4
3 · 2 · 2 · 3 · 2 · 2

108
4 · 27
2 · 2 · 3 · 3 · 3

■

Example 3. $4x^2$, $12xy$, and $8xy^2$

Solution *Step 1*

$$\begin{aligned} 4x^2 &= 2^2 \cdot \mathbf{x^2} \\ 12xy &= 2^2 \cdot \mathbf{3^1} \cdot x \cdot y \\ 8xy^2 &= \mathbf{2^3} \cdot x \cdot \mathbf{y^2} \end{aligned}$$

Step 2

$$\begin{aligned} &\mathbf{2^3 \cdot 3^1 \cdot x^2 \cdot y^2} \\ &= 24x^2y^2 \end{aligned}$$

■

Example 4. $x^2 - 4$, $x^2 + x - 6$, and $x^2 + 4x + 4$

Solution *Step 1*

$$\begin{aligned} x^2 - 4 &= (x-2)(x+2) \\ x^2 + x - 6 &= \mathbf{(x-2)} \quad \mathbf{(x+3)} \\ x^2 + 4x + 4 &= \quad \mathbf{(x+2)^2} \end{aligned}$$

Step 2

$$\mathbf{(x-2)(x+2)^2(x+3)}$$

■

As you can see from the examples, it takes two steps to find the LCD.

Procedure for Finding the LCD

1. Completely factor each of the given denominators, and write the factorization using exponents.
2. Find the product of the representatives of each factor, where the representative chosen is the one with the largest exponent.

Once you have the common denominator, multiply the terms by 1 (written in an appropriate form, as explained in the following examples) so that all of the terms have identical denominators.

In Examples 5–10, perform the indicated operations.

Example 5. $\dfrac{x}{24} + \dfrac{y}{36}$

Solution *First,* the LCD was found in Example 1.

$$\frac{x}{24}+\frac{y}{36}=\frac{x}{24}\cdot\frac{\mathbf{3}}{\mathbf{3}}+\frac{y}{36}\cdot\frac{\mathbf{2}}{\mathbf{2}}$$ *Next, multiply each fraction by 1 written as a fraction that will give the common denominator.*

$$=\frac{3x}{72}+\frac{2y}{72}$$ *Multiply the fractions.*

$$=\frac{3x+2y}{72}$$ *Complete the addition.* ∎

Example 6.

$$\frac{3y}{4x^2}-\frac{5}{12xy}=\frac{3y}{4x^2}\cdot\frac{\mathbf{3y}}{\mathbf{3y}}-\frac{5}{12xy}\cdot\frac{\boldsymbol{x}}{\boldsymbol{x}}$$

$$=\frac{9y^2}{12x^2y}-\frac{5x}{12x^2y}$$

$$=\frac{9y^2-5x}{12x^2y}$$ ∎

Example 7.

$$\frac{3}{x-1}+\frac{5}{1-x}=\frac{3}{x-1}+\frac{5}{1-x}\cdot\frac{\mathbf{-1}}{\mathbf{-1}}$$

$$=\frac{3}{x-1}+\frac{-5}{x-1}$$

$$=\frac{3-5}{x-1}$$

$$=\frac{-2}{x-1}$$ ∎

Example 8.

$$\frac{x^2+4}{x^2-4}+\frac{x+2}{2-x}=\frac{x^2+4}{(x-2)(x+2)}+\frac{x+2}{2-x}\cdot\frac{\mathbf{-1}}{\mathbf{-1}}$$

$$=\frac{x^2+4}{(x-2)(x+2)}+\frac{-x-2}{x-2}$$

$$=\frac{x^2+4}{(x-2)(x+2)}+\frac{(-x-2)(x+2)}{(x-2)\mathbf{(x+2)}}$$

$$=\frac{x^2+4}{(x-2)(x+2)}+\frac{-x^2-4x-4}{(x-2)(x+2)}$$

$$=\frac{x^2+4-x^2-4x-4}{(x-2)(x+2)}$$

$$=\frac{-4x}{x^2-4}$$ ∎

Example 9.
$$\frac{1}{x^2-3x}-\frac{1}{3x-9}=\frac{1}{x(x-3)}-\frac{1}{3(x-3)}$$
$$=\frac{\mathbf{3}\cdot 1}{\mathbf{3}x(x-3)}-\frac{\mathbf{x}\cdot 1}{3\mathbf{x}(x-3)}$$
$$=\frac{\overset{-1}{\cancel{3-x}}}{3x\cancel{(x-3)}}$$
$$=\frac{-1}{3x}$$
■

Example 10.
$$\frac{2}{x-2}-\frac{1}{x+2}-\frac{4}{x^2-4}$$
$$=\frac{2}{x-2}\cdot\frac{\mathbf{x+2}}{\mathbf{x+2}}-\frac{1}{x+2}\cdot\frac{\mathbf{x-2}}{\mathbf{x-2}}-\frac{4}{(x+2)(x-2)}$$
$$=\frac{2x+4-x+2-4}{(x-2)(x+2)}$$
$$=\frac{\cancel{x+2}}{(x-2)\cancel{(x+2)}}$$
$$=\frac{1}{x-2}$$
■

Problem Set 6.4

In Problems 1–30 find the LCD of the given sets of numbers or expressions that serve as denominators. See Examples 1–4.

1. 14, 21
2. 8, 12
3. 18, 27
4. 24, 36, 54
5. 18, 24, 72
6. 24, 36, 72
7. $3x^2$, $12xy$
8. $8xy^2$, $12x^2y$
9. $9ab$, $6a^2$
10. x, x^2, x^3
11. a, a^3, a^4
12. n^2, n^3, n^4
13. $2x$, $3y$, $9xy$
14. $5a$, $10ab$, $30b$
15. $4m^2$, $6mn$, $3n$
16. $x-2$, x^2-4
17. $x+3$, x^2-9
18. $1-x^2$, $2x+2$
19. $2x-4$, x^2-2x
20. $3x-2x^2$, $4x-6$
21. $3x^2+2x$, $4+6x$
22. x^2-1, $x-x^2$
23. $2n+4$, n^2-4
24. a^2-9, $2a+6$
25. x^2-4, x^2-5x+6
26. y^2-9, y^2-y-12
27. a^2-a-12, a^2-2a-8
28. $b^2+2b-15$, $b^2-7x+12$
29. x^2+3x-4, x^2-x-20
30. $y^2+3y-10$, y^2-y-30

Perform the indicated operations in Problems 31–70 and simplify. Assume that all values for the variables that cause the denominators to be zero are excluded. See Examples 5–10.

31. $\frac{x}{14}+\frac{y}{21}$

32. $\frac{a}{8}-\frac{b}{12}$

33. $\frac{m}{10}+\frac{n}{14}$

34. $\frac{1}{x}+\frac{2}{3}$

35. $\frac{3}{y}-\frac{4}{5}$

36. $\frac{3}{4}-\frac{1}{z}$

37. $a+\frac{1}{a}$

38. $b-\frac{2}{b}$

39. $\frac{1}{c}-c$

40. $\frac{2}{3m}-\frac{5}{2n}$

41. $\frac{7}{2s}+\frac{1}{3r}$

42. $\frac{3}{7x}-\frac{5}{y}$

43. $\frac{5}{3x^2}+\frac{3}{12xy}$

44. $\frac{4}{8xy^2}-\frac{2}{12x^2y}$

45. $\frac{2}{9ab}+\frac{5}{6a^2}$

46. $\frac{6}{x}-\frac{2}{x^2}+\frac{1}{x^3}$

47. $\frac{1}{a}+\frac{1}{a^3}+\frac{2}{a^4}$

48. $\frac{2}{n^2}-\frac{1}{n^3}+\frac{3}{n^4}$

49. $\frac{1}{x-2}-\frac{2}{x^2-4}$

50. $\frac{1}{x+3}+\frac{3}{x^2-9}$

51. $\frac{1}{1-x^2}+\frac{x}{2x+2}$

52. $\frac{3}{2x-4}+\frac{2}{x^2-2x}$

53. $\frac{2}{3x-2x^2}-\frac{x}{4x-6}$

54. $\frac{5}{3x^2+2x}-\frac{3}{4+6x}$

55. $\frac{x}{x^2-1}-\frac{4}{x-x^2}$

56. $\frac{3}{2n+4}+\frac{n}{n^2-4}$

57. $\frac{2}{a^2-9}-\frac{3a}{2a+6}$

58. $\frac{3}{2x-1}+\frac{x}{2x+2}$

59. $\frac{x}{2x+4}+\frac{1}{3x-3}$

60. $\frac{2}{4x-2}-\frac{3}{2x-4}$

61. $\frac{2}{3x}-\frac{1}{3x-5}+\frac{1}{5}$

62. $\frac{1}{2x}-\frac{2}{2x+3}+\frac{1}{3}$

63. $\frac{1}{2}-\frac{x}{2x-1}+\frac{1}{x}$

64. $\frac{-x}{2x-1}+\frac{1}{2}+\frac{1}{x-1}$

65. $\frac{5x}{x^2-y^2}+\frac{3}{x-y}-\frac{2}{x+y}$

66. $\frac{2b}{a^2-b^2}-\frac{5a}{a-b}+\frac{2b}{a+b}$

67. $\frac{3x}{2x+6}+\frac{5}{x^2-9}+\frac{2x}{9-3x}$

68. $\frac{x}{4-2x}-\frac{3x}{5x+10}-\frac{2}{x^2-4}$

69. $\frac{1}{2x-6}+\frac{1}{3}-\frac{x}{3x+6}$

70. $\frac{1}{3x+3}-\frac{3}{15}+\frac{x}{5x-10}$

6.5 Rational Equations

We have solved several types of equations, and now we can consider equations that involve rational expressions. For example, the equation

$$\frac{5}{x} - \frac{1}{3} = \frac{6}{x}$$

is called a ***rational equation,*** since it has at least one variable in the denominator. Essentially, the method of solution for such equations obtains an equivalent equation without fractions. However, the domain of the variable must exclude any value that will make a denominator zero. To clear the equation of fractions, we multiply both sides of the equation by the LCD of the denominators. Each of the denominators will divide evenly into the LCD, so the resulting equation will be similar to those we have solved earlier.

Example 1. Solve the rational equation:

$$\frac{5}{x} - \frac{1}{3} = \frac{6}{x}$$

Solution Note that $x \neq 0$, then multiply by the LCD, 3x.

$$(3x)\frac{5}{x} - (3x)\frac{1}{3} = (3x)\frac{6}{x}$$ *Multiply both sides by the LCD, noting that each of the denominators divides evenly into the LCD.*

$$3 \cdot 5 - x \cdot 1 = 3 \cdot 6$$ *Simplify.*

$$15 - x = 18$$ *Solve the resulting equation.*

$$-x = 3$$

$$\mathbf{x = -3}$$

Check by substituting $x = -3$ into the original equation.

$$\frac{5}{-3} - \frac{1}{3} = \frac{6}{-3}$$

$$\frac{-6}{3} = -2 \checkmark$$

■

Example 2. Solve the rational equation:

$$\frac{x}{x-2} = 2 - \frac{2}{2-x}$$

Solution Note that $x \neq 2$, then multiply both sides by the LCD, $x - 2$.

$$(x-2)\frac{x}{x-2} = (x-2)2 + (x-2)\frac{-2}{2-x}$$

$$x = (x-2)(2) + (-1)(-2)$$ Since $(x-2)\dfrac{-2}{2-x} = (-1)(-2)$

$$x = 2x - 4 + 2$$

$$-x = -2$$

$$x = 2$$

But this result is impossible. Recall that $x \neq 2$ was an initial restriction because it would cause division by zero. Thus, the solution set is empty. ■

Procedure for Solving Rational Equations

1. Exclude values that cause division by zero.
2. Multiply both sides by the LCD of the denominators.
3. Solve the resulting equation.
4. Check the solution set to make sure it is not one of the excluded values; that is, the values in the solution set must not cause division by zero.

Any value that is excluded in this manner is called an ***extraneous root.***

Example **3.** Solve the rational equation and note extraneous roots:

$$\frac{x}{x-1} = \frac{2x}{x^2-1} - \frac{x}{x+1}$$

Solution The LCD of $x-1$, x^2-1, and $x+1$ is $(x-1)(x+1)$, so multiply by that quantity, noting that $x \neq 1$, $x \neq -1$.

$$\mathbf{(x-1)(x+1)}\frac{x}{x-1} = \mathbf{(x-1)(x+1)}\frac{2x}{x^2-1} - \mathbf{(x-1)(x+1)}\frac{x}{x+1}$$

$$\begin{aligned} (x+1)x &= 2x - (x-1)x \\ x^2 + x &= 2x - x^2 + x \\ 2x^2 - 2x &= 0 \\ 2x(x-1) &= 0 \end{aligned}$$

$$\begin{aligned} 2x &= 0 \quad \text{or} \quad x - 1 = 0 \\ x &= 0 \qquad\qquad x = 1 \end{aligned}$$

But $x = 1$ is an extraneous root, so $\boldsymbol{x = 0}$ is the only solution. ■

Problem Set 6.5

Solve the equations in Problems 1–40 and note extraneous roots. See Examples 1–3.

1. $\frac{x}{3} + \frac{5x}{9} = 8$

2. $\frac{y}{2} + \frac{y}{4} = 6$

3. $\frac{2z}{5} + \frac{3z}{2} = 19$

4. $\frac{1}{2x} - \frac{2}{3x} = \frac{1}{6}$

5. $\frac{1}{3y} - \frac{1}{12} = \frac{1}{4y}$

6. $\frac{1}{2z} - \frac{1}{10} = \frac{2}{5z}$

7. $\frac{3}{x} + \frac{2}{x} = 5$

8. $\frac{6}{y} - \frac{4}{5y} = \frac{13}{5}$

9. $\frac{3}{4z} + \frac{6}{5z} = \frac{39}{20}$

10. $\frac{5}{2x} + 2 = \frac{2x}{x-1}$

11. $\frac{3y}{y-2} = \frac{2}{y} + 3$

12. $\frac{5z-7}{z-3} - 5 = \frac{2}{z}$

13. $\frac{1}{x} + \frac{1}{2x} - \frac{1}{3x} = \frac{7}{12}$

14. $\frac{1}{y} + \frac{2}{y} + \frac{3}{y} = 6$

15. $\frac{1}{2a} - \frac{1}{3a} - \frac{1}{4a} = \frac{1}{12}$

16. $\frac{1}{b} - \frac{2}{3b} - \frac{1}{5b} = \frac{1}{15}$

17. $\frac{4}{m+1} + 2 = \frac{6}{m+1}$

18. $\frac{9}{n-3} - 1 = \frac{6}{n-3}$

19. $\frac{6x}{2x-1} = 4 - \frac{3}{1-2x}$

20. $\frac{3}{y-4} - 2 = \frac{3y}{4-y}$

21. $\frac{x+2}{3x+6} = \frac{1}{3} + \frac{2-x}{2x+4}$

22. $\frac{y+3}{4y-4} = \frac{1}{12} + \frac{5-y}{6y-6}$

23. $\frac{4}{1-x} = \frac{3}{x-1} - 1$

24. $3 + \frac{1}{2-x} = \frac{2}{x-2}$

25. $\frac{2x+1}{x^2-x} = \frac{x+2}{x-1}$

26. $\frac{x-2}{x+2} = \frac{x+10}{x^2+2x}$

27. $\frac{x}{x^2-4} = \frac{3}{x+2}$

28. $\frac{x}{x^2-25} = \frac{2}{x-5}$

29. $\frac{x-2}{x+3} = \frac{x-5}{x+6}$

30. $\frac{2x+1}{x-5} = \frac{8x-1}{4x+5}$

31. $\frac{x+6}{x^2-7x+12} = \frac{6}{x-3}$

32. $\frac{4x-6}{x^2-9x+20} + \frac{6}{x-5} = 0$

33. $\frac{2x+4}{x^2-16} = \frac{1}{x+4} + \frac{3}{x-4}$

34. $\frac{4}{x-5} - \frac{16}{x+5} = \frac{3x-5}{x^2-25}$

35. $\frac{1}{a+2} - \frac{1}{2-a} = \frac{3a+8}{a^2-4}$

36. $\frac{5b+7}{b^2-25} + \frac{24}{5-b} + \frac{45}{b+5} = 0$

37. $\frac{2}{x-3} - \frac{20}{x^2+4x-21} = \frac{1-x}{x+7}$

38. $\frac{3}{x^2-5x-36} = \frac{x-3}{x-9} + \frac{1}{x+4}$

39. $\frac{2x}{3x+1} = \frac{1}{x-2} - \frac{4}{3x^2-5x-2}$

40. $\frac{x-2}{2x-3} + \frac{1}{x+2} = \frac{3x-1}{2x^2+x-6}$

6.6 Negative Exponents

The product $32 \cdot 8$ can be found by expressing each number as a power of 2 and using the First Law of Exponents.

$$\begin{aligned} 32 \cdot 8 &= 2^5 \cdot 2^3 \\ &= 2^{5+3} \\ &= 2^8 \quad \text{or} \quad 256 \end{aligned}$$

That is, to multiply two numbers with like bases, we add the exponents. Is there a similar property for division? Consider the quotient $\frac{32}{8}$.

$$\begin{aligned} \frac{32}{8} &= \frac{\not{2} \cdot \not{2} \cdot \not{2} \cdot 2 \cdot 2}{\not{2} \cdot \not{2} \cdot \not{2}} \\ &= 2 \cdot 2 \end{aligned}$$

$$\frac{2^5}{2^3} = 2^2 \quad \text{or} \quad 4$$

Notice that $2^{5-3} = 2^2$. Since division is the inverse operation of multiplication where we add exponents, we might reasonably expect division to result in subtracting exponents. That is, a property like $\frac{b^m}{b^n} = b^{m-n}$ is plausible. However, two cases need to be considered.

What if the numerator and denominator are the same? The quotient $\frac{8}{8}$ is clearly equal to 1, but,

$$\begin{aligned}\frac{8}{8} &= \frac{2^3}{2^3} \\ &= 2^{3-3} \\ &= 2^0\end{aligned}$$

A zero exponent has no meaning under our present definition of exponent; thus, if the property $\frac{b^m}{b^n} = b^{m-n}$ is to hold, then $2^0 = 1$. Consider the First Law of Exponents,

$$\begin{aligned}x^0 \cdot x^3 &= x^{0+3} \\ &= x^3 \\ &= 1 \cdot x^3\end{aligned}$$

This result can be true only if $x^0 = 1$. If these properties are to hold, we need the following definition.

Zero Exponent

$b^0 = 1$ for all nonzero b.

Examples

1. $10^0 = 1$
2. $\left(\frac{1}{2}\right)^0 = 1$
3. $(xy)^0 = 1$
4. $(-2)^0 = 1$
5. $-2^0 = -1$ ■

Thus, we have seen that

$$\frac{b^m}{b^n} = b^{m-n}$$

holds if $m = n$ and if $m > n$. What if $m < n$, as in the following example?

$$\begin{aligned}\frac{8}{32} &= \frac{2^3}{2^5} \\ &= 2^{3-5} \\ &= 2^{-2}\end{aligned} \qquad \begin{aligned}\frac{8}{32} &= \frac{\cancel{2}\cdot\cancel{2}\cdot\cancel{2}}{\cancel{2}\cdot\cancel{2}\cdot\cancel{2}\cdot 2\cdot 2} \\ &= \frac{1}{2\cdot 2} \\ &= \frac{1}{2^2}\end{aligned}$$

But our definition of exponents does not provide for negative exponents. If the division property is to hold,

$$2^{-2} = \frac{1}{2^2}$$

In general, the following definition is necessary.

Negative Exponents

$b^{-n} = \frac{1}{b^n}$ for all nonzero b and natural numbers n.

Example

6. $10^{-2} = \frac{1}{10^2} = \frac{1}{100} = .01$

7. $x^{-9} = \frac{1}{x^9}$

8. $\left(\frac{1}{2}\right)^{-3} = \frac{1}{\left(\frac{1}{2}\right)^3} = \frac{1}{\frac{1}{8}} = 8$

9. $x^{-3}y^{-2} = \frac{1}{x^3y^2}$

10. $\frac{1}{z^{-4}} = z^4$ ■

The definition of zero and negative exponents now allow the statement of a ***Fourth Law of Exponents.***

Fourth Law of Exponents

$\frac{b^m}{b^n} = b^{m-n}$ for all nonzero b and integers m and n.

To divide two numbers with like bases, subtract the exponents.

Examples

11. $\frac{81}{27} = \frac{3^4}{3^3}$

$= 3^{4-3}$

$= 3$

12. $\frac{5^2}{5^{-2}} = 5^{2-(-2)}$

$= 5^{2+2}$

$= 5^4$ or 625

13. $\frac{x^2y^5}{x^3y^3} = x^{2-3}y^{5-3}$

$= x^{-1}y^2$

$= \frac{y^2}{x}$

14. $\frac{2^5x^2}{2^3x^4} = 2^{5-3}x^{2-4}$

$= 2^2x^{-2}$

$= \frac{2^2}{x^2}$ or $\frac{4}{x^2}$ ■

A ***Fifth Law of Exponents*** can be derived from the other laws. Recall that $(ab)^m = a^m b^m$, and consider an analogous property for division.

$$\begin{aligned}
\left(\frac{a}{b}\right)^m &= \left(a\cdot\frac{1}{b}\right)^m && \textit{Definition of Division}\\
&= (a\cdot b^{-1})^m && \textit{Definition of Negative Exponent}\\
&= a^m(b^{-1})^m && \textit{Second Law of Exponents}\\
&= a^m(b^{-m}) && \textit{Third Law of Expressions}\\
&= a^m\cdot\frac{1}{b^m} && \textit{Definition of Negative Exponents}\\
&= \frac{a^m}{b^m} && \textit{Definition of Division}
\end{aligned}$$

Fifth Law of Exponents

$$\left(\frac{a}{b}\right)^m = \frac{a^m}{b^m},\ b \neq 0$$

To raise a quotient to a power, raise both the numerator and the denominator to that power.

We can combine these properties as shown in the following examples, in which all of the variables are nonzero real numbers.

Examples

15. $x^2(xy^2)^3 = x^2(x^{3\cdot 1}y^{3\cdot 2})$
$= x^2(x^3y^6)$
$= x^{2+3}y^6$
$= x^5y^6$

16. $\dfrac{8x^{-2}y^{-2}}{4x^3y^{-4}} = \dfrac{8}{4}x^{-2-3}y^{-2-(-4)}$
$= 2x^{-5}y^2$

17. $\left(\dfrac{x^{-3}}{y^2}\right)^{-2} = \dfrac{(x^{-3})^{-2}}{(y^2)^{-2}}$
$= \dfrac{x^6}{y^{-4}}$
$= x^6y^4$ ■

One application of positive and negative exponents that is particularly useful is called ***scientific notation.*** Any rational number can be written as the product of a number between 1 and 10 and a power of 10. When we write a number this way, we say that it is in scientific notation.

Examples *Write the following numbers in scientific notation.*

18. $579 = 5.79\cdot 10^2$

19. $4{,}800{,}000 = 4.8 \cdot 10^6$

20. $.0000217 = 2.17 \cdot 10^{-5}$

21. The earth is 93,000,000 miles from the sun.

$9.3 \cdot 10^7$ mi

22. The smallest living things visible with an ordinary microscope are bacteria as small as 0.00002 cm in length.

$2 \cdot 10^{-5}$ cm ■

Examples *Write the following numbers in decimal notation.*

23. $4.67 \cdot 10^{-8} = 0.0000000467$

24. $9.3 \cdot 10^{11} = 930{,}000{,}000{,}000$

25. The land area of the earth's surface is $1.49 \cdot 10^8$ km^2.

149,000,000 km^2 ■

Problem Set 6.6

Simplify Problems 1–40. Leave in exponent form without negative exponents. See Examples 1–17.

1. 3^0 **2.** 3^{-1} **3.** 2^{-1} **4.** 2^0

5. 3^{-2} **6.** 2^{-3} **7.** $(-3)^2$ **8.** $(-2)^3$

9. $5^3 \cdot 5^7$ **10.** $7^2 \cdot 7^4$ **11.** $(7^3)^4$ **12.** $(3^4)^5$

13. $(2 \cdot 3)^3$ **14.** $(5 \cdot 7)^2$ **15.** $(-2^3)^2$ **16.** $-(2^3)^2$

17. $\frac{2^7}{2^3}$ **18.** $\frac{3^3}{3^2}$ **19.** $\frac{5^7}{5^9}$ **20.** $\frac{2^{10}}{2^7}$

21. $\left(\frac{2}{3}\right)^{-1}$ **22.** $\left(\frac{2}{3}\right)^{-2}$ **23.** $\left(\frac{-2}{3}\right)^2$ **24.** $\left(\frac{-2}{3}\right)^{-2}$

25. $\frac{3^2 \cdot 3^3}{3}$ **26.** $\frac{2^5 \cdot 2^4}{2^3}$ **27.** $\frac{(5^2 \cdot 5)^3}{5^2}$ **28.** $\frac{(2^3 \cdot 2^2)^2}{2^3}$

29. $\frac{7^8}{7^2(7^3)^2}$ **30.** $\frac{5^9}{5^3(5^2)^3}$ **31.** $\frac{2^{-7}}{2^3}$ **32.** $\frac{3^3}{3^{-2}}$

33. $\frac{5^{-7}}{5^9}$ **34.** $\frac{2^{10}}{2^{-7}}$ **35.** $\frac{3^{-2}}{3^{-5}}$ **36.** $\frac{7^{-1}}{7^{-2}}$

37. $\frac{3^{-2} \cdot 3^3}{3}$ **38.** $\frac{2^{-5} \cdot 2^4}{2^3}$ **39.** $\frac{(5^{-2} \cdot 5)^3}{5^2}$ **40.** $\frac{(2^3 \cdot 2^{-2})^2}{2^{-3}}$

Simplify Problems 41–70. Leave variables in exponent form without negative exponents. See Examples 1–17.

41. $(2x)(3x^2)$ **42.** $(3y)(2y^3)$

43. $(2z^2)(z^5)$ **44.** $(3x^3)^2$

45. $(2y^2)^3$ **46.** $(3z^3)^3$

47. $\frac{x^5}{x^3}$ **48.** $\frac{y^7}{y^4}$

49. $\frac{z^3}{z^5}$

50. $\left(\frac{x^2}{y^3}\right)^3$

51. $\left(\frac{y^3}{z^2}\right)^3$

52. $\left(\frac{z^2}{x^3}\right)^3$

53. $(x^2y^{-3})^{-2}$

54. $(y^3z^{-2})^{-3}$

55. $(x^{-3}z^2)^{-3}$

56. $(2a^3b^{-5})^{-3}$

57. $(5b^{-4}c^3)^{-2}$

58. $(3a^{-2}c^{-3})^3$

59. $(2a^3b - 5b^{-4})^0$

60. $(3a^{-2}c + c^3)^0$

61. $\frac{x^{-3}x^3}{x^5}$

62. $\left(\frac{x^5y}{x^2y^2}\right)^3$

63. $\left(\frac{a^3b^2}{a^2b^3}\right)^2$

64. $\frac{12x^{-3}y^{-2}}{4x^5y^{-5}}$

65. $\frac{20(x^2y^{-1})^2}{(2xy^{-2})^3}$

66. $\frac{5a^5b^{-5}}{5^{-1}a^2b^3}$

67. $\left(\frac{5m^{-2}n}{15mn^3}\right)^{-1}$

68. $\left(\frac{x-1}{x}\right)\left(\frac{1-x}{x^2}\right)^{-1}$

69. $\left(\frac{4x-1}{x}\right)\left(\frac{1-4x}{x^2}\right)^{-1}$

70. $\left(\frac{5x-3}{x^3}\right)\left(\frac{3-5x}{x}\right)^{-1}$

Write the numbers in Problems 71–80 in scientific notation. See Examples 18–22.

71. 43,000,000
72. 63,000,000,000
73. .000681
74. 923,000,000,000,000
75. .06831
76. .000000000021
77. Pluto is about 3,666,000,000 miles from the sun.
78. 361,000,000 km^2 of the earth's surface is covered by oceans.
79. Wrought iron expands 0.00001166 in. for each degree Celsius of temperature increase.
80. The time required for a computer to do a certain arithmetic operation is 0.00000000235 s.

Write the numbers in Problems 81–90 in decimal notation. See Examples 23–25.

81. The velocity of sound is $3.3146 \cdot 10^2$ m/s in dry air.
82. An x-ray has a wavelength of approximately $1 \cdot 10^{-8}$ cm.
83. The mean distance of the planet Mercury from the sun is $5.787 \cdot 10^7$ km..
84. An atom of gold weighs approximately $3.25 \cdot 10^{-23}$ g.
85. A communications satellite can process $4.2 \cdot 10^{12}$ bits of information daily.
86. One second is approximately $2.778.\ 10^{-4}$ hours.
87. A single red cell of human blood contains $2.7 \cdot 10^8$ hemoglobin molecules.
88. The earth picks up approximately $1.3 \cdot 10^4$ tons of dust from the atmosphere every day.
89. The mass of a water molecule is $3 \cdot 10^{-23}$ g.
90. One cycle of a television broadcast signal takes $4.81 \cdot 10^{-9}$ s.

6.7 Complex Fractions

If an expression contains fractions in the numerator or denominator, it is called a ***complex fraction.*** The quotient of two rational expressions is a complex fraction. To simplify these expressions, we need to rewrite them as a single fraction. Such a process simply recognizes that the complex fraction may be rewritten as a division problem. The division is performed in the usual way.

$$\frac{\frac{a}{b}}{\frac{c}{d}} = \frac{a}{b} \div \frac{c}{d}$$

To divide complex fractions, we invert the divisor and multiply. This means multiply by the reciprocal of the divisor.

$$= \frac{a}{b} \cdot \frac{d}{c} \qquad b, c, d \neq 0$$

Simplify the following complex fractions.

Example 1. $\dfrac{\frac{5}{8}}{\frac{15}{16}}$

Solution $\dfrac{\frac{5}{8}}{\frac{15}{16}} = \dfrac{\overset{1}{\cancel{5}}}{\underset{1}{\cancel{8}}} \cdot \dfrac{\overset{2}{\cancel{16}}}{\underset{3}{\cancel{15}}}$ *Notice that you invert $\frac{15}{16}$ and then multiply.*

$$= \frac{2}{3}$$ ■

Example 2. $\dfrac{\frac{3a^2}{8b}}{\frac{9a}{2b^3}} = \dfrac{3a^2}{8b} \div \dfrac{9a}{2b^3}$ *This step is usually done mentally.*

$$= \frac{3a^2}{8b} \cdot \frac{2b^3}{9a}$$

$$= \frac{ab^2}{12}$$ ■

Example 3. $\dfrac{2 + \frac{1}{x}}{\frac{1}{x} - 1} = \dfrac{\frac{2x+1}{x}}{\frac{1-x}{x}}$ *The numerator and denominator must be written as single fractions before you can invert and multiply.*

$$= \frac{2x+1}{\cancel{x}} \cdot \frac{\cancel{x}}{1-x}$$ *Invert and multiply here.*

$$= \frac{2x+1}{1-x}$$ *Write in reduced form.* ■

Example 3 could have appeared with negative exponents.

$$\frac{2 + \frac{1}{x}}{\frac{1}{x} - x} = \frac{2 + x^{-1}}{x^{-1} - x}$$

Once written with positive exponents, such expressions can be simplified as in Example 3 in the following examples.

Example 4. $\dfrac{x^{-1}+y^{-1}}{x+y} = \dfrac{\frac{1}{x}+\frac{1}{y}}{x+y}$ *First rewrite with positive exponents.*

$= \dfrac{\frac{x+y}{xy}}{x+y}$ *Add fractions in the numerator.*

$= \dfrac{\cancel{x+y}}{xy}\cdot\dfrac{1}{\cancel{x+y}}$ *Invert and multiply.*

$= \dfrac{1}{xy}$ *Reduce and state answer.* ■

Example 5. $\dfrac{\frac{x^2-4}{x+3}}{\frac{x+2}{4x+12}} = \dfrac{x^2-4}{x+3}\cdot\dfrac{4x+12}{x+2}$

$= \dfrac{(x-2)\cancel{(x+2)}(4)\cancel{(x+3)}}{\cancel{(x+3)}\cancel{(x+2)}}$

$= 4(x-2)$ ■

Example 6. $\dfrac{5+\frac{1}{x}}{3+\frac{1}{x}} = \dfrac{\frac{5x+1}{x}}{\frac{3x+1}{x}}$

$= \dfrac{5x+1}{\cancel{x}}\cdot\dfrac{\cancel{x}}{3x+1}$

$= \dfrac{5x+1}{3x+1}$ ■

Problem Set 6.7

Simplify the expressions in Problems 1–40. See Examples 1–6.

1. $\dfrac{\frac{1}{5}}{\frac{2}{3}}$
2. $\dfrac{\frac{7}{8}}{\frac{3}{4}}$
3. $\dfrac{\frac{2}{3}}{\frac{15}{14}}$
4. $\dfrac{\frac{14}{20}}{\frac{7}{5}}$
5. $\dfrac{\frac{5x^2}{3}}{\frac{10x}{9}}$
6. $\dfrac{\frac{3}{5a^2}}{\frac{12}{a}}$
7. $\dfrac{\frac{x}{y^2}}{\frac{x^2}{y}}$
8. $\dfrac{\frac{a}{b}}{\frac{a^2}{b}}$

9. $\dfrac{\frac{ab}{c}}{\frac{a^2}{2c}}$ 10. $\dfrac{\frac{ab}{c}}{\frac{a^2b}{c^2}}$ 11. $\dfrac{\frac{2a^2}{9b}}{\frac{4a}{3b^3}}$ 12. $\dfrac{\frac{27a}{25b^3}}{\frac{3a^2}{5b}}$

13. $\dfrac{\frac{1}{2}+\frac{1}{3}}{\frac{3}{2}+\frac{2}{3}}$ 14. $\dfrac{-2+\frac{2}{3}}{1+\frac{1}{3}}$ 15. $\dfrac{\frac{1}{5}+\frac{2}{3}}{\frac{3}{5}-\frac{1}{3}}$ 16. $\dfrac{\frac{1}{2}-\frac{2}{5}}{\frac{1}{5}+\frac{1}{2}}$

17. $\dfrac{\frac{x-1}{x+1}}{\frac{x-1}{x+3}}$ 18. $\dfrac{\frac{x+1}{x}}{\frac{x-1}{x}}$ 19. $\dfrac{\frac{3x+1}{x-4}}{\frac{3x+1}{x+2}}$ 20. $\dfrac{\frac{2x-3}{x^2}}{\frac{2x-3}{x^3}}$

21. $\dfrac{1+\frac{1}{a}}{1-\frac{1}{b}}$ 22. $\dfrac{2-\frac{1}{x}}{1+\frac{1}{x}}$ 23. $\dfrac{x+\frac{1}{x}}{x-\frac{1}{3}}$ 24. $\dfrac{x+\frac{2}{x}}{3x-\frac{1}{2}}$

25. $\dfrac{x^{-1}+1}{x^{-1}-1}$ 26. $\dfrac{y^{-1}-2}{1-y^{-1}}$ 27. $\dfrac{a^{-2}+1}{a^{-1}-2}$ 28. $\dfrac{b^{-1}+1}{1-b^{-2}}$

29. $\dfrac{\frac{1}{a}+\frac{1}{b}}{\frac{1}{a+b}}$ 30. $\dfrac{a+\frac{1}{b}}{\frac{1}{a}+b}$ 31. $\dfrac{x-\frac{1}{x}}{x+\frac{1}{x}}$ 32. $\dfrac{y+\frac{1}{y}}{\frac{1}{y}-y}$

33. $\dfrac{a^{-1}+b^{-1}}{(a+b)^{-1}}$ 34. $\dfrac{a+b^{-1}}{a^{-1}+b}$ 35. $\dfrac{x-x^{-1}}{x+x^{-1}}$ 36. $\dfrac{y+y^{-1}}{y^{-1}-y}$

37. $\dfrac{\frac{x^2-9}{x-2}}{\frac{x+3}{2x-4}}$ 38. $\dfrac{\frac{x^2-16}{x+3}}{\frac{x-4}{x+3}}$ 39. $\dfrac{\frac{x^2+3x+2}{3x^2+8x+4}}{\frac{x^2-1}{3x^2-x-2}}$ 40. $\dfrac{\frac{x^2+3x+2}{x-3}}{\frac{x^2+3x+2}{x^2-6x+9}}$

6.8 Review

This chapter gave you additional practice with factoring as you applied it to rational expressions. You should now feel at ease simplifying, adding, subracting, multiplying, and dividing rational expressions. Negative exponents were developed in this chapter because they provide a compact, useful notation for writing certain rational expressions.

We also considered the solution of equations involving rational expressions. The important thing to remember about solving equations involving rational expressions is that all fractions and fractional forms can be eliminated by multiplying both sides of that equation by the appropriate expression. This cannot be done when simplifying rational equations or when an equation is not given.

Self-Test

Perform the indicated operations in Problems 1–7 and simplify without using negative exponents. Values for variables that cause division by zero are excluded.

[6.1] **1.** **a.** $\frac{7}{12} - \frac{4}{9} - \frac{5}{8}$ **b.** $\frac{5}{3}\left(\frac{1}{7} - \frac{2}{5}\right)$

[6.2] **2.** **a.** $\frac{12x^2y^3z}{16x^3yz^2}$ **b.** $\frac{2a+4}{a^2-4}$

[6.3] **3.** **a.** $\frac{5x^2y}{27z^2} \cdot \frac{9z}{125xy^2}$ **b.** $\frac{x-1}{x^2-4x+4} \div \frac{x^2-x}{(x-2)^2}$

[6.3],[6.4] **4.** **a.** $\frac{x^2+2x-15}{2x+1} \div (x^2-25)$ **b.** $\frac{5}{s-t} + \frac{5}{t-s}$

[6.4] **5.** **a.** $x - \frac{x-1}{x}$ **b.** $\frac{4}{3x+3y} - \frac{5}{2x+2y}$

[6.6] **6.** **a.** $(a^3b^{-2})^{-3}$ **b.** $\frac{9x^{-4}y^2}{15x^3y^{-5}}$

[6.6],[6.7] **7.** **a.** $4(m-n)^{-1} - 3(n-m)^{-1}$ **b.** $\frac{4 - \frac{a}{b}}{5 - \frac{a^2}{b^2}}$

[6.6] **8.** If $b \neq 0$, and m and n are integers:

a. $b^n =$ ____________ **b.** $b^0 =$ ____________

c. $b^{-n} =$ ____________ **d.** $b^mb^n =$ ____________

e. $(b^n)^m =$ ____________ **f.** $\frac{b^m}{b^n} =$ ____________

[6.6] **9.** **a.** Write in scientific notation.

i. 15,300,000 ii. .00613 iii. 10 million

iv. The mass of a molecule of water is 0.00000000000000000000003 g.

b. Write in decimal notation.

i. $3.01 \cdot 10^{-3}$ ii. $5.29 \cdot 10^5$ iii. 10^{-1}

iv. The volume of the earth's oceans is approximately $1.37 \cdot 10^9$ cubic kilometers.

[6.5] **10.** Solve the equations.

a. $\frac{4}{5} = \frac{7}{x}$ **b.** $\frac{4}{x} = \frac{5}{12}$

c. $\frac{5}{y} - \frac{1}{12} = \frac{3}{4y}$ **d.** $\frac{2x}{3x+1} + 2 = \frac{5x+11}{3x+1}$

Review Problems

Perform the indicated operations in Problems 1–80 and simplify without using negative exponents. Values for variables that cause division by zero are excluded.

1. $\frac{2}{3} + \frac{1}{2} - \frac{5}{6}$ **2.** $\frac{1}{4} + \frac{1}{2} - \frac{3}{4}$

3. $\frac{3}{4} + \frac{1}{6} - \frac{2}{3}$ **4.** $\frac{1}{4} + \frac{5}{6} - \frac{1}{3}$

5. $\frac{5}{12}+\frac{5}{9}-\frac{5}{18}$

6. $\frac{3}{8}-\frac{3}{32}+\frac{3}{4}$

7. $\frac{9}{11}\left(\frac{1}{9}-\frac{3}{5}\right)$

8. $\frac{7}{9}\left(\frac{2}{7}-\frac{1}{2}\right)$

9. $\frac{8}{3}\left(\frac{3}{4}-\frac{1}{2}\right)$

10. $\frac{3}{2}\left(\frac{2}{3}-\frac{5}{6}\right)$

11. $\frac{6xy}{4xz}$

12. $\frac{9xz}{6yz}$

13. $\frac{12x^2y}{15xy^2}$

14. $\frac{14x^3y}{21xy^2}$

15. $\frac{3a-3}{a^2-1}$

16. $\frac{b^2-1}{2b+2}$

17. $\frac{3x-6}{2x-4}$

18. $\frac{y^2+y}{3y+3}$

19. $\frac{x^2-x}{x^2-1}$

20. $\frac{4y+2}{4y^2-1}$

21. $\frac{2a}{b}\cdot\frac{b}{3}$

22. $\frac{b}{3a}\cdot\frac{6}{ab}$

23. $\frac{6x}{4y}\cdot\frac{2}{y}$

24. $\frac{10x}{15z}\cdot\frac{3z}{x^2}$

25. $\frac{3xy}{8yz^2}\cdot\frac{4z}{10x^2y}$

26. $\frac{2ab}{15b^2c}\cdot\frac{10c^3}{4ab}$

27. $\frac{2x^3}{25z^2}\div\frac{4x}{5yz}$

28. $\frac{3b^2}{14a^3c}\div\frac{b^3}{21ac^2}$

29. $\frac{a}{a-b}\cdot\frac{a-b}{b}$

30. $\frac{b}{a-b}\cdot\frac{b-a}{a}$

31. $\frac{2x-4}{3x+9}\cdot\frac{2x+6}{3x-6}$

32. $\frac{4-x}{3x+9}\cdot\frac{2x+6}{x-4}$

33. $\frac{x-1}{4x}\cdot\frac{2x+2}{x^2-1}$

34. $\frac{9y}{y+3}\cdot\frac{y^2-9}{3y^2-9y}$

35. $\frac{x^2-1}{1-x}\cdot\frac{x}{x+1}$

36. $\frac{a-a^2}{a^2-1}\div\frac{-a}{a^2+a}$

37. $\frac{x^2-5x+6}{2x-6}\cdot\frac{3x}{x-2}$

38. $\frac{3x+3}{6x}\cdot\frac{2x^2}{x^2-5x-6}$

39. $\frac{x^2-4x+4}{x^2-2x}\div\frac{x^2-4}{x^2+x-2}$

40. $\frac{x^2+3x}{x^2+6x+9}\div\frac{3x^2-9x}{x^2-9}$

41. $\frac{x}{6}+\frac{y}{4}$

42. $\frac{a}{6}-\frac{b}{9}$

43. $\frac{x}{6}+\frac{4}{y}$

44. $\frac{a}{6}-\frac{9}{b}$

45. $x+\frac{4}{y}$

46. $a-\frac{9}{b}$

47. $\dfrac{x}{6}+\dfrac{1}{x}$

48. $\dfrac{a}{6}-\dfrac{1}{a}$

49. $\dfrac{1}{x}+\dfrac{2}{y}$

50. $\dfrac{3}{a}-\dfrac{2}{b}$

51. $\dfrac{1}{x+1}+\dfrac{1}{x^2-1}$

52. $\dfrac{2}{y^2-4}-\dfrac{1}{y-2}$

53. $\dfrac{3}{2x-4}-\dfrac{4}{3x-6}$

54. $\dfrac{1}{5y+5}-\dfrac{2}{3y+3}$

55. $\dfrac{3}{2y+6}+\dfrac{2}{9+3y}$

56. 2^{-3}

57. $(-2)^3$

58. $(-2)^{-3}$

59. $(-3)^2$

60. -3^2

61. $(-3)^{-2}$

62. $\left(\dfrac{3}{2}\right)^{-1}$

63. $\left(\dfrac{-3}{2}\right)^{-1}$

64. $\left(\dfrac{-3}{2}\right)^{-2}$

65. $\dfrac{5^{-5}}{5^2}$

66. $\dfrac{7^{-3}}{7^{-5}}$

67. $\dfrac{9^{-4}}{9^{-3}}$

68. $\dfrac{a^7}{a^5}$

69. $\dfrac{b^5}{b^7}$

70. $\dfrac{c}{c^3}$

71. $(x^3y)^{-2}$

72. $(x^{-3}y)^2$

73. $(x^{-3}y)^{-2}$

74. $\left(\dfrac{a^2b}{ab^3}\right)^2$

75. $\left(\dfrac{a^3b^2}{ab^4}\right)^{-1}$

76. $\left(\dfrac{ab^3}{a^2b}\right)^{-2}$

77. $x(x^{-2}y^2)^{-2}$

78. $a^2(a^{-3}b)^{-2}$

79. $x^{-3}(x^2y^{-1})^{-3}$

80. $ab^2(a^2b^3)^{-1}$

Solve the equations in Problems 81–100 and note extraneous roots, if any.

81. $\dfrac{x}{2}+\dfrac{x}{3}=\dfrac{5}{6}$

82. $\dfrac{y}{5}-\dfrac{y}{10}=\dfrac{1}{5}$

83. $\dfrac{3n}{4}-\dfrac{5}{6}=\dfrac{n}{3}$

84. $\dfrac{2m}{3}+\dfrac{5}{6}=\dfrac{m}{2}$

85. $\dfrac{7x}{8}-2=\dfrac{3x}{4}$

86. $4+\dfrac{y}{6}=\dfrac{3y}{2}$

87. $\dfrac{a+1}{2}=\dfrac{2a+5}{6}$

88. $\dfrac{3-b}{8}=\dfrac{9-3b}{2}$

89. $\dfrac{5}{x}=-1$

90. $\dfrac{3}{y}=-9$

91. $\frac{5}{a} - \frac{3}{a} = 2$

92. $\frac{6}{b} - \frac{1}{2} = \frac{3}{b}$

93. $\frac{6}{x+3} = \frac{2}{x-1}$

94. $\frac{8}{2y+1} = \frac{6}{3y}$

95. $\frac{2x-1}{x+2} - 1 = \frac{x-3}{x-2}$

96. $\frac{y+4}{y+1} + 2 = \frac{3y-2}{y-1}$

97. $\frac{x+2}{x-2} + \frac{4}{x+2} = \frac{x^2+12}{x^2-4}$

98. $\frac{13}{y+4} - \frac{8}{y-4} = \frac{y-60}{y^2-16}$

99. $\frac{4}{x+2} + \frac{7}{x+3} = \frac{4}{x^2+5x+6}$

100. $\frac{y+1}{y-1} - \frac{y-1}{y+1} = \frac{1}{y^2-1}$

7

First-Degree Equations and Inequalities in Two Variables

For "is" and "is-not" though with rule and line
And "up-and-down" by logic I define,
Of all that one should care to fathom, I
Was never deep in anything but—wine.

Ah, but my computations, people say,
Reduced the year to better reckoning? Nay,
'Twas only striking from the calendar
Unborn to-morrow and dead yesterday.

The Rubáiyát of Omar Khayyám
trans. Edward Fitzgerald

7.1 Cartesian Coordinate System

Have you ever drawn a picture by connecting the dots or found a city or street on a map? If you have, then you have used some of the ideas discussed in this chapter. Suppose you wish to find Fisherman's Wharf in San Francisco. You look in the index for the map in Figure 7.1 and find Fisherman's Wharf listed as (6, D). Can you locate section (6, D) in Figure 7.1?

Figure 7.1
Map of San Francisco

In mathematics, the listing (6, D) is called an ***ordered pair.*** The first entry, 6, is called the ***first component*** and the second entry, D, is called the ***second component.***

Have you ever been given the location of a street and still not been able to find it easily? For example, suppose you want to find Main Street. Without the index the task is next to impossible, but even with the index listing Main at (7, B) the task of finding the street is not an easy one. How could we improve the map?

One way would be to create a smaller grid. However, a smaller grid would mean that a lot of letters are needed for the vertical scale, and it is possible to run out of letters. Therefore, suppose we use ordered pairs of numbers and relabel the vertical scale with numbers. Also change from labeling the spaces to labeling the lines as shown in Figure 7.2.

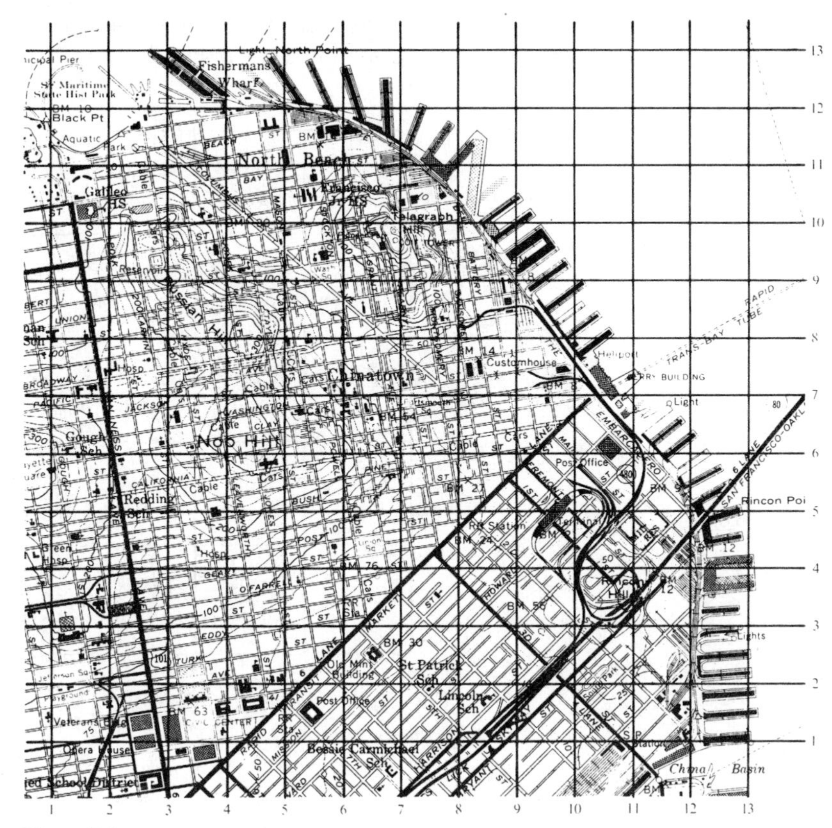

Figure 7.2
Map of San Francisco with coordinate grid

Now the location of any street on the map can be fixed quite precisely.

Notice that if we use an ordered pair of numbers (instead of a number and a letter), it is important to know which component of the ordered pair represents the horizontal direction and which component represents the vertical distance. Now, if you use Figure 7.2, you will see that the coordinates of Main Street are now (9.5, 6.3), or we can even say that Main Street runs from (9.5, 6.3) to (11.2, 4.5). When using pairs of numbers instead of numbers and letters, you must remember that the first component is measured horizontally and the second component is measured vertically.

Examples Find the following landmarks located in Figure 7.2 by the following coordinates.

	Answers
1. (3.5, 1.5)	**1.** Civic Center
2. (3.5, 12.5)	**2.** Fisherman's Wharf
3. (5, 11)	**3.** North Beach
4. (4, 6)	**4.** Nob Hill
5. (6, 7.3)	**5.** Chinatown ■

There are many ways to use ordered pairs to find particular locations. For example, a teacher may make out a seating chart like the one shown in Figure 7.3.

5	Terry Shell	Wayne Savick	Harvey Dunker	Richard Giles	Shannon Smith
4	Vicki Switzer	Harold Peterson	Bob Anderson	Josie Lee	Milt Hoehn
3	Jeff Atz	Carol Jones	Todd King	Jim Kintzi	Sharon Boschen
2	Warren Ruud	Cori Boyle	Steve Switzer	Clint Stevens	
1		Rose Cirner	Missy Smith	Niels Sovndal	Maureen Gradek
	1	2	3	4	5

Front

Figure 7.3
Seating chart

In the gradebook, the teacher records

Anderson (3, 4)
Atz (1, 3)
.
.
.

Anderson's seat is represented in column 3, row 4. Can you think of some other ways in which ordered pairs could be used to locate a position?

The idea of using an ordered pair to locate a certain position is associated with particular terminology. *Axes* are two perpendicular number lines, such as those in Figure 7.4a. The point of intersection of the axes is called the ***origin.*** We have associated positive numbers with the directions up and to the right. The arrows in Figure 7.4b are pointing in the positive direction. These perpendicular lines are usually drawn so that one is horizontal and the other is vertical. The horizontal axis is called the ***x-axis,*** and the vertical axis is called the ***y-axis.*** Notice that these axes divide the plane into four parts, which are called ***quadrants*** and are arbitrarily labeled as shown in Figure 7.4b.

You can now label points in the plane by using ordered pairs. The first component of the pair gives the horizontal distance (sometimes called the *ab-*

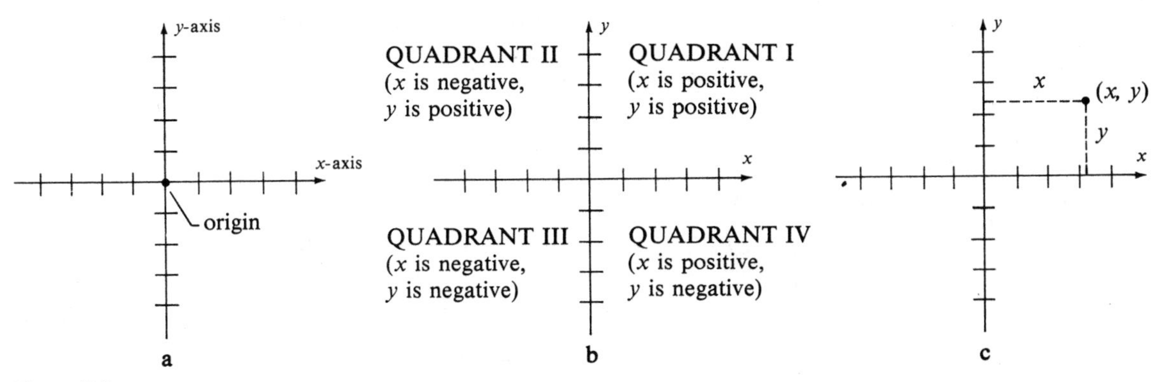

Figure 7.4
Coordinate system

scissa), and the second component gives the vertical distance (sometimes called the *ordinate*) (see Figure 7.4c). If x and y are components of the point, then this representation of the point (x, y) is called the *rectangular* or *Cartesian coordinates* of that point.

Examples *Plot the following points.*

6. $(5, 2)$ **7.** $(3, 5)$

8. $(-2, 1)$ **9.** $(0, 5)$

10. $(-6, -4)$ **11.** $(3, -2)$

12. 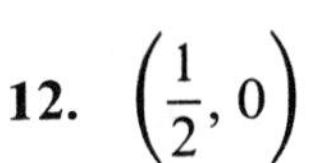 $\left(\frac{1}{2}, 0\right)$ **13.** $(0, 0)$

14.  $(5, 0)$

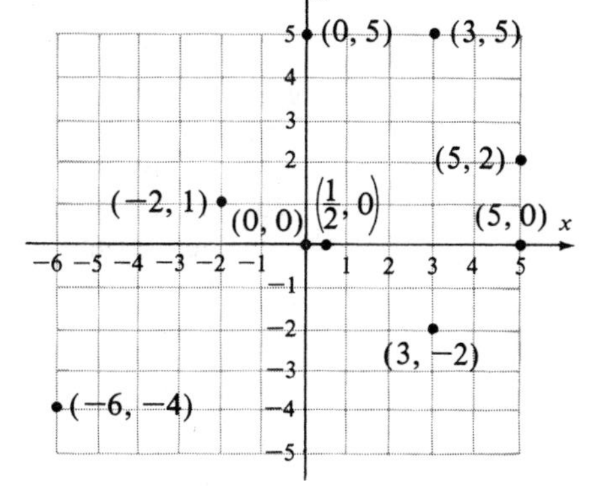

Answers: They are shown at the right. ■

The ordered pair $(3, -4)$ specifies a point in the plane *3* units to the *right* of the y-axis and *4* units *below* the x-axis. ((Notice that it is *not* -4 units below the x-axis.)

Examples *Each of the given ordered pairs specifies a point. Find the distance of that point from each of the coordinate axes.*

15. $(-5, 2)$; It is 5 units from the y-axis and 2 units from the x-axis.

16. $(5, -2)$; It is 5 units from the y-axis and 2 units from the x-axis.

17. $(w, -2)$; It is not correct to say that this specifies a point that is w units from the y-axis because what if $w = -5$ (as in Example 15)? Recall the notion of absolute value from Section 2.1. We need to specify undirected distance for this answer, so we use absolute value to say that it is $|w|$ units from the y-axis and 2 units from the x-axis. ■

Problem Set 7.1

Find the landmarks given by the coordinates in Problems 1–9. Use the map in Figure 7.2. See Examples 1–5.

1. (7, 10) **2.** (1, 2) **3.** (11, 6)
4. (12, 0.5) **5.** (11, 7.3) **6.** (1.5, 11.3)
7. (6.5, 2.5) **8.** (6.5, 4.5) **9.** (6.9, 9.8)

Using the seating chart in Figure 7.3, name the persons given by the coordinates in Problems 10–15.

10. (5, 1) **11.** (4, 1) **12.** (1, 2)
13. (4, 2) **14.** (2, 3) **15.** (3, 1)

Plot the points given in Problems 16–20. Use the indicated scale. See Examples 6–14.

16. (Scale: 1 square on your paper = 1 unit)
a. (1, 2) **b.** (3, −3) **c.** (−4, 3) **d.** (−6, −5) **e.** (0, 3)
18. (Scale: 1 square on your paper = 10 units)
a. (10, 25) **b.** (−5, 15) **c.** (0, 0) **d.** (−50, −35) **e.** (−30, −40)
19. (Scale: 1 square on your paper = 50 units)
a. (100, −225) **b.** (−50, −75) **c.** (0, 175) **d.** (−200, 125) **e.** (50, 300)
20. (Scale, x-axis: 1 square on your paper = 1 unit
Scale, y-axis: 1 square on your paper = 10 units)
a. (4, 40) **b.** (−3, 0) **c.** (2, −80) **d.** (−5, −100) **e.** (0, 0)

Find the distance between the point specified by the given ordered pair and each of the coordinate axes in Problems 21–23. See Examples 15–17.

21. a. (3, 7) **b.** (−3, 7) **c.** (−3, −7) **d.** (3, −7) **e.** (s, t)
22. a. (2, 5) **b.** (2, −5) **c.** (−2, −5) **d.** (−2, 5) **e.** (a, b)
23. a. (4, 8) **b.** (−4, −8) **c.** (−4, 8) **d.** (4, −8) **e.** (u, v)
24. Plot the following coordinates on graph paper, and connect each point with the preceding one: (−2, 2), (−2, 9), (−7, 2), (−2, 2). Start again: (−6, −6), (−6, −8), (10, −8), (10, −6), (6, −6), (6, 0), (10, 0), (10, 2), (6, 2), (6, 16), (−2, 14), (−11, 2), (−9, −1), (−2, −1), (−2, −6), (−6, −6).
25. Plot the following coordinates on graph paper, and connect each point with the preceding one: (9, 0), $(7, -\frac{1}{2})$, (6, −2), (8, −5), (5, −8), (−7, −10), (−1, −5), (0, −2), (−2, −1), (−6, −5), (−5, −3), (−6, −2), (−5, −1), (−6, 0), (−5, 1), (−6, 2), (−5, 3), (−6, 4), (−5, 5), (−6, 7), (−2, 3), (0, 4), (−1, 10), (−7, 15), (3, 10), (6, 5), (4, 3), (5, 2), (7, 2), (9, 0). Finally, plot a point at (7, 1).
26. Plot all the points for which $-10 \leq x \leq 10$ and for which the first and second coordinates are equal.
27. Connect the points (0, 2), (1, 5), (2, 8), (−1, −1), (−2, −4). What do you observe?
a. How many points lie on a line?
b. How many points do you need to plot in order to determine a line?
28. Plot five points whose first component is 2. Connect the plotted points.

29. Plot five points whose second component is 3. Connect the plotted points.
30. Plot five points whose second component is -4. Connect the plotted points.

7.2 Solving Equations

In Chapter 3 we solved first-degree equations with a single variable. In this chapter we will solve first-degree equations with two variables. Because there are *two variables,* solutions are represented as ordered pairs.

Suppose you want to solve

$$2x + y = 3$$

An ordered pair (x, y) that makes the equation true is called a ***solution*** or **root** of the equation. The variable associated with the first component is called the ***independent variable,*** because this value is chosen first—that is, independently from the value associated with the second component. The variable associated with the second component is called the ***dependent variable,*** because this value *depends* on the choice of the first component. The variable x is most often chosen as the independent variable and the variable y as the dependent variable. Thus, the ordered pairs involving x and y are almost always denoted by (x, y), not (y, x). In this book we will assume that when the variables x and y are used, x is the independent variable and y the dependent variable, and the form of the ordered pair is (x, y).

Examples *In Examples 1–3, find the value for the dependent variable y for the equation $2x + y = 3$ for each of the given values of the independent variable x, and state the result using ordered pair notation.*

1. $x = 1$: If $x = 1$, then $2(1) + y = 3$ *Now solve this equation for y.*
$y = 3 - 2$ *Subtract 2 from both sides.*
$y = 1$
The ordered pair $(1, 1)$ satisfies the equation.

2. $x = 2$: If $x = 2$, then $2(2) + y = 3$ *Subtract 4 from both sides.*
$y = -1$
The ordered pair $(2, -1)$ satisfies the equation.

3. $x = -1$: If $x = 1$, then $2(-1) + y = 3$
$-2 + y = 3$
$y = 5$
The ordered pair $(-1, 5)$ satisfies the equation. ■

In Examples 1–3 we have found three roots of the equation $2x + y = 3$. Can you find others? One of the most difficult steps for beginning students is finding *additional* ordered pairs satisfying a given equation. The reason is because any additional examples shown in a book makes it look like the book is selecting the choices for x, when in practice it is *you,* the student, who will be making those choices. Consider Example 4.

Example 4. Find three ordered pairs satisfying the equation $2x + y = 3$ in addition to those listed in Examples 1–3.

Solution Now *you* choose any x value (except those listed in Examples 1–3—namely $x = 1$, $x = 2$, and $x = -1$). What are you going to choose? You might have chosen $x = 0$: If $x = 0$, then

$$2(0) + y = 3$$
$$y = 3$$

Do you see why $x = 0$ was an *easy* choice for you? If you can pick any value, you might as well pick one that makes your work easy. Can you find two others? The work and the rest of the solution is left for you to fill in. One ordered pair is (0, 3); what are the other two you have found? ■

Have we now found the entire solution set for the equation $2x + y = 3$? We have found six ordered pairs satisfying this equation (see Examples 1–4). Are there others? Suppose you plot those points on a Cartesian coordinate system as shown in Figure 7.5. Do you notice anything about the arrangement of these points in the plane?

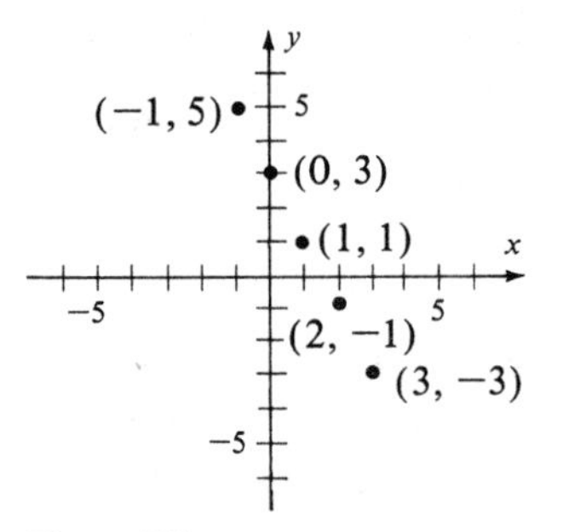

Figure 7.5

Suppose we draw a line passing through these points as shown in Figure 7.6.

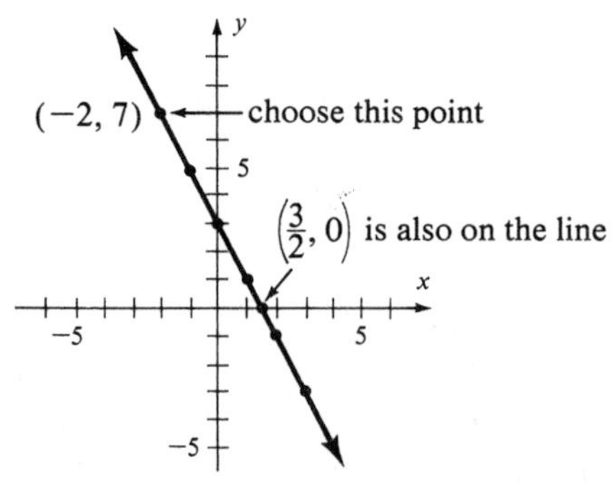

Figure 7.6

Now choose *any* other point on this line—say $(-2, 7)$. This point has coordinates that satisfy the equation.

$$2x + y = 3$$
$$2(-2) + 7 = 3$$

Also, if we find *any* other coordinates satisfying the equation and plot the point, it will fall on the line. For example, if $y = 0$, then $2x + (0) = 3$; $x = \frac{3}{2}$. The point for $(\frac{3}{2}, 0)$ also lies on the line.

This relationship between the solutions of a first-degree equation and the graph of a line always holds. That is, we can speak of ***the graph of an equation*** or ***the equation of a graph,*** and we mean a picture of the ordered pairs satisfying the equation.

Graph of an Equation

By the graph of an equation or the equation of a graph, we mean:

1. Every point on the graph has coordinates that satisfy the equation.
2. Every ordered pair satisfying the equation has coordinates that lie on the graph.

In particular, the graph of a first-degree equation is a line, and, conversely, the equation of a line is a first-degree equation of the form

$$Ax + By + C = 0$$

where A, B, and C are any specific constants (A and B not both zero). By using this fact, along with the fact from geometry that two distinct points determine a line, we can represent graphically all solutions of a first-degree equation with two variables by plotting two points and then drawing the line passing through these points. For this reason, when we want to solve a first-degree equation with two variables, we will talk about ***graphing the line.***

Procedure for Graphing a Line

1. Find two ordered pairs that are solutions of the equation. Find them either by inspection or by choosing an x value (because it is the independent variable) and then determining the corresponding y value.
2. Plot the points for the two pairs on a rectangular coordinate system. Be sure to label the axes and use an appropriate scale.
3. Draw the line passing through the two points. This line is a graph of all solutions of the equation.
4. Check by finding a third ordered pair that satisfies the equation and verifying that their point lies on the line.

Graph the given lines in Examples 5–9.

Example 5. $3x + 2y = 7$

Solution *Step 1* Find two points. Let $x = 1$ and $x = 2$.

If $x = 1$, then $3(1) + 2y = 7$

$2y = 4$

$y = 2$ Point $(1, 2)$

If $x = 2$, then $3(2) + 2y = 7$

$2y = 1$

$y = \frac{1}{2}$ Point $\left(2, \frac{1}{2}\right)$

Step 2 Plot the points as shown in Figure 7.7a.

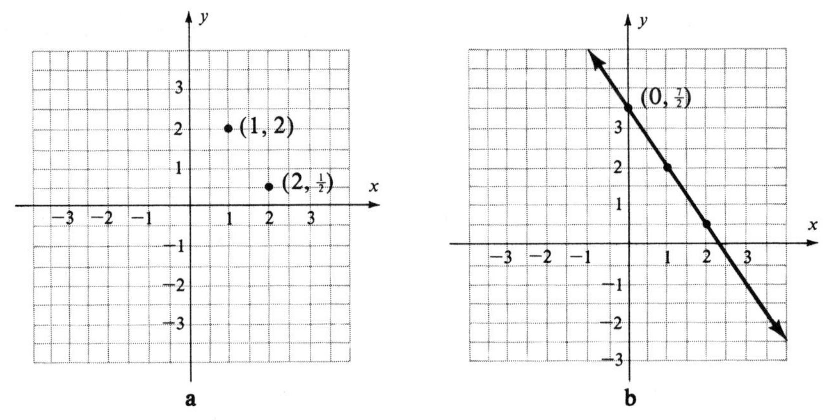

Figure 7.7

Step 3 Draw a line passing through the points as shown in Figure 7.7b.

Step 4 Find another point that satisfies the equation—say $x = 0$.

If $x = 0$, then $3(0) + 2y = 7$

$2y = 7$

$y = \frac{7}{2}$ Point $\left(0, \frac{7}{2}\right)$

Because this point lies on the line, the graph is correct. ■

Example 6. $x = y$

Solution *Step 1* $(0, 0)$ and $(1, 1)$ make the equation true.

Step 2 Plot the points; see Figure 7.8.

Step 3 Draw the line passing through the points as shown in Figure 7.8.

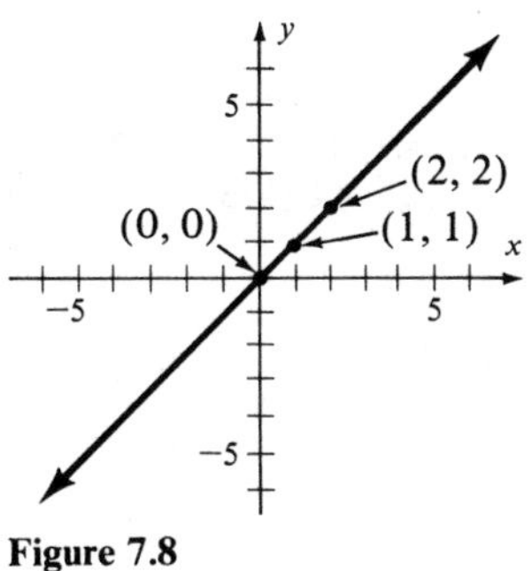

Figure 7.8
Graph of $x = y$

Step 4 Check; (2, 2) makes the equation true and also lies on the line. ■

Example 7. $4y = 3x$

Solution

Step 1 Points (0, 0) and (4, 3) make the equation true.

Step 2 Plot the points as shown in Figure 7.9.

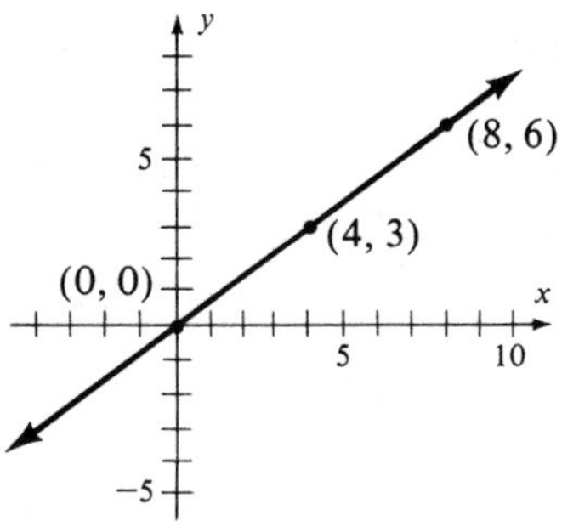

Figure 7.9
Graph of $4y = 3x$

Step 3 Draw the line passing through the points as shown in Figure 7.9.

Step 4 Check; (8, 6) satisfies the equation and also lies on the line. ■

Example **8.** $y = 3$

Solution The x-coordinate can be *any* value; the only restriction is that the second coordinate be 3. Thus, we can combine some of the steps by plotting three points: (0, 3), (2, 3), and (−2, 3). Next, draw the line passing through these points as shown in Figure 7.10.

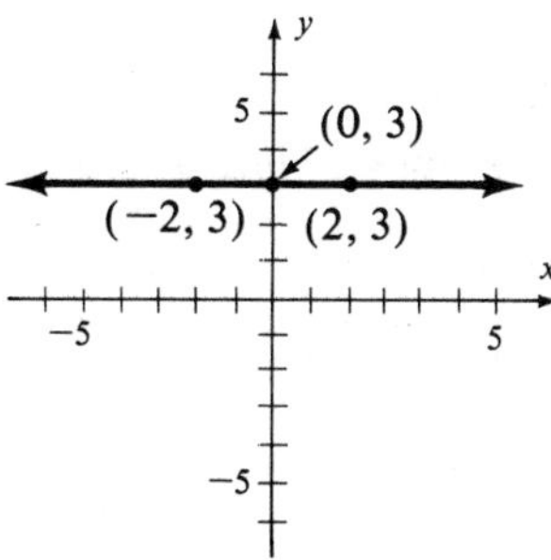

Figure 7.10
Graph of $y = 3$

■

Example 9. $x = -2$

Solution The y-coordinate can be any value, so plot the points $(-2, 0)$, $(-2, 4)$, and $(-2, -2)$ and draw the line passing through them as shown in Figure 7.11.

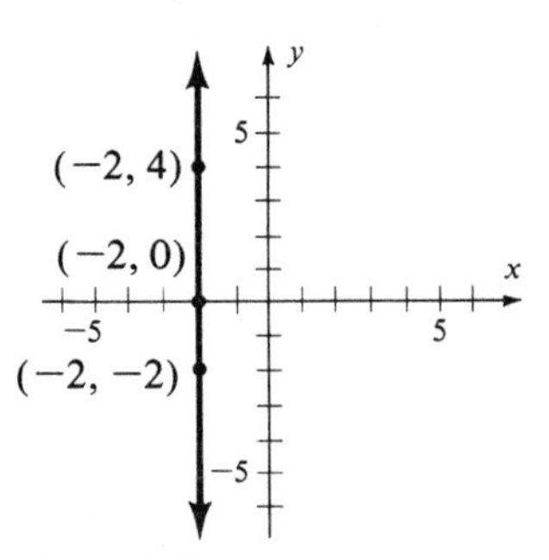

Figure 7.11
Graph of $x = -2$ ■

Look again at Figure 7.10. Notice that any equation of the form $y = k$, where k is any real number, is a ***horizontal line*** that intersects the y-axis at k. Look at Figure 7.11, and notice that an equation of the form $x = h$, where h is any real number, is a ***vertical line*** that intersects the x-axis at h.

In the next sections we will discuss more efficient ways of graphing lines than plotting points.

Problem Set 7.2

Write the set of ordered pairs generated for each of the equations in Problems 1–6. Let the values for the independent variable be 0, 1, 2, 3, and 4. See Examples 1–3.

1. $y = x + 3$, where the ordered pair is given by (x, y).
2. $y = x^2 + 1$, where the ordered pair is given by (x, y).
3. $d = 4t$, where the ordered pair is given by (t, d).
4. $d = 50t$, where the ordered pair is given by (t, d).
5. $C = \pi d$, where the ordered pair is given by (d, C).
6. $A = s^2$, where the ordered pair is given by (s, A).

In Problems 7–18 find three ordered pairs satisfying each equation. See Example 4.

7. $y = x + 3$
8. $y = 2x - 1$
9. $y = 2x + 5$
10. $y = x - 4$
11. $x - y = 1$
12. $y - x = 1$
13. $2x + y = 1$
14. $3x + y = 1$
15. $y - 2x = 1$
16. $y - 3x = 1$
17. $3x + 4y = 8$
18. $x + 2y = 4$

Graph the lines in Problems 19–39. See Examples 5–9.

19. $y = x + 3$
20. $y = x + 1$
21. $y = 2x + 1$
22. $y = 3x - 2$
23. $y = x - 1$
24. $y = 2x + 3$
25. $x + y = 0$
26. $x - y = 0$
27. $2x + y = 1$
28. $3x + y = 10$
29. $3x + 4y = 8$
30. $x + 2y = 4$
31. $x = 5$
32. $x = -4$
33. $x = -1$
34. $y = -2$
35. $y = 6$
36. $y = 4$
37. $x + y + 100 = 0$
38. $5x - 3y = 45$
39. $y = 50x$

40. The cost of a multivitamin depends on the weight purchased, according to the formula

$$C = 3w$$

where C is the cost in dollars and w is the weight in pounds. Give the cost for each of the following weights: $w = 2$; $w = 10$; $w = 100$. Write your answers in ordered-pair notation (w, C).

41. The admission price for a certain performance is $1.25 for adults and 50¢ for children. If 525 tickets are sold, the relationship

$$a + c = 525$$

represents the number of (a) adult admissions and the number of (c) children's admissions. Find the number of adult admissions for the following numbers of children's admissions: $c = 125$; $c = 200$; $c = 247$. Write your answers in the form (c, a).

42. The resistance (R) in an electric circuit is related to the voltage (E) by the formula

$$R = .07E$$

Find the resistance for the following voltages: $E = 120$; $E = 240$. Write your answers in the form (E, R).

43. The velocity v (in feet per second) of a rock dropped from a building is related to time t (in seconds) by the formula

$$v = 32t$$

Find the velocity of a rock at the time it is dropped, after 3 seconds, and when it hits at 6 seconds. Write your answers in the form (t, v).

7.3 The Slope of a Line

In the last section we discussed solving first-degree equations with two variables. You saw that the solution could be represented graphically as a line. All first-degree equations with two variables x and y can be written in what is called the ***standard form of the equation of a line*** by using constants A, B, and C (A and B are not both zero).

Standard Form of the Equation of a Line

$$Ax + By + C = 0$$

If $B \neq 0$, we can solve for y:

$$By = -Ax - C$$

$$y = -\frac{A}{B}x - \frac{C}{B}$$

Now let $m = -\frac{A}{B}$ and $b = -\frac{C}{B}$, and any first-degree equation with two variables x and y (where $B \neq 0$) can be written in the form

$$y = mx + b$$

This is a very useful way of writing the equation of a line because x is the independent variable. When working the problems in the last section, you may have discovered that it is often easier to solve for y *before* substituting in your values for x. You might want to review this process by looking at Examples 4–9 in Section 3.5.

In this section we will discuss the significance of m in the equation $y = mx + b$ and in the next section, the significance of b. In order to investigate the nature of m, let $b = 0$ and consider the equation $y = mx$ by choosing different values for m.

Given: $y = mx$

	Equation	Two Points	Check Point
Let $m = 0$;	$y = 0$	(0, 0), (1, 0)	(2, 0)
$m = 1$;	$y = x$	(0, 0), (1, 1)	(2, 2)
$m = 5$;	$y = 5x$	(0, 0), (1, 5)	(2, 10)
$m = -3$;	$y = -3x$	(0, 0), (1, −3)	(2, −6)
$m = -\frac{2}{3}$;	$y = -\frac{2}{3}x$	(0, 0), (3, −2)	(6, −4)

These graphs are shown in Figure 7.12. Notice that the steepness of these lines varies. The steepness of a line is called the ***slope*** of the line. But what do we mean by slope?

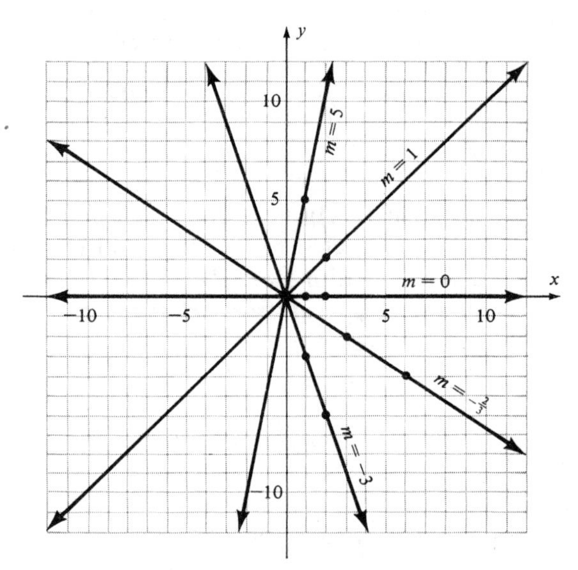

Figure 7.12
Graph of $y = mx$ for selected values of m

We have all noticed that certain hills or roads are steeper than others or that certain roofs slope more than others. How do we measure this steepness or slope? Consider two rooflines as shown in Figure 7.13.

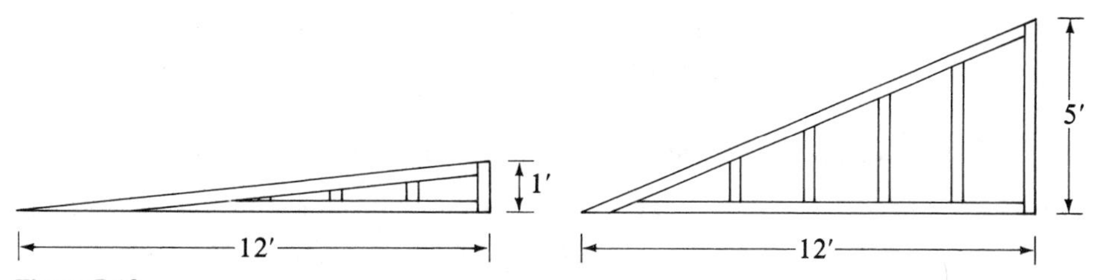

Figure 7.13
Cross sections of roof gables

The first roof rises 1 ft for every 12 ft in horizontal distance, and the second one rises 5 ft for every 12 ft in horizontal distance. It's obvious that the second one has a greater slope than the first one. The steepness is expressed as a ratio of vertical distance, called the ***rise,*** to the horizontal distance, called the ***run.*** That is, the slope is given by the equation:

$$\text{slope} = \frac{\text{RISE}}{\text{RUN}}$$

To relate the slope of a line to the line whose equation is

$$y = mx$$

you need to observe (Figure 7.12 may help) that for this equation, the rise divided by the run is always m. Remember, $y = mx$ has a graph that passes through the origin because (0, 0) makes the equation true regardless of the value of m.

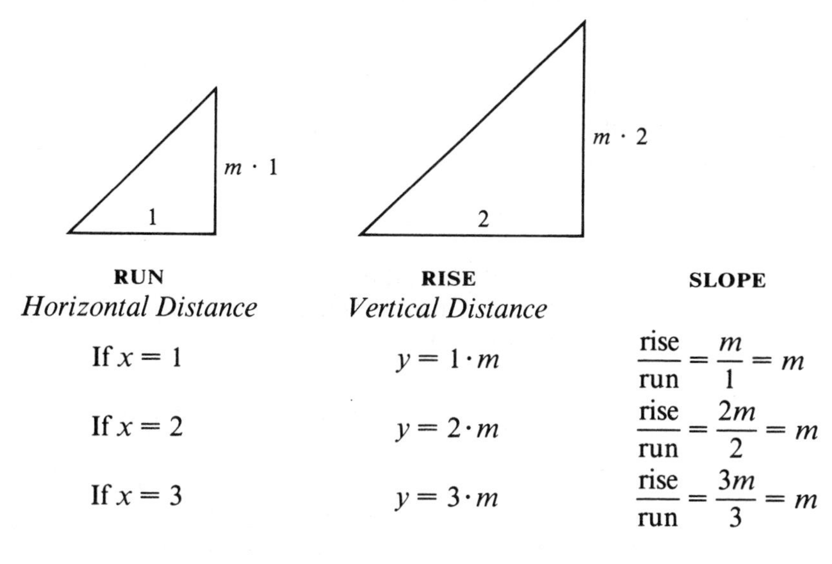

RUN *Horizontal Distance*	**RISE** *Vertical Distance*	**SLOPE**
If $x = 1$	$y = 1 \cdot m$	$\frac{\text{rise}}{\text{run}} = \frac{m}{1} = m$
If $x = 2$	$y = 2 \cdot m$	$\frac{\text{rise}}{\text{run}} = \frac{2m}{2} = m$
If $x = 3$	$y = 3 \cdot m$	$\frac{\text{rise}}{\text{run}} = \frac{3m}{3} = m$

Slope

The slope of the line $y = mx$ is m and is found by

$$m = \frac{\text{RISE}}{\text{RUN}} = \frac{\text{change in the vertical distance}}{\text{change in the horizontal distance}}$$

It is easy to remember *the rise over the run* because *rise* comes before *run* in alphabetical order.

Examples *Find the slopes by inspection from the given equations.*

1. $y = 4x$; *Compare with* $y = mx$ *and notice that* $m = 4$. *Thus, the slope is* 4.

2. $y = -2x$; *The slope is* -2.

3. $2x + 3y = 0$; *Solve for* y: $3y = -2x$

$$y = -\frac{2}{3}x$$

The slope is $-\frac{2}{3}$. ■

We can now use the idea of slope to graph an equation of the form $y = mx$ without randomly plotting points. First, remember that *this* line passes through the origin. Next, find m as in Examples 1–3 and write m as a fraction. Interpret this fraction as a ratio

$$\frac{\text{RISE}}{\text{RUN}}$$

as shown in Examples 4–6, and plot this point. The line passes through this point and the origin.

In Examples 4–5, graph the given equations using the slope method.

Example 4. $y = 4x$ $m = 4$; write this as a fraction:

$$m = \frac{4}{1}$$

Starting at the origin, count out a rise of 4 units and a run of 1 unit, marking the final point as shown in Figure 7.14. Draw the line through this point and the origin.

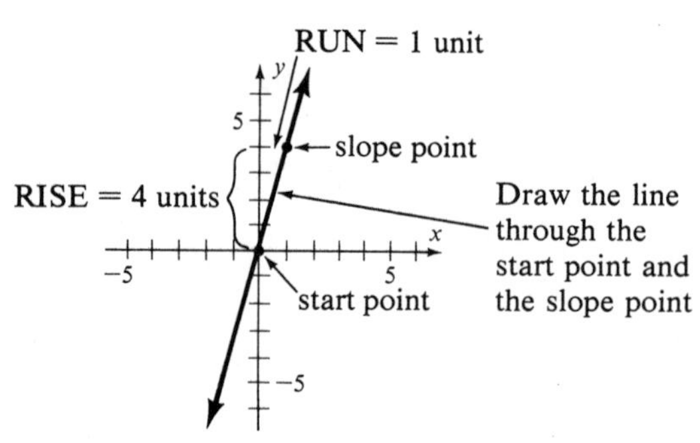

Figure 7.14
Graph of $y = 4x$ ■

Example 5. $y = -\frac{2}{3}x$ $m = -\frac{2}{3}$. Plot the start point, (0, 0), and the slope point (RISE $= -2$, RUN $= 3$). Draw the line as shown in Figure 7.15.

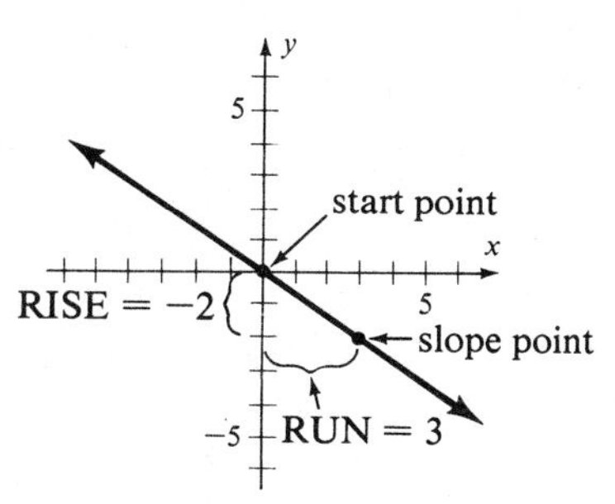

Figure 7.15
Graph of $y = -\frac{2}{3}x$ ■

In Example 4 we stated that to graph $y = 4x$ using the slope method you need to write the slope as a fraction. Is $m = \frac{4}{1}$ the only fraction that equals 4? Of course not; try $\frac{8}{2}$, $\frac{-4}{-1}$, or $\frac{400}{100}$. Would it make any difference if you use any of these fractions to find the slope point for the line in Figure 7.14?

Example 6. Rework Example 5 by finding two other slope points for $y = -\frac{2}{3}x$ to show that the graph in Figure 7.15 is unchanged.

Solution: Since $m = -\frac{2}{3}$, and in Example 5 we wrote RISE $= -2$ and RUN $= 3$, for this example we want to write the slope by writing the fraction $-\frac{2}{3}$ two other ways.

(1) Since $-\dfrac{2}{3} = \dfrac{2}{-3}$ we can write RISE $= 2$, RUN $= -3$.

(2) Since $-\dfrac{2}{3} = \dfrac{-4}{6}$ we can write RISE $= -4$, RUN $= 6$.

Both of these possibilities are plotted in Figure 7.16. Notice that the result is the same line that was graphed in Example 5 (Figure 7.15).

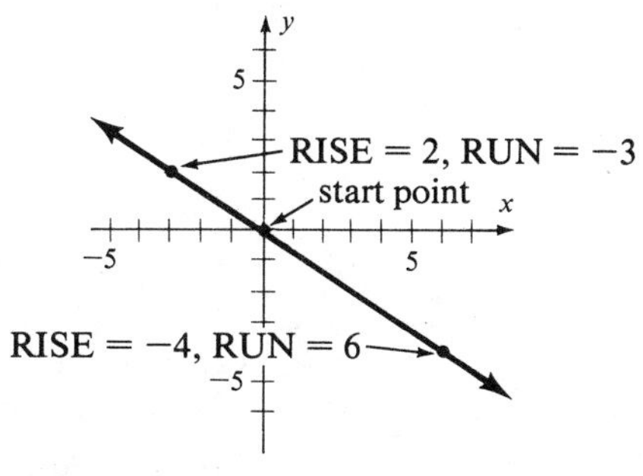

Figure 7.16
Graph of $y = -\frac{2}{3}x$

If we can use different ratios to find the slope point, it seems reasonable that we do not need to use the origin as a starting point. ■

Example 7. Draw the line passing through (2, 3) with slope $\frac{3}{5}$.

Solution First, plot the point (2, 3); then count out the slope ($m = \frac{3}{5}$; RISE $= 3$ and RUN $= 5$) to find the slope point as shown in Figure 7.17.

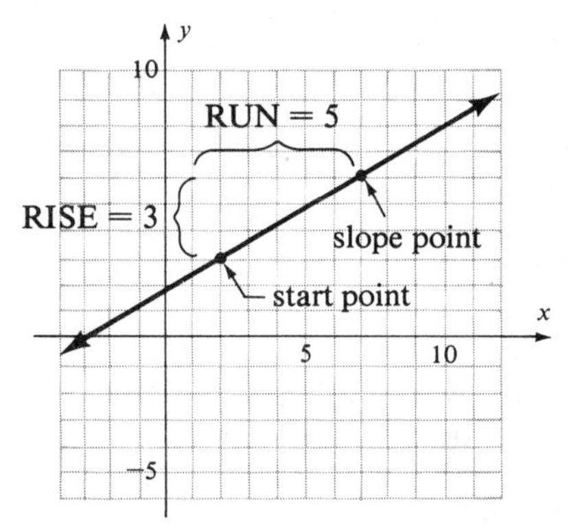

Figure 7.17
Graph of line passing through (2, 3) with slope $\frac{3}{5}$ ■

We conclude this section by stating a general formula for representing the slope of a line. You have seen that any ratio representing the same value of m will graph the same line. You have also seen that any point on the line can be used as a starting point. It seems reasonable, then, that you can represent the slope in terms of *any* two known points on the line. The notation mathematicians use for representing any two points on a line is $P_1(x_1, y_1)$ and $P_2(x_2, y_2)$. P_1 is read "P sub one"; P_2 is read "P sub two." Let $P_1(x_1, y_1)$ and $P_2(x_2, y_2)$ be any two points on the line as shown in Figure 7.18.

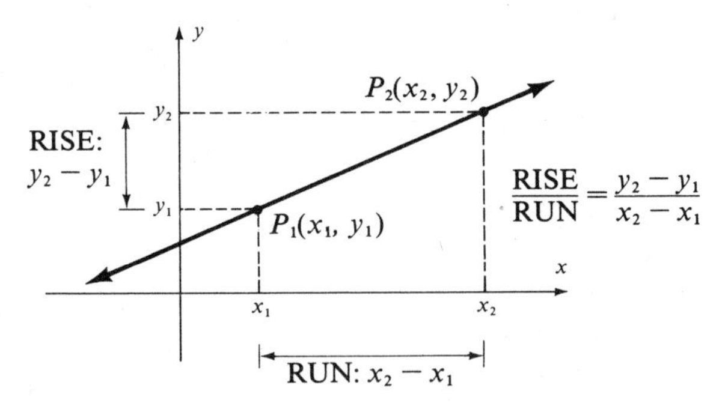

Figure 7.18
Slope of a general line

The RISE $= y_2 - y_1$ and the run $= x_2 - x_1$, so the slope of the line is given the general result summarized as follows.

Slope Formula

The slope, m, of a nonvertical line passing through $P_1(x_1, y_1)$ and $P_2(x_2, y_2)$ is given by

$$m = \frac{\text{RISE}}{\text{RUN}} = \frac{\text{change in vertical distance}}{\text{change in horizontal distance}} = \frac{y_2 - y_1}{x_2 - x_1}$$

Examples *Find the slopes of the lines passing through the given points.*

8. (2, 3) and (4, 5)

$$m = \frac{y_2 - y_1}{x_2 - x_1} = \frac{5-3}{4-2} = \frac{2}{2} = 1$$

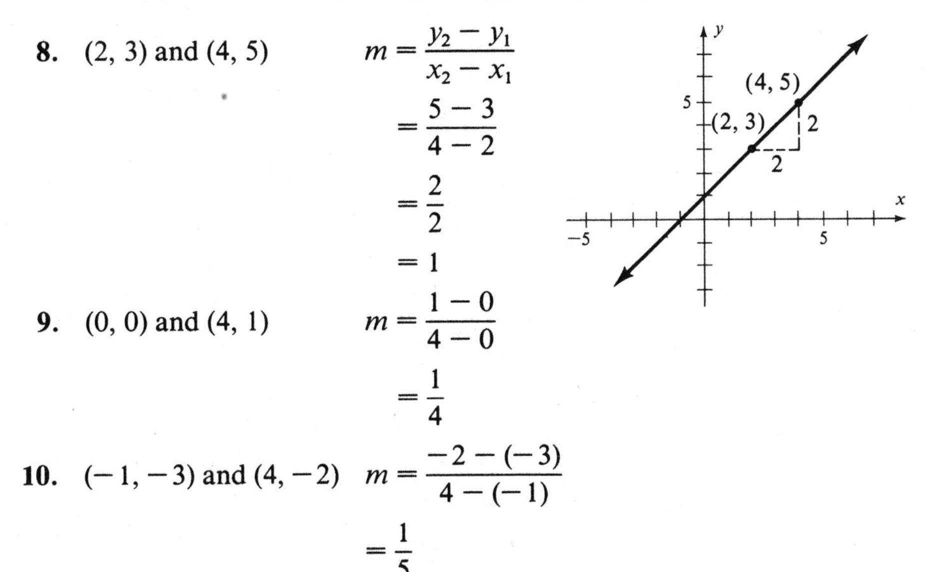

9. (0, 0) and (4, 1)

$$m = \frac{1-0}{4-0} = \frac{1}{4}$$

10. $(-1, -3)$ and $(4, -2)$

$$m = \frac{-2-(-3)}{4-(-1)} = \frac{1}{5}$$

■

Problem Set 7.3

Find the slopes in Problems 1–20 by inspection and then state each one as a ratio. See Examples 1–3.

1. $y = \frac{1}{2}x$
2. $y = \frac{1}{3}x$
3. $y = -\frac{2}{3}x$
4. $y = -\frac{4}{5}x$
5. $y = 7x$
6. $y = -9x$
7. $y = -11x$
8. $y = 24x$
9. $y = 1.5x$
10. $y = -2.5x$
11. $x + y = 0$
12. $x - y = 0$
13. $3x + y = 0$
14. $5x - y = 0$
15. $3x - 2y = 0$
16. $5x + 3y = 0$
17. $5x = 9y$
18. $4x = 11y$
19. $4y = -3x$
20. $6y = 4x$

Graph the equations given in Problems 21–41 using the slope method. See Examples 4–6.

21. $y = \frac{1}{2}x$
22. $y = \frac{1}{3}x$
23. $y = -\frac{2}{3}x$
24. $y = -\frac{4}{5}x$
25. $y = 7x$
26. $y = -9x$

27. $y = -11x$ **28.** $y = 24x$ **29.** $y = 1.5x$
30. $y = -2.5x$ **31.** $x + y = 0$ **32.** $x - y = 0$
33. $3x + y = 0$ **34.** $5x - y = 0$ **35.** $3x - 2y = 0$
36. $5x + 3y = 0$ **37.** $5x = 9y$ **38.** $4x = 11y$
39. $4y = -3x$ **40.** $6y = 4x$ **41.** $8y = 12x$

Graph the lines indicated in Problems 42–59. See Example 7.

42. $(2, 5), m = \frac{1}{3}$ **43.** $(-7, -3), m = \frac{1}{5}$ **44.** $(-4, -6), m = \frac{3}{4}$

45. $(5, 0), m = -\frac{2}{3}$ **46.** $(1, -1), m = -\frac{1}{7}$ **47.** $(6, 3), m = -\frac{2}{5}$

48. $(-1, 3), m = 1$ **49.** $(1, -1), m = 2$ **50.** $(-1, 3), m = 3$
51. $(2, 1), m = -8$ **52.** $(-2, 2), m = -2$ **53.** $(7, 2), m = -1$
54. $(2, 3), m = 0$ **55.** $(-7, -2), m = 0$ **56.** $(-3, 6), m = 0$
57. $(5, 0), m = -1.5$ **58.** $(-4, -6), m = 2.5$ **59.** $(-3, 6), m = -2.5$

Find the slope of the line passing through the points given in Problems 60–68, and also draw the line. See Examples 8–10.

60. $(2, 3), (5, 7)$ **61.** $(6, 3), (-2, 4)$ **62.** $(2, 3), (0, 0)$
63. $(-4, -1), (4, 3)$ **64.** $(-3, -1), (0, -2)$ **65.** $(2, 5), (-8, 1)$
66. $(3, -5), (8, 4)$ **67.** $(3, -5), (-2, -5)$ **68.** $(-2, -1), (-3, 5)$

7.4 Graphs of Linear Equations

In the last section the form

$$y = mx + b$$

where $b = 0$ was investigated in order to consider the notion of the slope of a line. In this section we consider the *general* equation $y = mx + b$.

A place where a graph crosses the y-axis is called a ***y-intercept.*** The first component of a y-intercept must be zero because a y-intercept requires no right or left movement. Thus, $x = 0$, and if we substitute this into the general equation

$$y = mx + b$$

we obtain

$$\begin{aligned} y &= m(0) + b \\ &= 0 + b \\ &= b \end{aligned}$$

Notice that it makes no difference what the slope, m, is since $m(0) = 0$ *for all* m.

This means that the y-intercept is b, and $(0, b)$ is a solution of $y = mx + b$.

Examples *Find the y-intercept (b) and slope (m) of the following lines.*

Answers

1. $y = 2x + 5$

1. y-intercept $b = 5$ (by inspection) means that the line passes through the point $(0, 5)$. $m = 2$ (by inspection)

2. $y = 4x - 7$ **2.** $b = -7$ (by inspection; do not forget the negative); $m = 4$ (by inspection)

3. $x + 2y = 1$ **3.** $2y = -x + 1$ *First solve for y.*

$$y = -\frac{1}{2}x + \frac{1}{2}$$

Thus, $b = \frac{1}{2}$ and $m = -\frac{1}{2}$ ■

Now use the ***y-intercept as the starting point*** for graphing a line as illustrated in the last section. The method of graphing a line by plotting the intercept as the starting point and then using the slope is called the ***slope-intercept method.*** Remember that a y-intercept of b means that the line passes through the point $(0, b)$.

Plot the lines in Examples 4–7 using the slope-intercept method.

Example 4. $y = -\frac{2}{3}x + 3$

Solution Notice that $b = 3$, which means $(0, 3)$ is the starting point, and $m = -\frac{2}{3}$. The graph is shown in Figure 7.19. Also notice

$$m = -\frac{2}{3} = \frac{-2}{3} = \frac{2}{-3}$$

so you obtain the same line if you use $m = \frac{-2}{3}$ or $m = \frac{2}{-3}$.

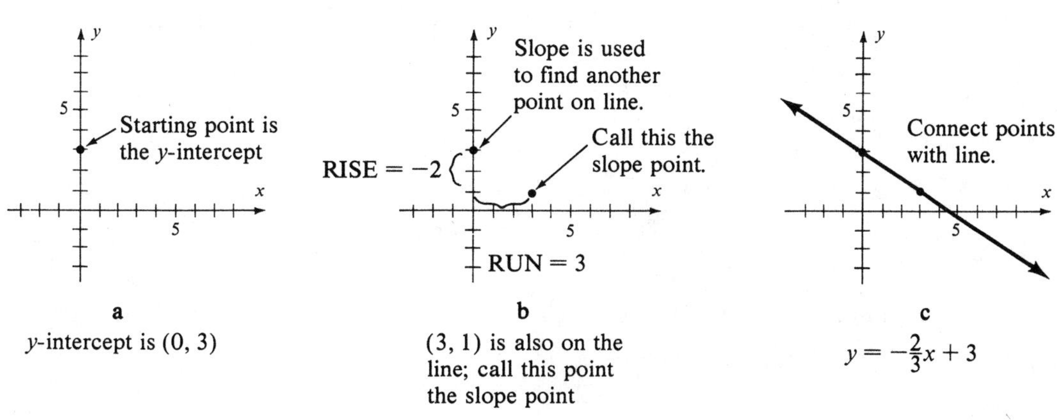

Figure 7.19

Steps in graphing $y = -\frac{2}{3}x + 3$

■

Example 5. $y = 4x - 7$

Solution Notice that $b = -7$ and $m = 4 = \frac{4}{1}$. The graph is shown in Figure 7.20. ■

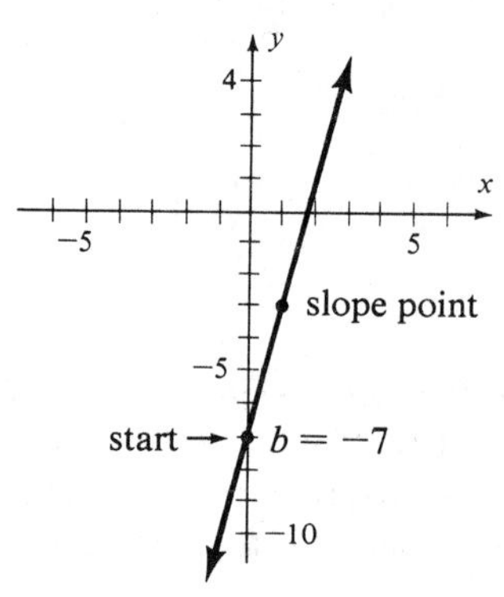

Figure 7.20
Graph of $y = 4x - 7$ ■

Example 6. $x + 2y = 1$

Solution First solve for y: $2y = -x + 1$

$$y = -\frac{1}{2}x + \frac{1}{2}$$

Notice that $b = \frac{1}{2}$ and $m = -\frac{1}{2}$. Do not let the fractions bother you; simply change the scale as shown in Figure 7.21.

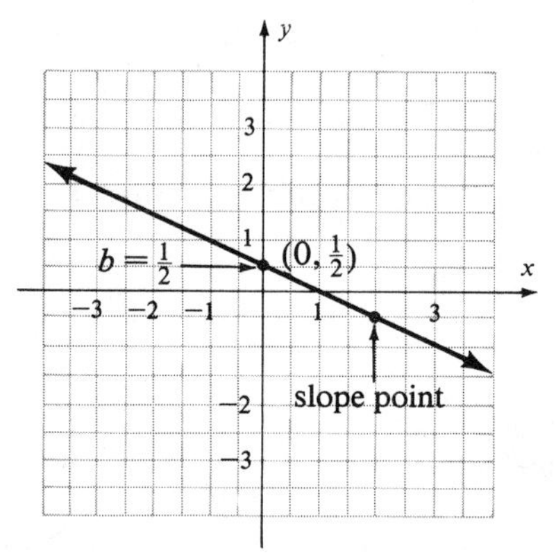

Figure 7.21
Graph of $x + 2y = 1$ ■

Example 7. $3y - 6 = 0$

Solution Solve for y: $3y = 6$

$y = 2$

This equation can be written in the form

$$y = mx + b \quad \text{as} \quad y = 0(x) + 2$$

Notice that $b = 2$ and $m = 0$; this equation graphs as a horizontal line as shown in Figure 7.22.

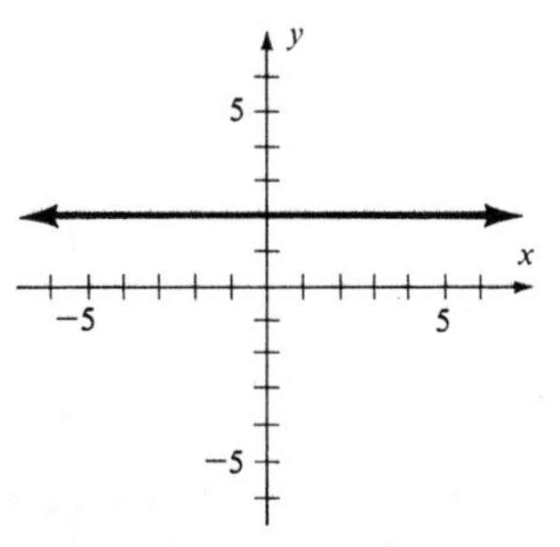

Figure 7.22
Graph of $3y - 6 = 0$ ■

Example 7 illustrates that a horizontal line has zero slope. What about a vertical line? In the discussion at the beginning of the last section about the line, the equation

$$Ax + By + C = 0$$

was called the *standard form* of the equation of a line. We made the restriction $B \neq 0$ so that we could solve for y. What if $B = 0$? Then

$$\begin{aligned} Ax + C &= 0 \\ Ax &= -C \\ x &= -\frac{C}{A} \end{aligned}$$

Since $A \neq 0$ (remember A and B cannot both be zero), this equation is of the form

$$x = h$$

for some constant h, where h represents $-C/A$. We have seen before that this equation forms a vertical line. For example,

$$x = -3$$

is shown in Figure 7.23. In general, this line has *no y-intercept* and *no slope.* This

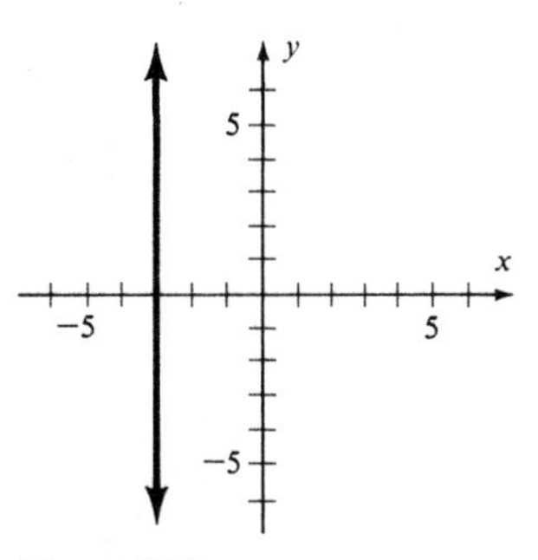

Figure 7.23
Graph of $x = -3$

line is a special case that is considered separately from those lines having the equation $y = mx + b$.

HORIZONTAL LINES	VERTICAL LINES
Horizontal lines have the form	Vertical lines have the form
$y = \text{constant}$	$x = \text{constant}$
and have zero slope.	and have no slope.

A general procedure may now be stated for graphing a line using the slope-intercept method.

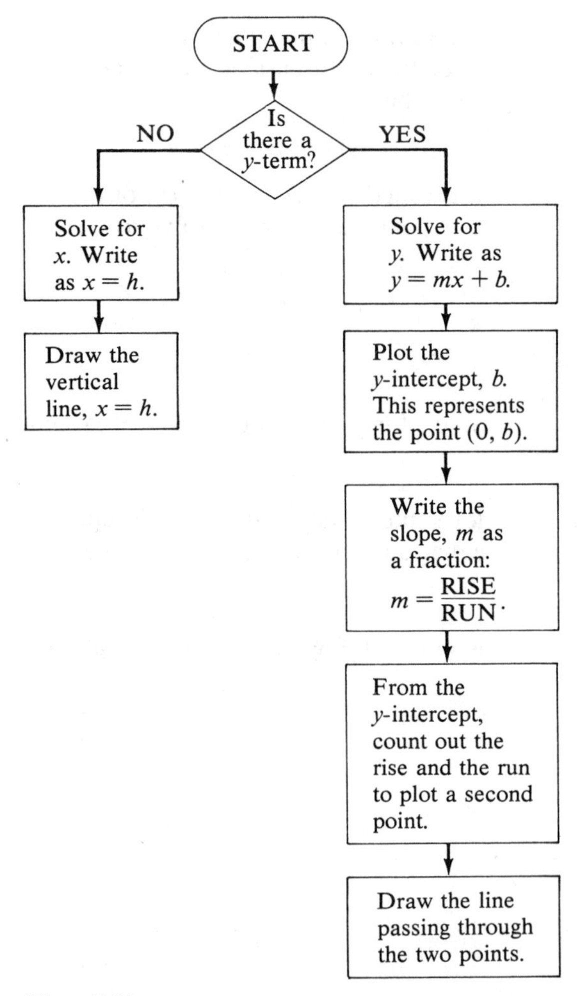

Figure 7.24
General procedure for graphing lines

Slope-Intercept Form

Every first-degree equation of the form $Ax + By + C = 0$, $B \neq 0$, can be written in the form

$$y = mx + b$$

where b is the y-intercept and m is the slope. This form is called the ***slope-intercept form*** of the equation of a line.

Problem Set 7.4

Find the slope and y-intercept of the lines given in Problems 1–21. See Examples 1–3.

1. $y = 3x + 2$ **2.** $y = -2x + 5$ **3.** $y = x - 4$
4. $y = x - 8$ **5.** $y = -3x - 2$ **6.** $y = -2x - 3$
7. $y = \frac{1}{2}x - \frac{3}{2}$ **8.** $y = -\frac{2}{3}x + \frac{4}{3}$ **9.** $y = \frac{2}{3}x - \frac{1}{3}$
10. $y = \frac{4}{5}x + \frac{2}{5}$ **11.** $y = -\frac{2}{3}x + \frac{1}{6}$ **12.** $y = -\frac{1}{2}x + \frac{1}{4}$
13. $y = \frac{4}{5}x + 8$ **14.** $y = -\frac{2}{3}x + 9$ **15.** $y = \frac{3}{4}x - 12$
16. $y = \frac{3}{5}x - 7$ **17.** $x = 6$ **18.** $x = -2$
19. $y = -3$ **20.** $y = 4$ **21.** $y = 0$

Graph the lines of the equations in Problems 22–51 using the slope intercept method. See Examples 4–7.

22. $y = 3x + 2$ **23.** $y = -2x + 5$ **24.** $y = x - 4$
25. $y = x - 8$ **26.** $y = -3x - 2$ **27.** $y = -2x - 3$
28. $y = \frac{1}{2}x - \frac{3}{2}$ **29.** $y = -\frac{2}{3}x + \frac{4}{3}$ **30.** $y = \frac{2}{3}x - \frac{1}{3}$
31. $y = \frac{4}{5}x + \frac{2}{5}$ **32.** $y = -\frac{2}{3}x + \frac{1}{6}$ **33.** $y = -\frac{1}{2}x + \frac{1}{4}$
34. $y = \frac{4}{5}x + 8$ **35.** $y = -\frac{2}{3}x + 4$ **36.** $y = \frac{3}{4}x - 12$
37. $y = \frac{3}{5}x - 7$ **38.** $x = 6$ **39.** $x = -2$
40. $y = -3$ **41.** $y = 4$ **42.** $y = 0$
43. $2x + y + 4 = 0$ **44.** $3x + y + 2 = 0$ **45.** $4x - y + 5 = 0$
46. $2x - y - 3 = 0$ **47.** $2x + 3y + 6 = 0$ **48.** $x - 3y + 2 = 0$
49. $3x - 2y + 5 = 0$ **50.** $3x + 4y - 5 = 0$ **51.** $5x - 4y - 8 = 0$

52. If an amount of money A results from investing a sum P at a simple interest rate of 7% for 10 years,

$$A = 1.7P$$

Graph this equation, where P is the independent variable.

53. An independent distributor bought a vending machine for $2000 with a probable scrap value of \$100 at the end of its expected 10-year life. The value V at the end of n years is given by

$$V = 2000 - 190n$$

Graph this equation, where n is the independent variable.

7.5 Linear Inequalities

In Section 3.7 we described a first-degree inequality with one variable. Consider the following inequalities with two variables.

$$y < x \qquad 2x + 3y \leq 6 \qquad 2x + 3y + 2 > 0 \qquad 5x \geq 4y$$

A ***linear inequality*** is an inequality statement involving $<$, $\leq$, $>$, or $\geq$, for which the variables are first-degree. In this section the linear inequalities will have two variables. We want to find the solution for such inequality statements. The techniques used are similar to the techniques used for solving first-degree equations with two variables.

First consider the relationship between a linear equation and a linear inequality. Look at the line

$$2x + y - 4 = 0$$

in Figure 7.25.

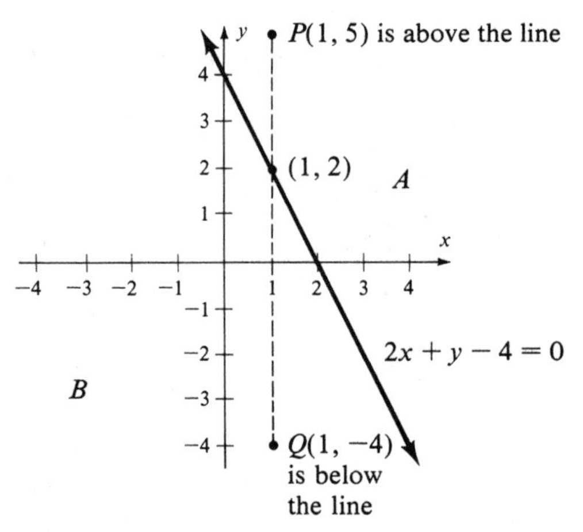

Figure 7.25

Notice that a line divides the plane into three parts. Two of the parts are the regions labeled A and B in Figure 7.25 and the third part is the set of points on the line. Regions A and B are called ***half-planes.*** Let (x_1, y_1) be a point on the line. Then another point (x_1, a) is said to be *above the line* if $a > y_1$ and *below the line* if $a < y_1$. In Figure 7.25 (1, 5) is above the line, and (1, −4) is below the line. Thus, A is the half-plane above the line, and B is the half-plane below the line.

Suppose we select some point not on the line, say (1, 5). Now, the half-planes are

$$2x + y - 4 > 0 \quad \text{and} \quad 2x + y - 4 < 0$$

In order to see which half-plane is above the line (Region A) and which is below the line (Region B), we substitute these coordinates into *either* inequality:

$$\begin{aligned} 2x + y - 4 &> 0 \\ 2(1) + 5 - 4 &\overset{?}{>} 0 \\ 3 &> 0 \end{aligned}$$

is ***true***
Thus, this inequality represents the half-plane ***containing*** the chosen point (1, 5). This is the half-plane above the line (Region A).

or

$$\begin{aligned} 2x + y - 4 &< 0 \\ 2(1) + 5 - 4 &\overset{?}{<} 0 \\ 3 &< 0 \end{aligned}$$

is ***false***
Thus, this inequality represents the half-plane ***not containing*** the chosen point (1, 5). This is the half-plane below the line (Region B).

In general, the solution set of a linear inequality is a half-plane. If the line is included in the solution set, as with

$$2x + y - 4 \geq 0$$

or

$$2x + y - 4 \leq 0$$

then we draw the line as shown in Figure 7.25. If the line is not included, we draw a dotted line.

Because we can pick any point that is not on the line—the point (1, 5) was used earlier—we usually pick (0, 0) if it is not on the line, because this choice makes the arithmetic easy.

Procedure for Graphing a Linear Inequality

Step 1. Replace the inequality symbol with an equality symbol.
Step 2. Graph the line. If the inequality symbol is $\leq$ or $\geq$, draw a solid line (it is included in the solution set); if the inequality symbol is $<$ or $>$, draw a dotted line (it is not included in the solution set).
Step 3. Choose any point in the plane that is not on the line (the origin is usually the simplest choice).
Step 4. If the chosen point satisfies the inequality, shade in that half-plane for the solution; if it does not satisfy the inequality, shade in the other half-plane for the solution.

Graph the given linear inequalities.

Example **1.** Graph $3x - y \geq 5$.

Solution Graph the (solid) line $3x - y = 5$ as shown in Figure 7.26a on page 188.

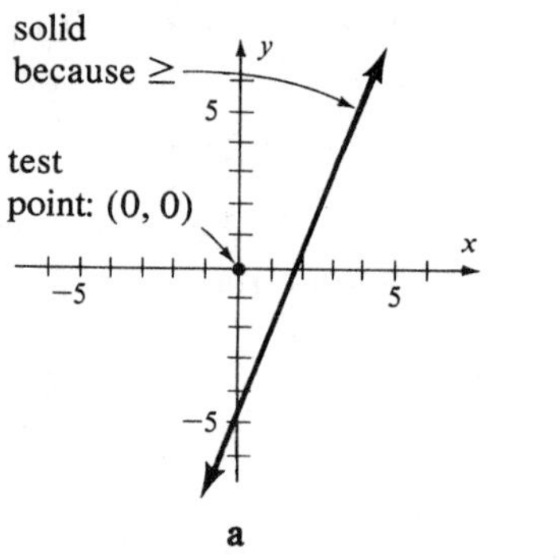

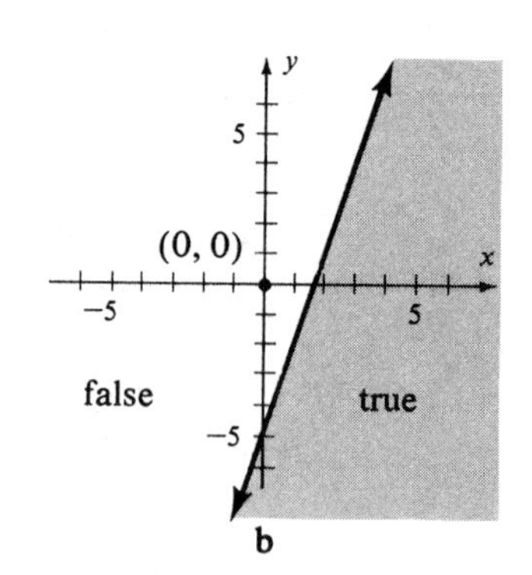

Figure 7.26

Check (0, 0): $3(0) - 0 \geq 5$; $0 \geq 5$ is false.

Thus, shade in the other half-plane as shown in Figure 7.26b. ■

Example 2. Graph $y < x$.

Solution Graph the (dotted) line $y = x$ as shown in Figure 7.27.

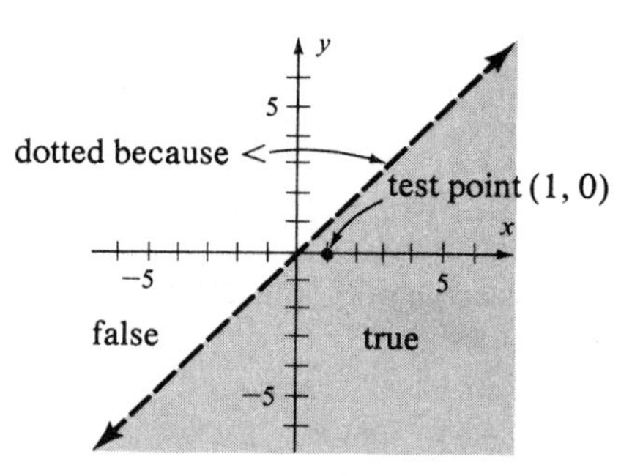

Figure 7.27

Check (1, 0): $0 < 1$ is true.

Thus, shade in the half-plane in which (1, 0) lies. ■

Problem Set 7.5

Graph the linear inequalities in Problems 1–30. See Examples 1 and 2.

1. $y \leq 2x + 1$
2. $y > 5x - 3$
3. $y > 3x - 3$
4. $y \geq -2x + 3$
5. $y > 3x - 3$
6. $y < 2x + 5$
7. $y > \frac{2}{3}x + 2$
8. $y \geq -\frac{1}{3}x - 3$
9. $y > \frac{4}{5}x - \frac{1}{5}$
10. $y < \frac{2}{3}x + \frac{5}{3}$
11. $3x \leq 2y$
12. $2x < 3y$
13. $x \geq y$
14. $y < \frac{4}{5}x$
15. $y \geq -\frac{2}{3}x$
16. $x \leq 3$
17. $y \leq 2$
18. $y \geq -1$
19. $x > -3$
20. $x \leq 0$
21. $y \geq 0$

22. $2x + 3y \leq 1$ **23.** $4x + 2y \geq -1$ **24.** $x + 4y > 8$

25. $x - 3y \geq 9$ **26.** $6x - 2y < 1$ **27.** $3x + 2y > 1$

28. $3x + y \leq -2$ **29.** $6x - 7y > 1$ **30.** $y \leq \frac{2}{3}x - \frac{3}{100}$

7.6 Review

In this chapter we have opened the door to a wide variety of problems by considering relationships between two sets of numbers. The ideas of a function (which relates sets of numbers), linear equations, and inequalities all are used to enable us to make observations about relationships between sets.

Self-Test

[7.1] **1.** Find the distance between the point specified by the given ordered pair and each of the coordinate axes.
a. $(-2, 4)$ **b.** $(6, -8)$ **c.** $(-5, -3)$ **d.** (m, n)

[7.2] **2.** Graph $3x - 4y = 12$ by plotting points. Label the points you have plotted.

[7.3] **3.** Find the slope of the line passing through $(4, 3)$ and $(-2, 1)$.

[7.4] **4.** **a.** What is the slope of a horizontal line?
b. What is the slope of a vertical line?

[7.4] *Graph the lines given in Problems 5–8.*

5. $y = \frac{4}{5}x - 2$ **6.** $x - y = 5$

7. $5x - 3y - 6 = 0$ **8.** $3x + 4y + 1200 = 0$

[7.5] *Graph the inequalities given in Problems 9 and 10.*

9. $y \geq 2x - 3$
10. $x - 5y < 125$

Review Problems

Find the slope of the line passing through the given pair of points in Problems 1–21.

1. $(5, 1)$ and $(6, 4)$ **2.** $(1, 3)$ and $(5, 8)$
3. $(2, 5)$ and $(7, 9)$ **4.** $(-1, -3)$ and $(1, 2)$
5. $(-2, -5)$ and $(3, 4)$ **6.** $(-1, 0)$ and $(6, 2)$
7. $(4, 3)$ and $(1, 1)$ **8.** $(6, 5)$ and $(3, 2)$
9. $(9, 5)$ and $(0, 1)$ **10.** $(-6, 4)$ and $(-3, -8)$
11. $(-4, -1)$ and $(-1, -3)$ **12.** $(6, -1)$ and $(-8, -2)$
13. $(4, -1)$ and $(-1, -1)$ **14.** $(6, -2)$ and $(-3, -2)$
15. $(-5, 1)$ and $(3, 1)$ **16.** $(8, 6)$ and $(8, 1)$
17. $(-5, -2)$ and $(-5, 3)$ **18.** $(6, -3)$ and $(6, 4)$
19. (a, b) and (c, d) **20.** (s, t) and (u, v)
21. (w, x) and (y, z)

Find the slope and intercept of the lines whose equations are given in Problems 22–54.

22. $y = 3x + 2$
23. $y = 2x - 5$
24. $y = -3x + 4$
25. $y = \frac{1}{2}x - 5$
26. $y = -\frac{3}{4}x + 2$
27. $y = \frac{1}{3}x - 1$
28. $y = 2x$
29. $y = -3x$
30. $y = x$
31. $x + y = 5$
32. $2x + y = 4$
33. $3x + y - 1 = 0$
34. $4x + y + 1 = 0$
35. $2x - y + 3 = 0$
36. $5x - y + 2 = 0$
37. $3x - y + 5 = 0$
38. $4x + 2y + 6 = 0$
39. $9x - 3y - 6 = 0$
40. $5x + 2y + 12 = 0$
41. $3x - 2y + 4 = 0$
42. $5x - 3y + 12 = 0$
43. $6x + 2y = 0$
44. $4x - 2y = 0$
45. $3x - 6y = 0$
46. $3x + 2y = 0$
47. $2x + 3y = 0$
48. $3x = 2y$
49. $5x = 3y$
50. $x = 5$
51. $x - 3 = 0$
52. $x + 4 = 0$
53. $y = -3$
54. $y + 6 = 0$

Graph the lines given in Problems 55–87.

55. $y = 3x + 2$
56. $y = 2x - 5$
57. $y = -3x + 4$
58. $y = \frac{1}{2}x - 5$
59. $y = -\frac{3}{4}x + 2$
60. $y = \frac{1}{3}x - 1$
61. $y = 2x$
62. $y = -3x$
63. $y = x$
64. $x + y = 5$
65. $2x + y = 4$
66. $3x + y - 1 = 0$
67. $4x + y + 1 = 0$
68. $2x - y + 3 = 0$
69. $5x - y + 2 = 0$
70. $3x - y + 5 = 0$
71. $4x + 2y + 6 = 0$
72. $9x - 3y - 6 = 0$
73. $5x + 2y + 12 = 0$
74. $3x - 2y + 4 = 0$
75. $5x - 3y + 12 = 0$
76. $6x + 2y = 0$
77. $4x - 2y = 0$
78. $3x - 6y = 0$
79. $3x + 2y = 0$
80. $2x + 3y = 0$
81. $3x = 2y$
82. $5x = 3y$
83. $x = 5$
84. $x - 3 = 0$
85. $x + 4 = 0$
86. $y = -3$
87. $y + 6 = 0$

Graph the inequalities in Problems 88–100.

88. $y > 2x - 1$
89. $y > 3x + 2$
90. $y > 2x - 3$
91. $y < 3x + 4$
92. $y < 5x - 3$
93. $y > 4x + 3$
94. $2x - y + 5 \leq 0$
95. $2x - 3y + 6 \leq 0$
96. $3x + 2y + 4 \leq 0$
97. $3y \geq 9$
98. $4x \geq 12$
99. $5x \geq 10$
100. $4y < 12$

8

Linear Systems

All animals are equal,
but some animals
are more equal than others.

Animal Farm
by George Orwell

8.1 Solving Systems of Equations by Graphing

In Chapter 7 we solved a linear equation with two variables. In this chapter we will consider solving *two linear equations* with two variables *simultaneously*—that is, together at the same time.

Definition of a System of Equations

> A ***system*** of equations is two (or more) equations with two (or more) variables.

A system is denoted by using a brace as shown in the following:

$$\begin{cases} 2x + 3y = 9 \\ 3x - y = 8 \end{cases}$$

A ***solution*** of a system is all of the values of the variables that satisfy both equations simultaneously. This means that a solution satisfies both (or all) the equations at the same time.

Determine if the given values are solutions to the systems given in Examples 1–3.

Example 1. $\begin{cases} 2x + 3y = 9 \\ 3x - y = 8 \end{cases}$ Is $x = 3$ and $y = 1$ a solution?

Check: $2x + 3y = 9$: $2(\mathbf{3}) + 3(\mathbf{1}) = 6 + 3 = 9$ ✓
$3x - y = 8$: $3(\mathbf{3}) - (\mathbf{1}) = 9 - 1 = 8$ ✓

Because they both check, $x = 3$ and $y = 1$ is a solution. The solution to a system is often stated using ordered-pair notation. For this example you might write the solution as $(3, 1)$ because it is understood that x is always the first component and y the second. ■

Example 2. $\begin{cases} 3a - b = 5 \\ 4a + b = 2 \end{cases}$ Is $a = 1$ and $b = -2$ a solution?

Check: $3a - b = 5$: $3(\mathbf{1}) - (\mathbf{-2}) = 3 + 2 = 5$ ✓
$4a + b = 2$: $4(\mathbf{1}) + (\mathbf{-2}) = 4 - 2 = 2$ ✓

Because they both check, $a = 1$ and $b = -2$ is a solution. In ordered-pair notation you would write $(a, b) = (1, -2)$. Notice that if variables other than x and y are used, you must show not only the numerical ordered pair, but also the variables. This is to show which variable represents which number because our only agreement about order is for the variables x and y. ■

Example 3. $\begin{cases} 5x - 2y = 10 \\ 3x + 2y = -2 \end{cases}$ Is $(2, 0)$ a solution?

Check: $5x - 2y = 10$: $5(2) - 2(0) = 10 - 0 = 10$ ✓
$3x + 2y = -6$: $3(2) + 2(0) = 6 + 0 = 6$ *This does not check.*

Be sure that you check the values in *both* equations of the system. ■

Now that you know how to check to see if some given values form a solution of a system, how can we go about finding that solution in the first place? Since *each* equation in a linear system has a solution that can be represented as a line, we can graph each solution separately as shown in Figure 8.1.

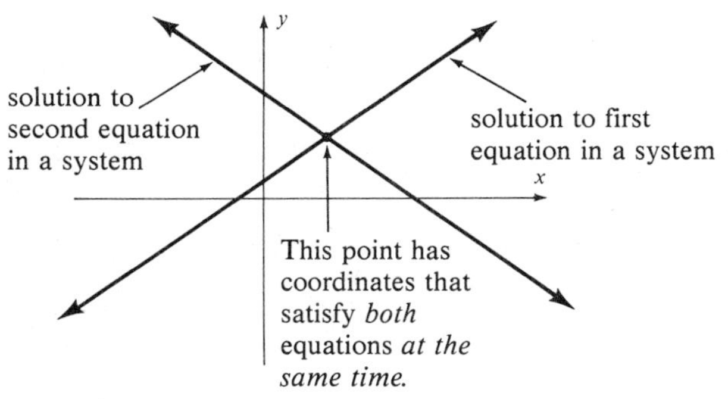

Figure 8.1
A typical solution to a linear system

The solution is the point of intersection of the lines. Remember from geometry that two *different* straight lines intersect in at most one point. You should therefore expect a single solution to a linear system of two equations with two unknowns. (There are exceptions to this expectation, and these will be discussed in Section 8.4.)

Solve the given systems in Examples 4–5.

Example 4. $\begin{cases} x + 2y = 10 \\ 3x - 2y = -2 \end{cases}$

Solution The most efficient way of graphing each of these equations is to use the slope-intercept method. Solve each equation for y:

$$y = -\frac{1}{2}x + 5;\ y\text{-intercept, } 5;\ \text{slope, } -\frac{1}{2}$$

$$y = \frac{3}{2}x + 1;\ y\text{-intercept, } 1;\ \text{slope, } \frac{3}{2}$$

The solution seems to be (2, 4).

Check:
$x + 2y = 10$: $(2) + 2(4) = 2 + 8 = 10$ ✓
$3x - 2y = -2$: $3(2) - 2(4) = 6 - 8 = -2$ ✓

The solution is (2, 4). ■

Example 5. $\begin{cases} 5s + 3t = 9 \\ 2s - 3t = 12 \end{cases}$

Solution If variables other than x and y are used, you must choose one of the variables to be the independent variable. In most cases, you can choose either one (it will not matter). It is customary, however, to write the ordered pairs in alphabetical order. For this example, let the ordered pairs be represented as (s, t). This means that s is the independent variable and t the dependent variable. Thus, when using the slope-intercept method for graphing we will solve for t.

$t = -\frac{5}{3}s + 3$; y-intercept, 3; slope, $-\frac{5}{3}$

$t = \frac{2}{3}s - 4$; y-intercept, -4; slope, $\frac{2}{3}$

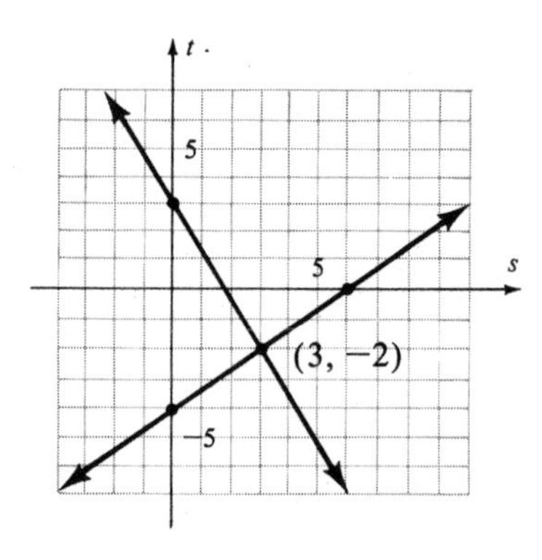

The solution seems to be $(s, t) = (3, -2)$. Can you carry out the check? ■

Word Problems: Supply and Demand (Optional)

Word problems have been emphasized throughout the book. It is important to continue your practice with solving applied problems. Fortunately, systems considerably simplify the process because they allow you to deal with more than a single unknown.

In this chapter we will present one new type of application in each section, and then summarize and develop them all in Section 8.5. The application type presented will be solved most conveniently by the method presented in that section. By concentrating on one new application in each section it is hoped that you progress toward mastering the solution of word problems.

One type of application problem has to do with supply and demand. If supply greatly exceeds demand, then money will be lost because of unsold items. On the other hand, if demand greatly exceeds supply, then money will be lost because of insufficient inventory. The most desirable situation is when the supply and demand are equal. If we assume that supply and demand are linear, then the solution to the system is called the *equilibrium point* and represents the point for which supply and demand are equal.

Example 6. *Suppose you have a small product that is marketable to the students on your campus. A little research shows that only 200 people would buy it at \$10, but 2000 would buy it at \$1.*

This information represents the ***demand.***

Let p = PRICE
n = NUMBER OF ITEMS

Then, because the demand, n, is determined by the price, let the price be the independent variable. That is, let the ordered pairs be (p, n).

From the given information: (10, 200)

↑ price ↑ number ↓ ↓

(1, 2000)

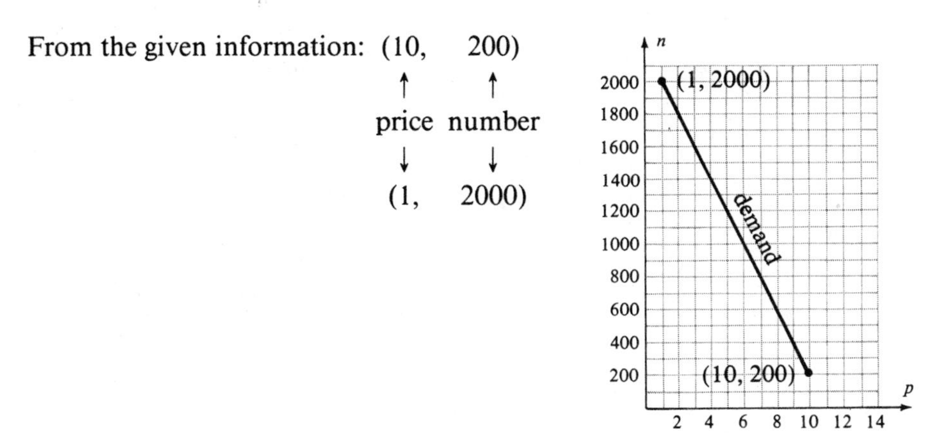

Draw a line passing through these points, and label the line "demand."

A local shop can make the product during slack time and could supply 100 items at a price that allows you to sell the item for $1. To supply more, the shop must use overtime. If the shop supplies 1500 items, you would have to charge $6.

This information represents the ***supply.*** Use (p, n) as defined earlier.

From the given information: (1, 100)

↑ price ↑ number ↓ ↓

(6, 1500)

Draw a line passing through these points on the same axes used for the demand. Label this line "supply."

What is the price you should charge for this item to maximize your profit?

Your profit is maximized at the equilibrium point. This is the point of intersection, which for this example is the point (5, 1200). This means that the price charged should be \$5. It also says that you should expect to sell 1200 items.

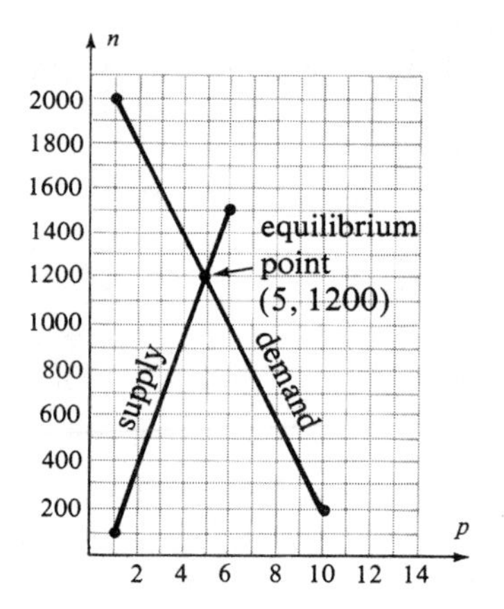

Notice that this question in Example 6 was interrupted several times to show the solution. In the problem set, this question would have shown only the part in italic type.

Problem Set 8.1

In Problems 1–12, decide whether the given ordered pair is a solution of the given system. See Examples 1–3.

1. $\begin{cases} x + 3y = 5 \\ 4x + 5y = 7 \end{cases}$ $(-2, 3)$

2. $\begin{cases} 5x - 6y = 1 \\ 3x - 4y = -3 \end{cases}$ $(11, 9)$

3. $\begin{cases} x - y = 9 \\ x + y = 33 \end{cases}$ $(12, 21)$

4. $\begin{cases} 3a + b = 17 \\ a - 2b = 1 \end{cases}$ $(a, b) = (5, 2)$

5. $\begin{cases} D + B = 7 \\ D + 4B = 13 \end{cases}$ $(D, B) = (5, 2)$

6. $\begin{cases} W + F = 6 \\ 3W - 2F = -2 \end{cases}$ $(W, F) = (4, 2)$

7. $\begin{cases} 2L + 2W = 80 \\ L = 20 + W \end{cases}$ $(L, W) = (10, 30)$

8. $\begin{cases} m + n = 800 \\ m + 100 = n \end{cases}$ $(m, n) = (350, 450)$

9. $\begin{cases} n = 500p + 800 \\ n = 3200 - 300p \end{cases}$ $(p, n) = (3, 2300)$

10. $\begin{cases} y = -3x + 2 \\ y = -1 \end{cases}$ $(1, -1)$

11. $\begin{cases} y = -\dfrac{2}{3}x - 4 \\ y = -8 \end{cases}$ $(3, -6)$

12. $\begin{cases} y = \dfrac{1}{2}x - 5 \\ y = 1 \end{cases}$ $(4, -1)$

In Problems 13–24, solve each system graphically. See Examples 4 and 5.

13. $\begin{cases} x + y = 5 \\ 2x - y = 1 \end{cases}$

14. $\begin{cases} 2x + y = 5 \\ x - y = 4 \end{cases}$

15. $\begin{cases} 2x + y = 1 \\ x - y = 5 \end{cases}$

16. $\begin{cases} 2x = y - 2 \\ 3x + y = 7 \end{cases}$

17. $\begin{cases} 2x = 6 + y \\ 4x + y = 6 \end{cases}$

18. $\begin{cases} y = -4x - 3 \\ 2x + y = 1 \end{cases}$

19. $\begin{cases} 3x + 2y = 4 \\ x + y = 1 \end{cases}$

20. $\begin{cases} x + 2y = 2 \\ 3x + y = -4 \end{cases}$

21. $\begin{cases} 4x + y = -3 \\ 3x + 2y = 4 \end{cases}$

22. $\begin{cases} 2x - 3y = 6 \\ x + 2y = 10 \end{cases}$

23. $\begin{cases} x - 2y = -2 \\ 3x - y = 4 \end{cases}$

24. $\begin{cases} x + y = 2 \\ 3x + 2y = -4 \end{cases}$

Read the supply and demand applications in Problems 25–30. Let p = PRICE and n = NUMBER OF ITEMS. Give two coordinates for the "demand" and two coordinates for the "supply." You do not need to solve the systems, simply give the ordered pairs. For Example 6, your answer would be:

Demand: (10,200) and (1,2000)
Supply: (1,100) and (6,1500)

25. The demand for a product varies from 200 at a price of \$6 to 800 at a price of \$2. Also, 800 could be supplied at a price of \$7, whereas only 200 could be supplied for \$1. What price maximizes the profit?

26. The demand for a product varies from 1000 at a price of \$9 to 7000 at a price of \$3. Also, 6000 could be supplied at a price of \$10, whereas only 2000 could be supplied for \$2. What price maximizes the profit?

27. The demand of a commodity is 5000 at a dollar price down to 1000 at a seven dollar price. It is possible to obtain 2000 at a two dollar price but 5000 can be available at eight dollars. Find the equilibrium point of the system.

28. The demand of a commodity is twenty at \$2000 but only five at \$8000. It is possible to obtain nine at \$2000 but eleven can be available at \$10,000. Find the equilibrium point of the system.

29. California Instruments, manufacturers of electronic calculators, find by test marketing their calculators at UCLA that they can sell 180 calculators when they are priced at \$10, but can sell only 20 calculators when they are pried at \$40. On the other hand, they find that 20 calculators can be supplied at \$10 each. If they were supplied at \$40 each, overtime shifts could be used to raise the supply to 180 calculators. What is the optimum price for the calculators?

30. A manufacturer of class rings test-marketed a new item at Bowling Green University. They found that they could sell 900 items when they are priced at \$1, but can sell only 300 items if the price is raised to \$7. On the other hand, they find that 600 items can be supplied at \$1 each. If they were supplied at \$9 each, overtime shifts could be used to raise the supply to 1,000 items. What is the optimum price for the items?

Answer the questions asked in Problems 31–36. See Example 6.

31. What price maximizes the profit in Problem 25?

32. What price maximizes the profit in Problem 26?

33. Find the equilibrium point for the system in Problem 27.

34. Find the equilibrium point for the system in Problem 28.

35. What is the optimum price for the calculators in Problem 29?

36. What is the optimum price for the rings in Problem 30?

8.2 Solving Systems of Equations by Adding

In the last section we sketched the graphs of two equations in a system and read a solution from the graph. This solution is an approximation, because it depends on the accuracy of the drawing and the keenness of our eyesight. Actually, it is impossible for the eye to differentiate among (1, 1), (0.99, 1.01), and (1.001, 0.985) on an ordinary scale. We need an algebraic method to complement the geometric one that we have studied.

All of our techniques for equation solving deal with a single equation that has one variable. These techniques essentially reduce the equation to a more familiar form so that the equation can be solved easily. Thus, the system of equations must somehow be reduced from two equations to a single equation—an equation to which the usual methods can be applied.

Consider Example 4 from the previous section.

$$\begin{cases} x + 2y = 10 \\ 3x - 2y = -2 \end{cases}$$

How can you "get rid of" one equation and one unknown so that you have a single equation in one unknown that can be solved? Well, in each equation there are two equal quantities. So, if you add equal quantities, the results should be equal. That is, because the members of one equation are equal, you can use the addition property of equality to add them to the members of the second equation.

$$+\begin{cases} x + 2y = 10 \\ 3x - 2y = -2 \end{cases}$$
$$4x + 0 = 8$$
$$x = 2$$

We indicate what we are doing by noting the operation to the left of the brace.

If $x = 2$, what is y? Consider the original equations with $x = 2$.

$$\begin{aligned} (2) + 2y &= 10 \\ 2y &= 8 \\ y &= 4 \end{aligned} \qquad \textbf{or} \qquad \begin{aligned} 3(2) - 2y &= -2 \\ 6 - 2y &= -2 \\ -2y &= -8 \\ y &= 4 \end{aligned}$$

Notice that $x = 2$ may be substituted into either of the two original equations to determine y.

Thus, the solution is (2, 4).

We found the common solution for the system easily, because the y-terms were eliminated. Will this always happen? Consider another example.

$$\begin{cases} 3x - y = 24 \\ x - 2y = -7 \end{cases}$$

If you add the members of the two equations in this example, neither x nor y will be eliminated in the resulting equation.

$$+\begin{cases} 3x - y = 24 \\ x - 2y = -7 \end{cases}$$
$$4x - 3y = 17$$

What was it in the first example that allowed the elimination of the y-terms? You added $2y$ and $-2y$ and ***they were opposites.*** Can you arrange it so that you have opposites in this example? The multiplication property allows you to multiply both sides of an equation by any nonzero number, so:

$$-2\begin{cases} 3x - y = 24 \\ x - 2y = -7 \end{cases}$$

We indicate to the left of the brace what we are doing to the equation.

$$+\begin{cases} \mathbf{-6x + 2y = -48} \\ x - 2y = -7 \end{cases}$$
$$-5x = -55$$
$$x = 11$$

Now replace x with 11 in the second of the original equations.

$$\begin{aligned} (\mathbf{11}) - 2y &= -7 \\ -2y &= -18 \\ y &= 9 \end{aligned}$$

We arbitrarily choose the second equation, but it does not matter which of the original equations you choose.

The solution is (11, 9).

In Examples 1–3, solve for x and y.

Example 1. $\begin{cases} x + 2y = 2 \\ x - 2y = 4 \end{cases}$

Solution The coefficients of y are opposites, thus you can eliminate y by adding.

$$+\begin{cases} x + 2y = 2 \\ \underline{x - 2y = 4} \end{cases}$$

$$\begin{aligned} 2x \quad &= 6 \\ x &= 3 \end{aligned}$$

$$\begin{aligned} \mathbf{3} + 2y &= 2 \\ 2y &= -1 \\ y &= -\frac{1}{2} \end{aligned}$$

Replace x = 3 in one of the original equations and solve for y.

The solution is $\left(3, -\frac{1}{2}\right)$. ■

Example 2. $\begin{cases} 2s + 3t = 5 \\ 4s + 5t = 7 \end{cases}$

Solution To eliminate one variable, multiply the first equation by -2.

$$\mathbf{-2}\begin{cases} 2s + 3t = 5 \\ 4s + 5t = 7 \end{cases}$$

$$+\begin{cases} \mathbf{-4s - 6t = -10} \\ \underline{4s + 5t = 7} \end{cases}$$

Now the coefficients of s are opposites.

$$\begin{aligned} -t &= -3 \\ t &= 3 \end{aligned}$$

$$\begin{aligned} 4s + 5(\mathbf{3}) &= 7 \\ 4s + 15 &= 7 \\ 4s &= -8 \\ s &= -2 \end{aligned}$$

Once the value of t is found, use that value to find s from one of the original equations.

The solution is $(s, t) = (-2, 3)$. Remember if variables other than x and y are used, you must show which number represents the first variable and which represents the second. ■

Example 3. $\begin{cases} 5x = 1 + 6y \\ 3 = 4y - 3x \end{cases}$

Solution In the addition method it is necessary to align the variables, so we first rewrite the equations in the usual form.

$$\begin{cases} 5x - 6y = 1 \\ 3x - 4y = -3 \end{cases}$$

To eliminate one variable, you want coefficients that are opposites.

$$\begin{array}{rl} 3 & 5x - 6y = 1 \\ -5 & 3x - 4y = -3 \end{array}$$

$$\begin{array}{rr} & \mathbf{15x - 18y = 3} \\ + & \underline{\mathbf{-15x + 20y = 15}} \\ & 2y = 18 \\ & y = 9 \end{array}$$

$$\begin{aligned} 3x - 4(9) &= -3 \\ 3x - 36 &= -3 \\ 3x &= 33 \\ x &= 11 \end{aligned}$$

Replace $y = 9$ in one of the original equations, and solve for x.

The solution is (11, 9). ■

Addition Method

The addition method, which we have developed by example, can be outlined by the following steps.

1. *Multiply* one or both of the equations by constants so that the coefficients of one of the variables are opposites.
2. *Add* corresponding members of the equations to obtain a new equation in a single variable.
3. *Solve* the derived equation for that variable.
4. *Substitute* the value of that variable into either of the original equations, and solve for the second variable.
5. *State* the solution.

Word Problems: Combining Rates (Optional)

We conclude this section with some examples of systems applied to another type of application problem, namely that of ***combining rates.*** Imagine a conductor walking toward the rear of a slowly moving train. Better still, picture a chase scene atop the freight cars of a moving train. We have all seen films of canoeists shooting the rapids while paddling furiously through the white water. Think of those canoeists paddling upstream. All of these situations and many more involve a common principle. When you move in air, in water, on a treadmill, or on some other medium that is also moving, you combine rates. *If you move in the same direction, your rate is added to the rate of the medium. If you move against the movement of the medium, the rates are subtracted.* Consider the following examples.

Example 4. A deep-sea fishing boat travels at 20 mph going out with the tide. Sometimes it comes in against the tidal current and is able to make only 15 mph. What is the boat's rate without the current, and what is the rate at which the tide moves the water?

Solution

$$\begin{cases} \text{rate with the tide is 20 mph} \\ \text{rate against the tide is 15 mph} \end{cases}$$

$$\begin{cases} (\text{RATE OF BOAT}) + (\text{RATE OF TIDE}) = 20 & \textit{Same direction: add rates} \\ (\text{RATE OF BOAT}) - (\text{RATE OF TIDE}) = 15 & \textit{Opposite directions: subtract rates} \end{cases}$$

There are two unknowns: let b = RATE OF BOAT and t = RATE OF TIDE.

$$\begin{cases} b + t = 20 \\ b - t = 15 \end{cases}$$

$$+\begin{cases} b + t = 20 \\ \underline{b - t = 15} \end{cases}$$

$$2b = 35$$

$$b = 17.5 \qquad (17.5) + t = 20$$

$$t = 2.5$$

The boat's speed is 17.5 mph, and the tide moves at 2.5 mph. ■

Example 5. On a certain day a large bird was clocked at 80 mph flying with the wind, but could fly only 10 mph against the wind. What would be the speed of the bird in still air?

Solution

$$\begin{cases} \text{rate with the wind is 80 mph} \\ \text{rate against the wind is 10 mph} \end{cases}$$

$$\begin{cases} \text{BIRD'S RATE} + \text{WIND'S RATE} = 80 \\ \text{BIRD'S RATE} - \text{WIND'S RATE} = 10 \end{cases}$$

Let b = BIRD'S RATE and w = WIND'S RATE.

$$+\begin{cases} b + w = 80 \\ \underline{b - w = 10} \end{cases}$$

$$2b = 90$$

$$b = 45$$

The bird's rate in still air would be 45 mph. ■

Problem Set 8.2

Solve each system in Problems 1–30 by the addition method. See Examples 1–3.

1. $\begin{cases} x + y = 36 \\ x - y = 2 \end{cases}$

2. $\begin{cases} x + y = 18 \\ x - y = 4 \end{cases}$

3. $\begin{cases} x + y = -8 \\ x - y = 18 \end{cases}$

4. $\begin{cases} g - h = 36 \\ g + h = -10 \end{cases}$

5. $\begin{cases} m + n = 14 \\ m - n = 6 \end{cases}$

6. $\begin{cases} a + b = -8 \\ a - b = 8 \end{cases}$

7. $\begin{cases} x - 2y = 5 \\ x + 2y = 7 \end{cases}$ **8.** $\begin{cases} x - 3y = 4 \\ x + 2y = -6 \end{cases}$ **9.** $\begin{cases} x + 5y = 10 \\ x - 3y = 2 \end{cases}$

10. $\begin{cases} 2c - d = 9 \\ c + 4d = 9 \end{cases}$ **11.** $\begin{cases} 2r + 3s = 5 \\ r - 2s = 6 \end{cases}$ **12.** $\begin{cases} u - 2v = 1 \\ 2u - 5v = 0 \end{cases}$

13. $\begin{cases} x = 1 + 2y \\ 2x - 3y = 2 \end{cases}$ **14.** $\begin{cases} 2x + 7 = y \\ 3x + 2y = 7 \end{cases}$ **15.** $\begin{cases} 3y = x - 18 \\ x + 2y + 2 = 0 \end{cases}$

16. $\begin{cases} 5x + y = 0 \\ 2x - 3y = 17 \end{cases}$ **17.** $\begin{cases} 9x + 2y = 1 \\ 4x + 6y = 3 \end{cases}$ **18.** $\begin{cases} 5x + 4y = 4 \\ 3x + 2y = 0 \end{cases}$

19. $\begin{cases} 5p + 2q = 3 \\ 3p + 4q = -1 \end{cases}$ **20.** $\begin{cases} 2w + 3y = 25 \\ 3w + 2y = 20 \end{cases}$ **21.** $\begin{cases} 13r + 3s = -1 \\ 15r + 4s = 1 \end{cases}$

22. $\begin{cases} 2x - 3y = 43 \\ 3x - 2y = 47 \end{cases}$ **23.** $\begin{cases} 3x + 5y = 3 \\ 5x + 7y = 1 \end{cases}$ **24.** $\begin{cases} 4x - 3y = 8 \\ 7x - 4y = 9 \end{cases}$

25. $\begin{cases} 7x - 9y = 3 \\ 5x - 6y = 3 \end{cases}$ **26.** $\begin{cases} 2x + 3y = 6 \\ 7x + 5y = -1 \end{cases}$ **27.** $\begin{cases} 3x + 5y = 8 \\ 4x - 3y = 1 \end{cases}$

28. $\begin{cases} 4W + 46R = 79 \\ W + R = 4 \end{cases}$ **29.** $\begin{cases} 180F + 140S = 1100 \\ F + S = 7 \end{cases}$ **30.** $\begin{cases} 65T + 45B = 230 \\ T + B = 4 \end{cases}$

In Problems 31–40 write equations to solve the stated problems, solve the equations, state the answers in words, and check your solutions. See Examples 4 and 5.

31. Two numbers yield 100 when added and 8 when subtracted. Find the numbers.

32. When you add two numbers you get 29, and when you subtract them you get 5. What are the numbers?

33. One number is three more than three times the second number. If their sum is 31, what are the numbers?

34. The first number is four less than four times the second. What are the numbers if their sum is 31?

35. Four times the smaller number reduced by three times the larger is 15. Three times the smaller less twice the larger is 29. What is the smaller number?

36. Twice the larger number subtracted from three times the smaller number is 1. Twice the smaller number reduced by the larger number is 12. What is the larger number?

37. A motorboat pulls a skier on a certain stretch of river. The boat travels 30 mph (relative to the shore) on the downstream run and is slowed to 26 mph on the way upstream. What is the rate of the current?

38. An airplane is able to travel 370 mph with the wind. On a return flight, flying into the wind, the plane's speed is only 320 mph. What is its speed in still air?

39. A canoeist rows downstream in $1\frac{1}{2}$ hours and back upstream in 3 hours. What is the rate of the current if he or she rows 9 miles in each direction?

40. A plane makes a 630-mile flight with the wind in $2\frac{1}{2}$ hours. Returning against the wind takes 3 hours. Find the wind speed.

8.3 Solving Systems of Equations by Substitution

It is often helpful to replace one expression with another in order to be understood. In algebra this replacement is called ***substitution.*** A substitution changes not the meaning, but the appearance or form. If you have two equations of lines in the slope-intercept form, then, in that form, each is an expression of y in terms of x. To solve the system, substitute to obtain a single equation with a single unknown as shown in Example 1.

Example 1. Solve for x and y.

$$\begin{cases} y = 2x + 7 \\ y = -3x + 2 \end{cases}$$

Solution Since $\mathbf{y = 2x + 7}$

and $\mathbf{y} = -3x + 2$

substitute

$$\begin{aligned} \mathbf{2x + 7} &= -3x + 2 && \textit{Substitution eliminates y.} \\ 5x + 7 &= 2 && \textit{Solve for x.} \\ 5x &= -5 \\ x &= -1 \end{aligned}$$

Once x is known, substitute back to find y (you can use either of the given equations):

$$\begin{aligned} y &= 2\mathbf{x} + 7 \\ &= 2(\mathbf{-1}) + 7 && \textit{Since } \mathbf{x = -1} \\ &= 5 \end{aligned}$$

The solution is $(-1, 5)$. ■

We now have a third method for solving systems of linear equations. Although we will continue to use the addition method and on occasion the graphing method to get approximations, there will also be occasions when the ***substitution method*** will be more convenient for a particular system. Moreover, with systems that are not linear, the addition method may not work, but the substitution method will always apply.

Example 2. Solve for x and y.

$$\begin{cases} y = 3x - 1 \\ 2x + 3y = 19 \end{cases}$$

Solution Since the first equation is solved for y, we consider substitution. That is, substitute $3x - 1$ from the first equation for the value of y in the second equation:

$$\mathbf{y = 3x - 1}$$

$$2x + 3\mathbf{y} = 19$$

substitute

$$\begin{aligned} 2x + 3(\mathbf{3x - 1}) &= 19 && \textit{Solve this equation for x.} \\ 2x + 9x - 3 &= 19 \\ 11x - 3 &= 19 \\ 11x &= 22 \\ x &= 2 \end{aligned}$$

Now substitute $\mathbf{x} = \mathbf{2}$ into $y = 3\mathbf{x} - 1$ and determine the value for y.

$$\begin{aligned} y &= 3(\mathbf{2}) - 1 \\ &= 5 \end{aligned}$$

The solution is (2, 5). ■

Let us summarize this new procedure by outlining the method.

Substitution Method

1. *Solve* one of the equations for one of the variables.
2. *Substitute* the expression that you obtain into the other equation for that variable.
3. *Solve* the resulting single equation for the value of that variable.
4. *Substitute* that value into either of the original equations to determine the value of the other variable.
5. *State* an answer.

In the next example, we will number these steps as we have in the preceding outline.

Example 3. Solve by substitution.

$$\begin{cases} x + 2y = 10 \\ 5x + 3y = 1 \end{cases}$$

Solution

Step 1 $\begin{cases} x + 2y = 10 \\ 5x + 3y = 1 \end{cases}$ $x = \mathbf{10 - 2y}$ *Solve for one variable.*

Step 2 $5(\mathbf{10 - 2y}) + 3y = 1$ *Substitute for x.*

Step 3 *Solve for y.*

$$\begin{aligned} 50 - 10y + 3y &= 1 \\ 50 - 7y &= 1 \\ -7y &= -49 \\ y &= 7 \end{aligned}$$

Step 4 *Substitute for y.*

$$\begin{aligned} x &= 10 - 2(\mathbf{7}) \\ x &= -4 \end{aligned}$$

Step 5 $(\mathbf{-4, 7})$ *State the solution.* ■

Word Problems: Coin Problems (Optional)

Substitution is one of the primary tools used in solving word problems. Many word problems are given in a form that indicates *more than two variables.* There are at least three ways you can reduce the number of variables in an applied problem:

1. Substitute numbers for variables. Use this type of substitution when the value of a variable is known.

2. Substitute variables for other variables by using a known formula.
3. Substitute variables for other variables when relationships between those variables are given in the problem.

Substitution, therefore, can allow your problem-solving ability to take a giant leap forward because now you can write down the statements as given in the problem without regard to the number of variables you are using. As you use substitution to eliminate variables, the algebraic equation or equations should become obvious.

The following example focuses on a common word problem type—money-value problems. You must distinguish between the *number of coins* and the *value of coins* because, for example, the number of nickels is not the same as the value of those same nickels. The formulas you will need for money problems are:

VALUE OF QUARTERS = 25(NUMBER OF QUARTERS)
VALUE OF DIMES = 10(NUMBER OF DIMES)
VALUE OF NICKELS = 5(NUMBER OF NICKELS)
VALUE OF PENNIES = NUMBER OF PENNIES

Notice that the values in the preceding formulas are in terms of cents, not dollars.

Example 4. A box of coins has dimes and nickels totaling \$4.20. If there are 3 more nickels than dimes, how many of each type of coin is there in the box?

Solution Restate the problem in equation form. It does not matter if you use more than two unknowns.

$$\begin{cases} \text{VALUE OF DIMES} + \text{VALUE OF NICKELS} = \text{TOTAL VALUE} \\ \text{NUMBER OF NICKELS} = \text{NUMBER OF DIMES} + 3 \end{cases}$$

(We used 5 different unknowns in this statement.)

Substitute known numbers:

VALUE OF DIMES + VALUE OF NICKELS = TOTAL VALUE

↑ ***substitute***

***TOTAL VALUE* = 420**

NUMBER OF NICKELS = NUMBER OF DIMES + 3

Substitute using a formula:

VALUE OF DIMES + VALUE OF NICKELS = **420**

↑ ***substitute*** ↑

10(NUMBER OF DIMES) ***5(NUMBER OF NICKELS)***

NUMBER OF NICKELS = NUMBER OF DIMES + 3

$$\begin{cases} \mathbf{10(\text{NUMBER OF DIMES})} + \mathbf{5(\text{NUMBER OF NICKELS})} = 420 \\ \text{NUMBER OF NICKELS} = \text{NUMBER OF DIMES} + 3 \end{cases}$$

Let d = NUMBER OF DIMES and n = NUMBER OF NICKELS.

$$\begin{cases} 10d + 5n = 420 \\ n = d + 3 \end{cases}$$

Solve *this* system by substitution:

$$\begin{aligned} 10d + 5(\boldsymbol{d + 3}) &= 420 & &\textit{Substitute } d + 3 \textit{ for } n \textit{ in first equation.} \\ 10d + 5d + 15 &= 420 & &\textit{Solve for } d. \\ 15d + 15 &= 420 \\ 15d &= 405 \\ d &= 27 \end{aligned}$$

Then $n = 27 + 3$
$= 30$

There are 27 dimes and 30 nickels. ■

Problem Set 8.3

Solve each system in Problems 1–24 by substitution. See Examples 1–3.

1. $\begin{cases} y = 3x + 7 \\ y = 2 - 2x \end{cases}$
2. $\begin{cases} y = 3x + 11 \\ y = 1 - 2x \end{cases}$
3. $\begin{cases} y = 2x + 15 \\ y = 3 - 2x \end{cases}$
4. $\begin{cases} y = 2x \\ y = 4 - 2x \end{cases}$
5. $\begin{cases} y = 3x - 11 \\ y = 1 - 3x \end{cases}$
6. $\begin{cases} y = 2x - 8 \\ y = 2 - 3x \end{cases}$
7. $\begin{cases} x + y = -2 \\ y = x + 2 \end{cases}$
8. $\begin{cases} 3x + 9 = y \\ 4y + 5x = 2 \end{cases}$
9. $\begin{cases} y = 3x - 4 \\ 5x - 2y - 5 = 0 \end{cases}$
10. $\begin{cases} x = 2y - 5 \\ 4x + 3y = 24 \end{cases}$
11. $\begin{cases} 2y - 3 = x \\ 2x - 3y + 2 = 0 \end{cases}$
12. $\begin{cases} 3y + 2 = x \\ 5x - 6y - 1 = 0 \end{cases}$
13. $\begin{cases} 3x + y = 1 \\ 5x + 2y = 14 \end{cases}$
14. $\begin{cases} x - 4y = 13 \\ 3x + 7y = 1 \end{cases}$
15. $\begin{cases} x + y = 2 \\ 2x - 3y = 19 \end{cases}$
16. $\begin{cases} x - y = 2 \\ 3x - 4y = 1 \end{cases}$
17. $\begin{cases} x + 5y + 2 = 0 \\ 2x + 3y = 3 \end{cases}$
18. $\begin{cases} 4x + y = 16 \\ x + 2y = 18 \end{cases}$
19. $\begin{cases} x - 3y = 21 \\ 2x + y = 14 \end{cases}$
20. $\begin{cases} 3y = 21 - 2x \\ 3x = y + 15 \end{cases}$
21. $\begin{cases} 2x = y + 9 \\ 4y = 9 - x \end{cases}$
22. $\begin{cases} 2x + 3y - 5 = 1 \\ x + 2y - 2 = -1 \end{cases}$
23. $\begin{cases} x - 2y - 1 = 0 \\ 2x - 5y = 2 \end{cases}$
24. $\begin{cases} x - 2y - 1 = 0 \\ 2x = 5y \end{cases}$

Answer the questions in Problems 25–27. See Example 4.

25. What is the value of 12 quarters?
26. What is the value of 37 dimes?
27. What is the value of 17 nickels?

Write a verbal expression in equation form for each the statements given in Problems 28–36. See Example 4.

28. A box contains $7.15 in dimes and nickels.
29. A box contains $3.75 in dimes and quarters.
30. A box contains $6.12 in nickels, dimes, and pennies.
31. A box contains $12.14 in nickels, dimes, and quarters.
32. The number of quarters in a box is twice the number of dimes.

33. The number of dimes in a box is 1 less than twice the number of nickels.
34. The number of dimes is three less than twice the number of quarters.
35. The number of dimes is six less than twice the number of pennies.
36. The number of nickels is four more than twice the number of dimes.

Solve each of Problems 37–40, and state an answer in words. See Example 4.

37. A box contains $8.40 in quarters and dimes. The number of quarters is twice the number of dimes. How many of each type of coin is in the box?
38. A box containing only dimes and quarters has three less than twice as many dimes as quarters. If the total value of the coins is $22.65, how many dimes are in the box?
39. A box contains $8.40 in nickels, dimes, and pennies. How many of each type of coin is there in the box if the number of dimes is six less than twice the number of pennies, and there is an equal number of dimes and nickels in the box.
40. Sherlock Holmes was called in as a consultant to solve the Great Bank Robbery. He was told that the thief had made away with a bag of money containing $5, $10, and $20 bills totaling $1390. When the bankers were checking serial numbers to see how many of each denomination were taken, Holmes said "It is elementary, since there were five times as many $10 bills as $5 bills and three more than twice as many $20 bills as $5 bills." How many of each denomination were taken?

8.4 Solving Systems

You now have three methods for solving linear systems—graphing, adding, and substitution. Sometimes, as in previous problem sets, you are told which method to use. Other times you must decide which method is best. Even though a particular problem may be solved by more than one method, we offer the following suggestions:

METHOD	WHEN TO USE
graphing	This is the least accurate and therefore least used of the methods. You would use it if you know points on intersecting lines. Use this method for supply and demand applications.
adding	This is the easiest of the three methods. You would use this method if graphing or substitution is not appropriate. Use this method for combining rates applications.
substitution	You would use this if one (or both) of the equations is solved for one of the variables, or if it is easy to solve one of the equations for one of the variables. This is the most useful method for eliminating variables when solving word problems.

Examples *Name the most appropriate method for solving each of the following systems.*

1. $\begin{cases} 2x + 3y = 6 \\ 3x - 6y = 30 \end{cases}$ *Method: addition*

2. $\begin{cases} y = \dfrac{2}{3}x + 5 \\ x + 3y = 33 \end{cases}$ *Method: substitution*

3. The intersection of the following lines: passing through (4, 3) and (− 1, 2), and passing through (3, −2) and (−4, 0). *Method: graphing*
4. Combining rates application *Method: adding*
5. Supply and demand application *Method: graphing*
6. Money-value application *Method: substitution* ■

Just as there are three ways to solve a system, there are three distinct results that you may obtain when solving a system. As we have said before, you should generally expect a single solution to a system of linear equations. However, there are two other possibilities as summarized in the following table.

Graph		Algebraic Equations	Solution
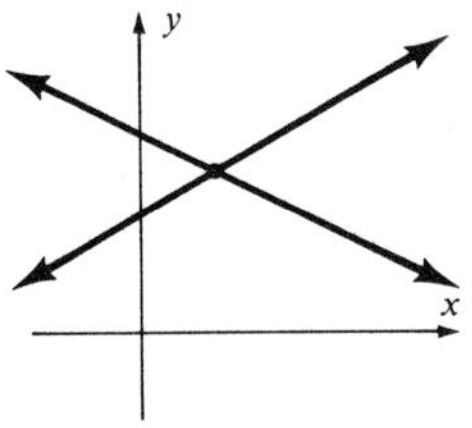	represent two *intersecting lines* a *single point* of intersection	You obtain a single value for each of the variables.	State the point of intersection as an ***ordered pair.***
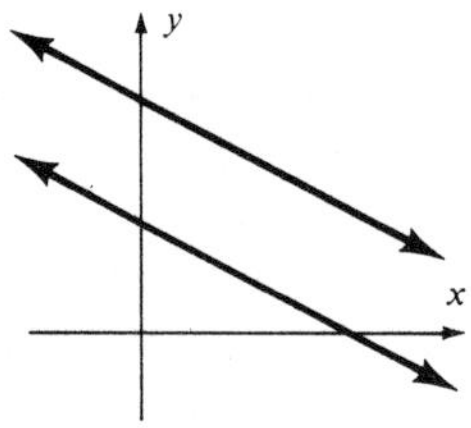	represent two *parallel lines* *no point* of intersection	The variable drops out of the equations, and you obtain an equation that is false, like 2 = 3.	Describe the system as ***inconsistent.***
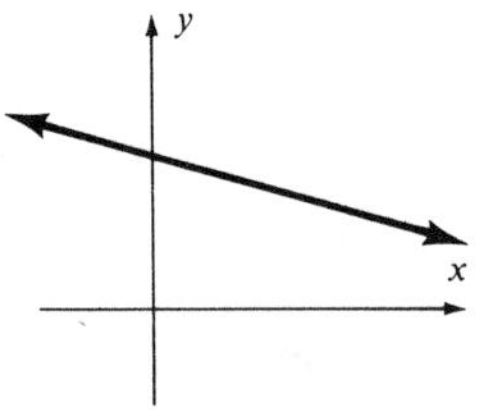	represent the *same line* *unlimited number* of solutions	The variable drops out of the equations, and you obtain an equation that is true, like 2 = 2.	Describe the system as ***dependent.***

Solve the systems given in Examples 7–9.

Example 7. $\begin{cases} 2x - 3y = 5 \\ y = \dfrac{2}{3}x + 1 \end{cases}$ *Solve this system by substitution.*

Solution

$$2x - 3\left(\frac{2}{3}x + 1\right) = 5$$
$$2x - 2x - 3 = 5$$
$$-3 = 5 \qquad \textit{This is false.}$$

Inconsistent system ■

Example 8. $\begin{cases} 3x + 4y = 9 \\ 6x - 3y = -4 \end{cases}$ *Solve this system by adding.*

Solution

$$-2\begin{cases} 3x + 4y = 9 \\ 6x - 3y = -4 \end{cases}$$

$$+\begin{cases} -6x - 8y = -18 \\ \underline{6x - 3y = -4} \end{cases}$$
$$-11y = -22$$
$$y = 2 \qquad \text{Thus, } 3x + 4(2) = 9$$
$$3x = 1$$
$$x = \frac{1}{3}$$

$\left(\frac{1}{3}, 2\right)$ *The equations are satisfied by a single set of values for x and y. The lines intersect in a single point.* ■

Example 9. $\begin{cases} x - 2y = 6 \\ y = \frac{1}{2}x - 3 \end{cases}$ *Solve this system by substitution.*

Solution

$$x - 2\left(\frac{1}{2}x - 3\right) = 6$$
$$x - x + 6 = 6$$
$$6 = 6 \qquad \textit{This is true.}$$

Dependent system ■

Problem Set 8.4

Solve Problems 1–30 using any appropriate method. See Examples 1–9.

1. $\begin{cases} x - y = 2 \\ x + y = 8 \end{cases}$ **2.** $\begin{cases} 2x - 3y = 1 \\ 2x + 3y = -17 \end{cases}$ **3.** $\begin{cases} 3x - 2y = -12 \\ 2x + 2y = 2 \end{cases}$

4. $\begin{cases} 2x + 5y = -7 \\ y = 2x + 1 \end{cases}$ **5.** $\begin{cases} 3x - 2y = 6 \\ y = -2x + 4 \end{cases}$ **6.** $\begin{cases} 4x + 3y = -11 \\ y = -x - 4 \end{cases}$

7. The intersection of the following lines: passing through (1, 1) and (4, 2), and passing through (3, 1) and (0, −2).

8. The intersection of the following lines: passing through (−1, 2) and (2, −1), and passing through (0, 4) and (6, 0).

9. The intersection of the following lines: passing through $(-6, 7)$ and $(8, 1)$, and passing through $(5, 8)$ and $(-2, 1)$.

10. $\begin{cases} y = \frac{3}{7}x + 2 \\ 5x - 7y + 14 = 0 \end{cases}$

11. $\begin{cases} x + 3y = -2 \\ 2x + 6y = 11 \end{cases}$

12. $\begin{cases} 4x + 5y = 2 \\ y = -\frac{4}{5}x + 1 \end{cases}$

13. $\begin{cases} 3x + 2y = 5 \\ 6x - 4y = 2 \end{cases}$

14. $\begin{cases} y = -\frac{2}{3}x + 3 \\ 2x + 3y - 9 = 0 \end{cases}$

15. $\begin{cases} x - 2y = 4 \\ 3x - 6y - 12 = 0 \end{cases}$

16. $\begin{cases} x + 3y = 0 \\ x - 3y = -12 \end{cases}$

17. $\begin{cases} x + y = 2 \\ y = 3x - 14 \end{cases}$

18. $\begin{cases} y = 9 - 2x \\ x + 4y = 8 \end{cases}$

19. $\begin{cases} x - 2y = 5 \\ x + 2y = 7 \end{cases}$

20. $\begin{cases} x = y + 1 \\ x = 2y - 6 \end{cases}$

21. $\begin{cases} y = -5x \\ 2x - 3y = 17 \end{cases}$

22. $\begin{cases} x = 2y + 1 \\ 2x - 5y = 2 \end{cases}$

23. $\begin{cases} 3y = 6 - 2x \\ 5y + 2 = -2x \end{cases}$

24. $\begin{cases} 4x = 4 - 5y \\ 3y = -2x \end{cases}$

25. $\begin{cases} 2x - 3y + 18 = 0 \\ 5x + 3y + 24 = 0 \end{cases}$

26. $\begin{cases} 4x - 6y + 10 = 0 \\ 2x - 3y + 5 = 0 \end{cases}$

27. $\begin{cases} 5x + 2y = 23 \\ 2x + 7y = 34 \end{cases}$

28. $\begin{cases} 2x - 3y - 15 = 0 \\ y = \frac{2}{3} - 5 \end{cases}$

29. $\begin{cases} 3x - 4y = 32 \\ y = \frac{1}{4}x - 6 \end{cases}$

30. $\begin{cases} 2x - 5y = 5 \\ y = \frac{3}{5}x - 2 \end{cases}$

31. If a plane flies 400 mph with the wind, but only 100 mph against the wind, what is the rate of the plane in still air?

32. A boat is able to make 16 knots with the current but only 10 knots against the current. What is the rate of the current?

33. In counting change, you find you have \$24 in quarters and dimes. If there are 147 coins, how many are quarters?

34. Suppose your piggy bank contains \$4.05 and contains dimes, nickels, and quarters. How many dimes do you have if there are twice as many nickels as dimes, the same number of dimes and quarters, and a total of 36 coins?

35. Suppose you want to sell pens to students on your campus. You need to decide on a price and conduct a survey to find that 800 students would buy a pen for \$1, but none would buy one for \$7. You also find a supplier who can supply 200 pens if they are sold for \$1, but will supply 600 if they are sold for \$7. What is the price for which the supply equals the demand?

8.5 Solving Word Problems with Systems

There is no surefire, step-by-step procedure for solving all word problems. However, by systematically approaching word problems, you can gain success. We have looked at three types of word problems in this chapter: supply and demand, combining rates, and money-value applications. In this section we will review these types as well as strengthen your problem-solving ability in general. We will also consider three additional types of application problems: age problems, rate problems, and mixture problems.

Let us begin by reviewing the steps to follow when solving word problems.

1. *Read the problem.* Note what it is all about. Focus on processes rather than numbers. You cannot work a problem you do not understand.
2. *Restate the problem.* Rephrase it as many times and in as many ways as necessary to clarify the facts and relationships. Look for equality. If you cannot find equal quantities, you will never find an equation.
3. *Write the equation* or equations that evolve from the facts of the problem. Do not be concerned with the number of variables or unknowns that you use in this step.
4. *Use substitution* to reduce the number of variables. Reduce to one variable if you have one equation or to two variables if you have two equations.
5. *Solve the equation or equations.* This is the easy step. Be sure they were the right equations by checking the result in the original problem. Your solutions should make sense.
6. *State an answer* to the original problem. There were no variables defined when you started, so $x = 3$ is not the answer. Pay attention to units of measure and other details of the problem.

Example 1. Alan Alda is 6 years older than Barbra Streisand. If the sum of their birth years is 3878, when was Streisand born?

Solution Let $x =$ Alda's birth year
$y =$ Streisand's birth year

The older person has the smaller birth year. Thus,

$$+\begin{cases} y - x = 6 \\ y + x = 3878 \end{cases}$$
$$2y = 3884 \qquad \textit{Solve by using the addition method.}$$
$$y = 1942$$

Barbra Streisand was born in 1942. ■

The second application we will consider in this section involves the rate formula:

$$\text{distance} = (\text{rate})(\text{time}) \quad \text{or} \quad d = rt$$

It is often helpful to draw a figure or a diagram to help you understand the problem.

Example 2. Once upon a time (about 450 B.C.), a Greek named Zeno made up a word problem about a race between Achilles and a tortoise. The tortoise has a 100-meter head start. Achilles runs at a rate of 10 meters per second, whereas the tortoise runs 1 meter per second (it is an extraordinarily swift tortoise). How long does it take Achilles to catch up with the tortoise?

Solution

Step 1 Reread the problem.

Step 2 Write down a verbal description of the problem. We use a diagram to help us.

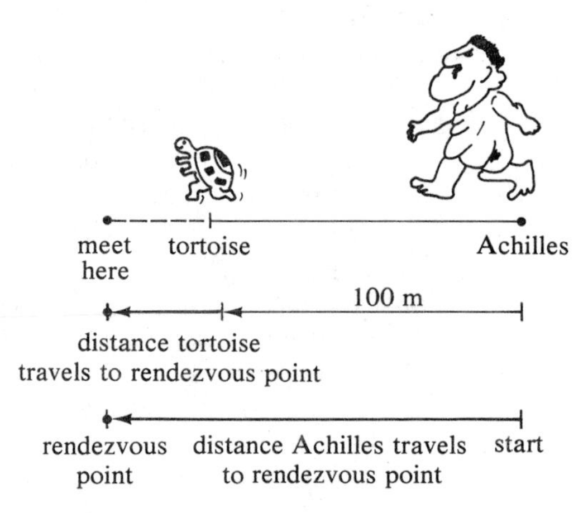

Step 3 Write down an equation.

$$\begin{pmatrix}\text{ACHILLES' DISTANCE}\\ \text{TO RENDEZVOUS}\end{pmatrix} = \begin{pmatrix}\text{TORTOISE'S DISTANCE}\\ \text{TO RENDEZVOUS}\end{pmatrix} + \begin{pmatrix}\text{HEAD}\\ \text{START}\end{pmatrix}$$

Step 4 Substitute.

First, substitute known numbers:

$$\begin{pmatrix}\text{ACHILLES' DISTANCE}\\ \text{TO RENDEZVOUS}\end{pmatrix} = \begin{pmatrix}\text{TORTOISE'S DISTANCE}\\ \text{TO RENDEZVOUS}\end{pmatrix} + \underset{\substack{\uparrow\\ \mathbf{100}}}{\begin{pmatrix}\text{HEAD}\\ \text{START}\end{pmatrix}}$$

Next, substitute using known formulas:

$$\text{ACHILLES' DISTANCE} = \begin{pmatrix}\text{RATE OF}\\ \text{ACHILLES}\end{pmatrix}\begin{pmatrix}\text{TIME TO}\\ \text{RENDEZVOUS}\end{pmatrix}$$

$$\text{TORTOISE'S DISTANCE} = \begin{pmatrix}\text{RATE OF}\\ \text{TORTOISE}\end{pmatrix}\begin{pmatrix}\text{TIME TO}\\ \text{RENDEZVOUS}\end{pmatrix}$$

Thus,

$$\underset{\begin{pmatrix}\textbf{\textit{RATE OF}}\\ \textbf{\textit{ACHILLES}}\end{pmatrix}\begin{pmatrix}\textbf{\textit{TIME TO}}\\ \textbf{\textit{RENDEZVOUS}}\end{pmatrix}}{\underset{\uparrow}{\begin{pmatrix}\text{ACHILLES' DISTANCE}\\ \text{TO RENDEZVOUS}\end{pmatrix}}} = \underset{\begin{pmatrix}\textbf{\textit{RATE OF}}\\ \textbf{\textit{TORTOISE}}\end{pmatrix}\begin{pmatrix}\textbf{\textit{TIME TO}}\\ \textbf{\textit{RENDEZVOUS}}\end{pmatrix}}{\underset{\uparrow}{\begin{pmatrix}\text{TORTOISE'S DISTANCE}\\ \text{TO RENDEZVOUS}\end{pmatrix}}} + 100$$

Now the equation has new variables, so see if any of these are known numbers; if so, substitute again:

$$\underset{\substack{\uparrow\\ \mathbf{10}}}{\begin{pmatrix}\text{RATE OF}\\ \text{ACHILLES}\end{pmatrix}}\begin{pmatrix}\text{TIME TO}\\ \text{RENDEZVOUS}\end{pmatrix} = \underset{\substack{\uparrow\\ \mathbf{1}}}{\begin{pmatrix}\text{RATE OF}\\ \text{TORTOISE}\end{pmatrix}}\begin{pmatrix}\text{TIME TO}\\ \text{RENDEZVOUS}\end{pmatrix} + 100$$

The equation now has a single unknown:

$$10\begin{pmatrix}\text{TIME TO}\\ \text{RENDEZVOUS}\end{pmatrix} = \begin{pmatrix}\text{TIME TO}\\ \text{RENDEZVOUS}\end{pmatrix} + 100$$

Let t = TIME TO RENDEZVOUS.

$$10t = t + 100$$

Step 5 Solve the equation.

$$\begin{aligned} 10t &= t + 100\\ 9t &= 100\\ t &= \frac{100}{9}\\ &= 11\frac{1}{9} \end{aligned}$$

Step 6 ***It takes Achilles $11\frac{1}{9}$ seconds to catch up with the tortoise.*** ■

Example 3. On a business trip a businesswoman logs an hour and a half in a jet airliner and four hours in a rent-a-car to reach her destination. If the total trip is a distance of 1100 miles, what is the average speed of each of the means of transportation if the jet travels twelve times as fast as the car?

Solution

$$\begin{cases} \begin{pmatrix}\text{DISTANCE}\\ \text{BY PLANE}\end{pmatrix} + \begin{pmatrix}\text{DISTANCE}\\ \text{BY CAR}\end{pmatrix} = \underset{\substack{\uparrow\\ \mathbf{1100}}}{\begin{pmatrix}\text{TOTAL}\\ \text{DISTANCE}\end{pmatrix}}\\[2ex] \begin{pmatrix}\text{PLANE}\\ \text{SPEED}\end{pmatrix} = 12\begin{pmatrix}\text{CAR}\\ \text{SPEED}\end{pmatrix} \end{cases}$$

Now, $\text{DISTANCE BY PLANE} = \begin{pmatrix}\text{PLANE}\\ \text{SPEED}\end{pmatrix}\begin{pmatrix}\text{PLANE}\\ \text{TIME}\end{pmatrix}$

$$\text{DISTANCE BY CAR} = \begin{pmatrix}\text{CAR}\\ \text{SPEED}\end{pmatrix}\begin{pmatrix}\text{CAR}\\ \text{TIME}\end{pmatrix}$$

Substitute these formulas into the first of the two equations:

$$\begin{cases} \begin{pmatrix}\text{DISTANCE}\\ \text{BY PLANE}\end{pmatrix} + \begin{pmatrix}\text{DISTANCE}\\ \text{BY CAR}\end{pmatrix} = 1100 \\ \quad\uparrow \qquad\qquad\qquad\qquad \uparrow \\ \begin{pmatrix}\textit{PLANE}\\ \textit{SPEED}\end{pmatrix}\begin{pmatrix}\textit{PLANE}\\ \textit{TIME}\end{pmatrix} \qquad \begin{pmatrix}\textit{CAR}\\ \textit{SPEED}\end{pmatrix}\begin{pmatrix}\textit{CAR}\\ \textit{TIME}\end{pmatrix} \\ \begin{pmatrix}\text{PLANE}\\ \text{SPEED}\end{pmatrix} = 12\begin{pmatrix}\text{CAR}\\ \text{SPEED}\end{pmatrix} \end{cases}$$

Finally, substitute known numbers to new variables:

$$\begin{cases} \begin{pmatrix}\text{PLANE}\\ \text{SPEED}\end{pmatrix}\underset{\substack{\uparrow\\ \mathbf{1.5}}}{\begin{pmatrix}\text{PLANE}\\ \text{TIME}\end{pmatrix}} + \begin{pmatrix}\text{CAR}\\ \text{SPEED}\end{pmatrix}\underset{\substack{\uparrow\\ \mathbf{4}}}{\begin{pmatrix}\text{CAR}\\ \text{TIME}\end{pmatrix}} = 1100 \\ \begin{pmatrix}\text{PLANE}\\ \text{SPEED}\end{pmatrix} = 12\begin{pmatrix}\text{CAR}\\ \text{SPEED}\end{pmatrix} \end{cases}$$

This gives a system with two equations and two unknowns:

$$\begin{cases} 1.5(\text{PLANE SPEED}) + 4(\text{CAR SPEED}) = 1100 \\ \text{PLANE SPEED} = 12(\text{CAR SPEED}) \end{cases}$$

Let $P =$ PLANE SPEED and $C =$ CAR SPEED.

$$\mathbf{-1.5}\begin{cases} 1.5P + 4C = 1100 \\ P - 12C = 0 \end{cases}$$

$$+\begin{cases} 1.5P + 4C = 1100 \\ -1.5P + 18C = 0 \end{cases}$$

$$22C = 1100 \qquad P = 12(50)$$
$$C = 50 \qquad P = 600$$

The plane averages 600 mph, and the car averages 50 mph. ■

The third application of this section is called a ***mixture problem.***

Example 4. How many liters of water must be added to 3 liters of an 80% acid solution (80% acid and 20% water by volume) to obtain a solution that is 30% acid?

Solution Just as with money-value problems, there are two amounts you need to distinguish:

AMOUNT OF SOLUTION
AMOUNT OF ACID

These amounts are related by the following formula:

$$\begin{pmatrix}\text{AMOUNT}\\ \text{OF}\\ \text{ACID}\end{pmatrix} = \begin{pmatrix}\text{PERCENT}\\ \text{ACID}\end{pmatrix} \times \begin{pmatrix}\text{AMOUNT}\\ \text{OF}\\ \text{SOLUTION}\end{pmatrix}$$

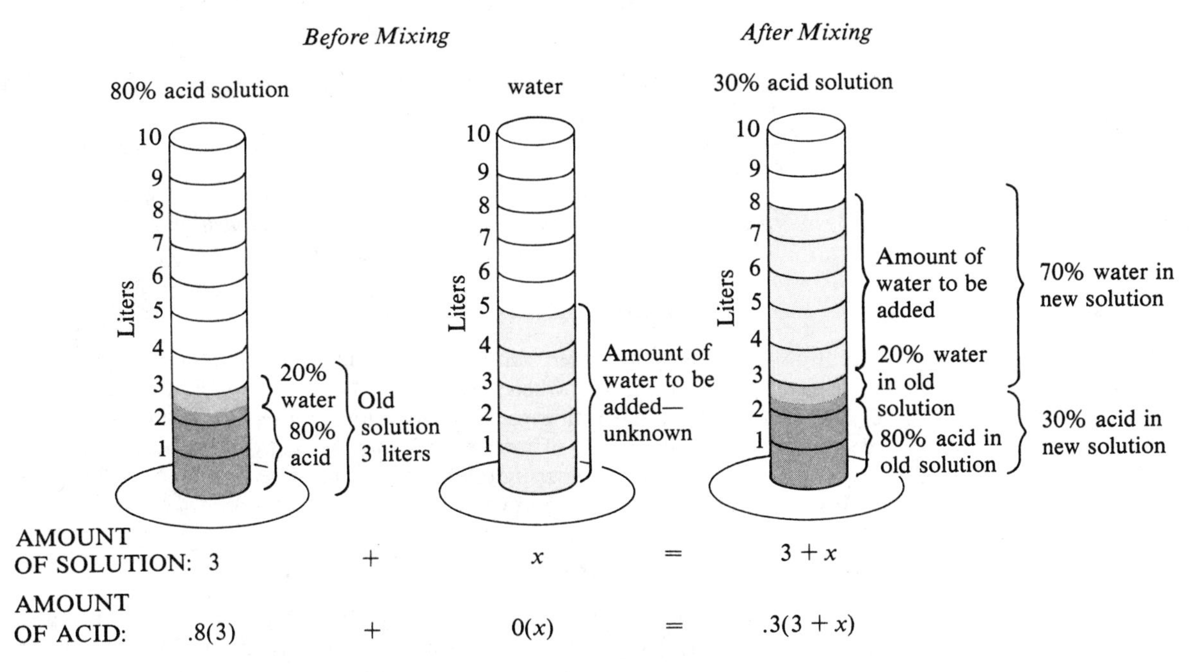

Let x = AMOUNT OF WATER TO BE ADDED

Since there is only one unknown in this problem, we can use the second equation only:

$$2.4 = .3(3 + x)$$
$$2.4 = .9 + .3x$$
$$1.5 = .3x$$
$$5 = x$$

The amount of water that should be added is 5 liters. ■

Example 5. Milk containing 10% butterfat and cream containing 80% butterfat are mixed to produce half-and-half, which is 50% butterfat. How many gallons of each must be mixed to make 140 gallons of half-and-half?

Solution Let m = AMOUNT OF MILK

c = AMOUNT OF CREAM

	Before Mixing		*After Mixing*
AMOUNT OF SOLUTION:	$m + c$	$=$	140
AMOUNT OF MIXTURE:	$.1m + .8c$	$=$	$.5(140)$

In this problem there are two unknowns so we use both equations:

$$-10\begin{cases} m + c = 140 \\ .1m + .8c = 70 \end{cases}$$

$$+\begin{cases} m + c = 140 \\ -m - 8c = -700 \end{cases}$$

$$-7c = -560$$

$$c = 80 \qquad \text{If } c = 80\text{, then } m + \mathbf{80} = 140$$

$$m = 60$$

You must add 60 gallons of milk and 80 gallons of cream. ■

Problem Set 8.5

Carefully interpet each problem, restate the information and relationships until you are able to write the equations, solve the equations, and state an answer.

Problems 1–3 are patterned after Example 1.

1. Mel Brooks is seven years older than fellow funnyman Dom DeLuise. If the sum of their birth years is 3859, then when was DeLuise born?
2. Charles Bronson was born nine years before another movie hard-guy, Clint Eastwood. If the sum of their years of birth is 3853, then in what year was Eastwood born?
3. Ray Charles is fifteen years older than fellow musician Jose Feliciano. If the sum of their years of birth is 3875, then in what year was Feliciano born?

Problems 4–6 are patterned after Example 2.

4. Two persons are to run a race, but one can run 10 meters per second, whereas the other can run 6 meters per second. If the slower runner has a 50-meter head start, how long will it be before the faster runner catches the slower runner if they begin at the same time?
5. If the rangefinder on Starbuck's *Viper* shows the enemy Cylon 4500 km away, how long will it take to catch up to the Cylon if Starbuck's *Viper* travels at 15,000 kph and the Cylon travels at 12,000 kph?
6. If a speeding car is traveling at 80 mph when a police car starts pursuit at 100 mph, how long will it take the police car to catch up to the speeding car? Assume that the speeding car has a 2-mile head start and that the cars travel at constant rates.

Problems 7–9 are patterned after Example 3.

7. A commuter rides a train into the city and catches a taxicab to the office, traveling 55 miles to work each day. The train averages 45 mph and the taxi 20 mph. How much time does the commuter spend in a cab each morning if the total trip to work averages an hour and a half?
8. Rich is hitchhiking back to school from his home, which is 79 miles away. He makes 4 mph walking, until he gets a ride, and 46 mph riding. If this trip takes 4 hours, how long does Rich walk before getting a ride to his dorm?
9. A plane averages 180 mph on the first leg of its flight but its speed is cut to 140 mph on the second leg. If the total trip of 1100 miles takes 7 hours, then what is the distance on each leg?

Problems 10–12 are patterned after Example 4.

10. How many ounces of a base metal must be alloyed with 100 ounces of 21% alloy to obtain an alloy that is 15% silver?

11. An after-shave lotion is 50% alcohol. If you have 6 fluid ounces of the lotion, how much water must be added to reduce the mixture to 20% alcohol?

12. How much pure silver must be alloyed with a 24% silver alloy to obtain 100 ounces of 43% silver?

Problems 13–15 are patterned after Example 5.

13. Milk containing 20% butterfat and cream containing 60% butterfat are mixed to produce half-and-half, which is 50% butterfat. How many gallons of each must be mixed to make 180 gallons of half-and-half?

14. A chemist has two solutions of sulfuric acid. One is a 50% solution, and the other is a 75% solution. How many liters of each does the chemist mix to get 10 liters of a 60% solution?

15. A dairy has cream containing 23% butterfat and milk that is 3% butterfat. How much of each must be mixed to obtain 30 gallons of a richer milk containing 4% butterfat?

Problems 16–33 provide a mixture of types. Each problem should match up with some example in this chapter.

16. The combined area of New York and California is 204,192 square miles. The area of California is 108,530 square miles more than that of New York. Find the land area of each state.

17. The area of Texas is 208,044 square miles greater than that of Florida. Their combined area is 316,224 square miles. What is the area of each state?

18. The combined height of the World Trade Center and the Empire State Building is 2600 ft. The Trade Center is 100 ft taller. What is the height of each of these New York City skyscrapers?

19. An airplane takes off one day after a ship sails, yet both arrive at their common destination simultaneously. If the ship travels at 20 mph and the plane at 260 mph, how long is the plane's flight?

20. Two joggers set out at the same time from their homes 18 miles apart. They agree to meet at a point in between in two hours. If the rate of one is 1 mile per hour faster than the other, find the rate of each.

21. The Standard Oil and the Sears buildings have a combined height of 2590 ft. The Sears Building is 318 ft taller. What is the height of each of these Chicago towers?

22. *Fiddler on the Roof* and *Man of La Mancha* ran for a total of 5570 performances. *Fiddler,* the all-time longest running play in Broadway history, ran 914 performances longer than *La Mancha.* How many times did *Fiddler* play on Broadway?

23. *Hello Dolly!* and *My Fair Lady* are fourth and fifth on the all-time Broadway list with a total of 5561 performances. If *Dolly* ran 127 performances longer, then how many times was *My Fair Lady* performed?

24. The combined length of the Golden Gate and San Francisco Bay bridges is 6510 ft. If the Golden Gate is 1890 ft longer, what is its length?

25. End-to-end the Verrazano-Narrows and the George Washington bridges would span 7760 ft. If the Verrazano-Narrows is the longer of the two New York structures by 760 ft, how long is it?

26. How many gallons of 24% butterfat cream must be mixed with 500 gallons of 3% butterfat milk to obtain a 4% butterfat milk?

27. The radiator of a car holds 17 quarts of water. If it now contains 15% antifreeze, how many quarts must be replaced by antifreeze to give the car a 60% solution in its radiator?
28. Sterling silver conains 92.5% pure silver. How many grams of pure silver and sterling silver must be mixed to get 100 grams of a 94% alloy?
29. Noxin Electronics has investiated the feasibility of introducing a new line of magnetic tape. The study shows that both supply and demand are linear. The supply can increase from 1000 items at \$2 each to 5000 units at \$4 each. The demand ranges from 1000 items at \$4 to 7000 at \$3. What is the equilibrium point of this supply-and-demand system?
30. Noxin (Problem 29) is also considering producting a small cassette line. Research shows a linear demand for from 40,000 cassettes at \$2 to 100,000 at \$1. Similarly, the supply goes from 20,000 at 50¢ to 80,000 at \$5. What is the equilibrium point?

8.6 Solving Systems of Inequalities by Graphing

Just as we solved systems of equations, we can solve ***systems of linear inequalities.*** You might wish to review Section 7.4 because, for this section, you will need to know how to graph a linear inequality. Graphically, the solution of a system of two linear equations is the intersection of the lines. For inequalities, it is also the intersection of the individual solution sets as shown in Example 1.

Example 1. Graph the solution of

$$\begin{cases} x + y \leq 2 \\ 2x - y > 10 \end{cases}$$

Solution Graph $x + y \leq 2$; the boundary is $y = -x + 2$ and is graphed as shown in Figure 8.2a.

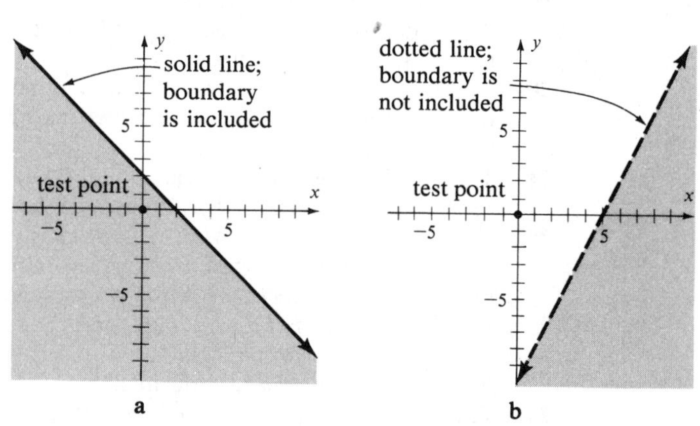

Figure 8.2

Next, graph $2x - y > 10$; the boundary is $y = 2x - 10$ and is graphed as shown in Figure 8.2b. Of course, in your work you will show both of these on the same coordinate axes because you are interested in the intersection as shown in Figure 8.3a.

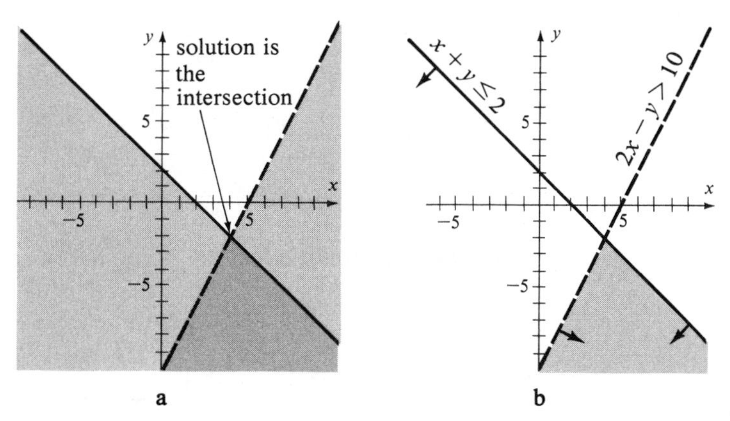

Figure 8.3

In fact, in your work you will probably not completely shade in the individual parts as shown in Figure 8.3a, but only the intersection as shown in Figure 8.3b. Notice the little arrows on the boundaries to remind you of the half-planes being considered. ■

With systems of linear equations we were limited to two equations, but we can consider the intersection of more than two linear inequalities.

Example **2.** Graph the solution of the following system.

$$\begin{cases} x + y \leq 6 \\ 3x + y \leq 12 \\ x > 0 \\ y > 0 \end{cases}$$

Solution *Step 1* $x > 0$ and $y > 0$ is the first quadrant—that is, everything to the right of the y-axis and above the x-axis (see Figure 8.4a).

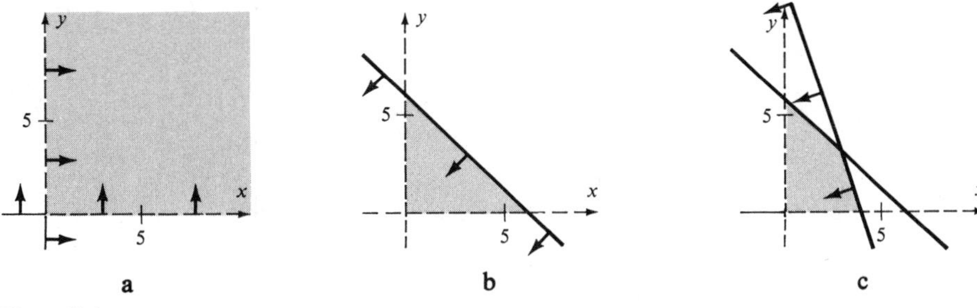

Figure 8.4

Step 2 Find the intersection of $x + y \leq 6$ and the region shown in Figure 8.4a (see Figure 8.4b).

Step 3 Find the intersection of $3x + y \leq 12$ and the region shown in Figure 8.4b (see Figue 8.4c). ■

Example 3. Graph the solution of

$$\begin{cases} y \le \frac{1}{2}x + 3 \\ 2 < x < 6 \\ y > 1 \end{cases}$$

Solution Apply the inequalities one at a time to obtain the region specified as shown in Figure 8.5.

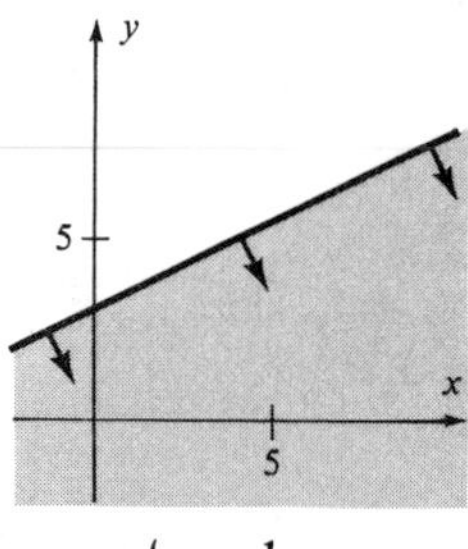

a. $\left\{ y \le \frac{1}{2}x + 3 \right.$

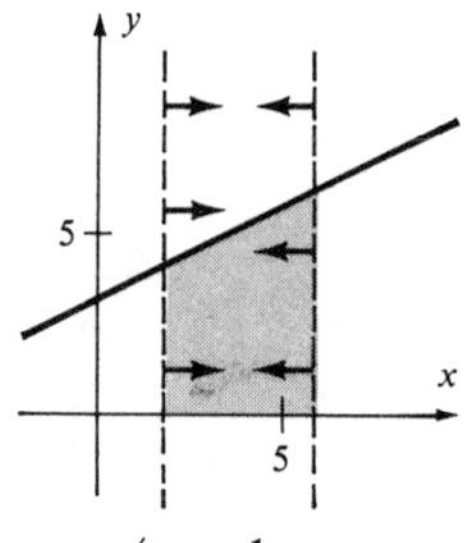

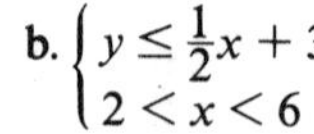

b. $\begin{cases} y \le \frac{1}{2}x + 3 \\ 2 < x < 6 \end{cases}$

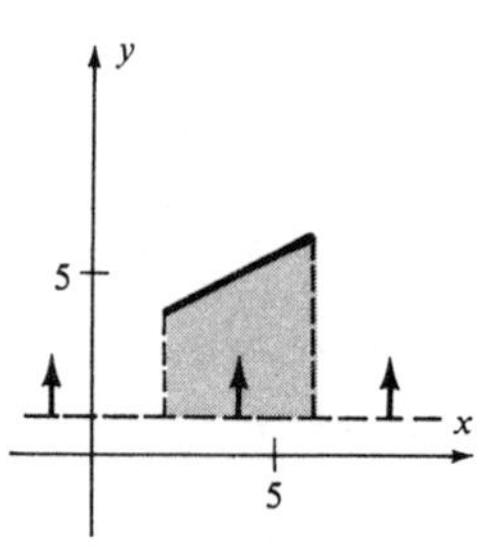

c. $\begin{cases} y \le \frac{1}{2}x + 3 \\ 2 < x < 6 \\ y > 1 \end{cases}$ ■

Figure 8.5

Problem Set 8.6

For Problems 1–29 sketch the solution to the system of inequalities given. See Examples 1–3.

1. $\begin{cases} y \ge x + 2 \\ y \le 2x + 5 \end{cases}$

2. $\begin{cases} y \le 3x + 4 \\ y \ge -2x - 1 \end{cases}$

3. $\begin{cases} y \le -2x + 1 \\ y \le 3x + 7 \end{cases}$

4. $\begin{cases} 3x + y < 4 \\ x + y > 3 \end{cases}$

5. $\begin{cases} x - y < 5 \\ y + 2x > -3 \end{cases}$

6. $\begin{cases} 2x - 3y < 6 \\ x + 2y \le -4 \end{cases}$

7. $\begin{cases} 0 \le x \le 5 \\ 0 \le y \le 3 \end{cases}$

8. $\begin{cases} -3 \le x \le 0 \\ 0 \le y \le 5 \end{cases}$

9. $\begin{cases} 2 \le x \le 5 \\ -3 \le y \le 4 \end{cases}$

10. $\begin{cases} y \le x \\ x \le 6 \\ y \ge 0 \end{cases}$

11. $\begin{cases} y > x \\ x > 0 \\ y < 8 \end{cases}$

12. $\begin{cases} y < 6 - x \\ x > 0 \\ y > 0 \end{cases}$

13. $\begin{cases} y \le 8 - x \\ x \ge 0 \\ y \ge 0 \end{cases}$

14. $\begin{cases} y \le x + 6 \\ x \le 0 \\ y \ge 0 \end{cases}$

15. $\begin{cases} y > x - 8 \\ x > 0 \\ y < 0 \end{cases}$

16. $\begin{cases} y > 2x - 6 \\ y > 2 \\ y < 6 \end{cases}$

17. $\begin{cases} y \le 2x + 8 \\ y > 2 \\ y \le 7 \end{cases}$

18. $\begin{cases} y \ge \frac{1}{2}x + 1 \\ 2 \le x \le 6 \\ y \le 6 \end{cases}$

19. $\begin{cases} y > \frac{1}{3}x + 2 \\ 3 < x < 9 \\ y < 7 \end{cases}$

20. $\begin{cases} y \geq \frac{2}{5}x + 2 \\ -2 < x \leq 5 \\ y < 5 \end{cases}$

21. $\begin{cases} y \leq -\frac{3}{5}x + 6 \\ -1 < x \leq 6 \\ y > 1 \end{cases}$

22. $\begin{cases} x \geq 0 \\ y \geq 0 \\ 2x + 3y \leq 15 \\ 3x + y \leq 12 \end{cases}$

23. $\begin{cases} x \geq 0 \\ y \geq 0 \\ 2x + 3y \leq 12 \\ 2x - 3y \geq -6 \end{cases}$

24. $\begin{cases} y < -x - 6 \\ -7 \leq x \leq -5 \\ y \geq -7 \end{cases}$

25. $\begin{cases} y \geq -x - 6 \\ -1 \leq y \leq 1 \\ x \leq -3 \end{cases}$

26. $\begin{cases} y \geq \frac{2}{3}x \\ y \leq -\frac{2}{3}x \\ x \geq -3 \end{cases}$

27. $\begin{cases} x \geq 0 \\ x + y \leq 10 \\ x - y \leq -2 \end{cases}$

28. $\begin{cases} y - 7 \leq 0 \\ 7x + 4y \geq 28 \\ 7x - 4y \leq 28 \end{cases}$

29. $\begin{cases} y \geq -4 \\ 4y \leq 7x + 8 \\ 4y \leq 56 - 7x \end{cases}$

30. Write a system of inequalities whose graph is the indicated quadrant.

a. Quadrant I

b. Quadrant II

c. Quadrant III

d. Quadrant IV

8.7 Review

In this chapter you have extended the solution of linear equations and inequalities to problems with two variables. The graphic methods of the previous chapter are employed to find solutions to these systems, and two other methods, addition and substitution, are developed for equations. The ability to work with two variables simplifies somewhat the solution of word problems, because the facts of the problem need not be evolved to a single variable.

Self-Test

[8.1] **1.** Solve by graphing.

a. $\begin{cases} y = 2x - 10 \\ x + 2y = 5 \end{cases}$

b. $\begin{cases} x + y = 14 \\ y = x + 4 \end{cases}$

[8.2] **2.** Solve by adding.

a. $\begin{cases} 2x - 5y = 9 \\ 3x + 4y = 2 \end{cases}$

b. $\begin{cases} 8x - 18y = 7 \\ 2y = 3 - x \end{cases}$

[8.3] **3.** Solve by substitution.

a. $\begin{cases} y = 7 - 2x \\ 4x + 7y + 1 = 0 \end{cases}$

b. $\begin{cases} 2x + 2y = 10 \\ y - x = 1 \end{cases}$

[8.4] *Solve Problems 4 and 5 using any method, but solve at least one by graphing, one by adding, and one by substitution.*

4. **a.** $\begin{cases} y = 4x - 5 \\ 5x - 4y + 2 = 0 \end{cases}$ **b.** $\begin{cases} 2x + 5y - 3 = 0 \\ y = -\frac{2}{5}x + 3 \end{cases}$

5. **a.** $\begin{cases} y = 2x - 7 \\ 2x - y - 2 = 0 \end{cases}$ **b.** $\begin{cases} 3x - 2y = 6 \\ 2x + 4y = 4 \end{cases}$

[8.5] **6.** Merrywell is considering producing a small, inexpensive cassette line. Research shows a linear demand for from 40,000 cassettes at $2 to 100,00 at $1. Similarly, the supply goes from 20,000 at 50¢ to 80,000 at $5. What is the equilibrium point for this supply-and-demand system?

[8.5] **7.** A boat is able to make 33 knots with the current. But, going upriver, its speed is cut to 17 knots. What is the speed of the river's current?

[8.5] **8.** Holiday Park admitted a total of 1017 people over the weekend. Adult tickets are $2.25 and children's are a dollar. How many children were there if the total receipts were $1392?

[8.5] **9.** Two concentrations of an acid are available, 12.5% and 65%. How much of each must be mixed to form 60 ml of a 30% acid concentration?

[8.6] **10.** Sketch the solution of each system.

a. $\begin{cases} 1 \le x \le 5 \\ 1 \le y \le 5 \\ 2x + 3y < 18 \end{cases}$ **b.** $\begin{cases} 3y - x > 15 \\ 2x + 3y < 15 \\ 3y - 4x < 30 \end{cases}$

Review Problems

Solve the systems of equations in Problems 1–50.

1. $\begin{cases} x + 2y = 13 \\ 3x - 2y = -1 \end{cases}$ **2.** $\begin{cases} x + 3y = 11 \\ 2x - 3y = -5 \end{cases}$ **3.** $\begin{cases} x + 2y = 0 \\ -x + 3y = 5 \end{cases}$

4. $\begin{cases} x + y = -1 \\ -x + 3y = -7 \end{cases}$ **5.** $\begin{cases} 3x - 4y = -5 \\ 2x + 4y = -10 \end{cases}$ **6.** $\begin{cases} 2x - 5y = -9 \\ 3x + 5y = -1 \end{cases}$

7. $\begin{cases} 3x + 5y = 1 \\ -3x + 2y = -8 \end{cases}$ **8.** $\begin{cases} 2x + 3y = 0 \\ -2x + 5y = -16 \end{cases}$ **9.** $\begin{cases} 2x - 3y = 1 \\ -6x + 3y = 3 \end{cases}$

10. $\begin{cases} 4x - 5y = -10 \\ -6y + 5y = 10 \end{cases}$ **11.** $\begin{cases} -3x + 8y = -18 \\ 3x - 10y = 24 \end{cases}$ **12.** $\begin{cases} -4x + 5y = 2 \\ 4x - 7y = -6 \end{cases}$

13. $\begin{cases} x + 2y = 6 \\ 2x + 5y = 15 \end{cases}$ **14.** $\begin{cases} 3x + 2y = 0 \\ x - 3y = 22 \end{cases}$ **15.** $\begin{cases} 3x + y = -6 \\ 5x - 2y = -10 \end{cases}$

16. $\begin{cases} 3x - 2y = 1 \\ 5x + y = 19 \end{cases}$ **17.** $\begin{cases} 2x + 3y = 11 \\ 6x - 6y = -12 \end{cases}$ **18.** $\begin{cases} 2x - 5y = 9 \\ 6x + 10y = 2 \end{cases}$

19. $\begin{cases} 3x + 4y = 19 \\ x - y = 4 \end{cases}$ **20.** $\begin{cases} 2x + 5y = 7 \\ x - y = 7 \end{cases}$ **21.** $\begin{cases} 2x - 3y = -13 \\ x + y = -4 \end{cases}$

22. $\begin{cases} 3x - 2y = -8 \\ x + y = -6 \end{cases}$ **23.** $\begin{cases} 3x - 2y = 0 \\ x + y = -5 \end{cases}$ **24.** $\begin{cases} 3x - y = 5 \\ x + 2y = -3 \end{cases}$

25. $\begin{cases} y = 3x + 5 \\ x + y = 9 \end{cases}$ **26.** $\begin{cases} y = 2x + 1 \\ x + y = 7 \end{cases}$ **27.** $\begin{cases} y = 2x + 8 \\ 2x + y = 0 \end{cases}$

28. $\begin{cases} y = 3x - 6 \\ 3x + y = 0 \end{cases}$ **29.** $\begin{cases} y = 2x + 7 \\ 5x - y = 2 \end{cases}$ **30.** $\begin{cases} y = 3x + 2 \\ 4x - y = -1 \end{cases}$

31. $\begin{cases} y = 3x + 1 \\ 4x - y = 0 \end{cases}$

32. $\begin{cases} y = 2x + 2 \\ 3x - y = -1 \end{cases}$

33. $\begin{cases} y = -2x \\ 3x + 2y = 1 \end{cases}$

34. $\begin{cases} y = -3x - 6 \\ 3x - 2y = 3 \end{cases}$

35. $\begin{cases} y = 5x + 11 \\ 2x - 3y = -7 \end{cases}$

36. $\begin{cases} y = 4x - 3 \\ 2x - 3y = -1 \end{cases}$

37. $\begin{cases} y = 5x + 1 \\ y = 2x + 4 \end{cases}$

38. $\begin{cases} y = 2x + 5 \\ y = 3x + 6 \end{cases}$

39. $\begin{cases} y = -x - 1 \\ y = 2x - 7 \end{cases}$

40. $\begin{cases} y = -x + 2 \\ y = 3x - 14 \end{cases}$

41. $\begin{cases} y = x \\ y = -3x + 24 \end{cases}$

42. $\begin{cases} y = x \\ y = 2x - 5 \end{cases}$

43. $\begin{cases} 3x - 2y = 0 \\ 5x - 3y - 1 = 0 \end{cases}$

44. $\begin{cases} 5x + 2y + 10 = 0 \\ 4x + 3y - 1 = 0 \end{cases}$

45. $\begin{cases} 2x - 3y + 6 = 0 \\ 3x - 2y + 1 = 0 \end{cases}$

46. $\begin{cases} 3x + 2y - 6 = 0 \\ 5x + 3y - 11 = 0 \end{cases}$

47. $\begin{cases} 4x - 3y - 5 = 0 \\ 3x + 2y - 8 = 0 \end{cases}$

48. $\begin{cases} 3x - 4y + 7 = 0 \\ 2x - 5y + 14 = 0 \end{cases}$

49. $\begin{cases} 5x - 2y + 40 = 0 \\ 3x + 3y + 3 = 0 \end{cases}$

50. $\begin{cases} 3x + 4y = 0 \\ 4x + 5y + 1 = 0 \end{cases}$

9

Square Roots

Everyone is ignorant,
only about different things.

Will Rogers

We boil at different degrees.

Ralph Waldo Emerson

9.1 Real Numbers

Find a number whose square is 2. Notice that $1^2 = 1$ and $2^2 = 4$, so the number whose square is 2 is somewhere between 1 and 2. You can try 1.5, but $1.5^2 = 2.25$, which is too big. By continuing to guess, you might develop the following sequence.

$$\begin{aligned} 1.4^2 &= 1.96 \\ 1.41^2 &= 1.9881 \\ 1.414^2 &= 1.999396 \\ 1.4142^2 &= 1.99996164 \\ 1.41421^2 &= 1.999989924 \\ 1.414213^2 &= 1.999998409 \\ 1.414214^2 &= 2.000001238 \end{aligned}$$

The number for which you are searching is somewhere between 1.414213 and 1.414214, and that number is the square root of 2. It does not have a decimal representation that terminates or repeats so it is written $\sqrt{2}$.

Definition of Square Root

For nonnegative numbers a

$$\sqrt{a} = b \text{ only if } b^2 = a.$$

The symbol $\sqrt{\ }$ is called a ***radical.*** The number under the radical is called the ***radicand.***

These nonrational numbers have many applications. Consider an historical application from ancient Egypt. In the time of the Pharaohs and pyramid building, a simple device was used to aid surveying. Their device was a rope with 12 equal divisions marked by knots (see Figure 9.1). When the rope was stretched

Figure 9.1

and staked so that a triangle was formed with sides 3, 4, and 5, the angle formed by the shorter sides was a right angle, 90° as shown in Figure 9.2, thus allowing the surveyors to make square corners on the land. This method was extremely useful

Figure 9.2

in Egypt, where the Nile flooded the rich lands close to the river each year. The lands had to be resurveyed when the waters subsided. This method used the ***Pythagorean Theorem,*** which is stated in the following box.

Pythagorean Theorem

For a right triangle ABC, with sides a, b, and c,

$$a^2 + b^2 = c^2$$

where c is the side opposite the right angle and is called the ***hypotenuse.***

The knotted rope formed a right triangle because $3^2 + 4^2 = 5^2$ or $9 + 16 = 25$. Consider a seemingly simpler case. Let the two shorter sides be one unit long. How long must the hypotenuse be? If we use the Pythagorean Theorem, $1^2 + 1^2 = c^2$, and $2 = c^2$. That side, c, must be a number that when squared is 2. We are back to $\sqrt{2}$. We know its value approximately, so we use the symbol $\approx$ instead of $=$. Some other examples are also listed.

$$\begin{aligned} \sqrt{2} &\approx 1.414214 & \sqrt{6} &\approx 2.449490 \\ \sqrt{3} &\approx 1.732051 & \sqrt{7} &\approx 2.645751 \\ \sqrt{4} &= 2.000000 & \sqrt{8} &\approx 2.828427 \\ \sqrt{5} &\approx 2.236068 & \sqrt{9} &= 3.000000 \end{aligned}$$

As the number increases, its square root correspondingly increases, and numbers in between natural numbers show similar behavior.

$$\begin{aligned} \sqrt{6.25} &= 2.50000 \\ \sqrt{6.3} &\approx 2.509980 \end{aligned}$$

Some numbers in the preceding list are rational, namely $\sqrt{4}$ and $\sqrt{9}$ and $\sqrt{6.25}$; the others are not. Numbers that are not rational are called ***irrational.***

In Examples 1–2, determine if the given numbers represent the sides of a right triangle.

Example 1. 16, 30, 34 Using the Pythagorean Theorem,

$$\begin{aligned} 16^2 + 30^2 &\stackrel{?}{=} 34^2 \\ 256 + 900 &\stackrel{?}{=} 1156 \\ 1156 &= 1156 \end{aligned}$$

Thus, ***16, 30, and 34 are the sides of a right triangle.*** ■

Example **2.** 47, 56, 73 Using the Pythagorean Theorem,

$$47^2 + 56^2 \stackrel{?}{=} 73^2$$
$$2209 + 3136 \stackrel{?}{=} 5329$$
$$5345 \neq 5329$$

Thus ***47, 56, and 73 are not sides of a right triangle.*** ■

One square root of 16 is 4, since

$$4^2 = 16$$

but another square root of 16 is -4, since

$$(-4)^2 = 16$$

Every positive number has *two* square roots. The symbol we use to represent one of them is the radical symbol $\sqrt{a}$. This expression is called *the* ***square root*** *of a* and denotes the *positive* square root of the number a. For example,

$$\sqrt{16} = 4, \textit{ not } -4$$
$$-\sqrt{16} = -(\sqrt{16}) = -(4) = -4$$

Real Numbers

> The set of ***real numbers*** is the set containing all the rationals and all the irrationals, and it is denoted by the letter R.

- R Real Numbers
  - Q Rationals
    - I Integers
      - N Natural Numbers
      - Zero
      - Opposites of N
    - Fractions
  - Irrationals

An irrational number you may be familiar with is usually represented by the Greek letter π; π is found in many common formulas—usually involving a circle. (See Section 1.4.) This important number is the ratio of the circumference of a circle to its diameter. That is, if you had a circle with a diameter of 1 and you rolled your circle one revolution, it would go π units. An old wagon wheel, for example, stands about 1 meter tall. If we took a band of iron and strapped it around the wheel, the band would have to be π meters long. The real number π is about 3.141592654—like $\sqrt{2}$, its decimal repesentation does not terminate and does not repeat—it is an irrational number. You may have used $3\frac{1}{7}$ or $\frac{22}{7}$ for π. This approximation is only a rough estimate that has the advantage of a simple fraction form.

$$\pi = 3.141592654\ldots$$

$$\frac{22}{7} = 3.142857142857\overline{142857}$$

The bar over 142857 is used to indicate the part that repeats. If zero is the repeating part, the bar is usually left off.

Notice that they agree in only two decimal places and thus represent different numbers. But more importantly the approximation is rational, and π is irrational. Recall that to find the decimal representation of a fraction, you simply perform the indicated division.

Example 3. Express the fraction $\frac{61}{495}$ in decimal form.

```
        .12323···
495)61.000000
    49 5
    11 50   <——
     9 90
     1 600
     1 485
      1150  <——
       990
      1600
      1485
       1150 <——
```

Once the remainder is repeated, the division will repeat.

$$\frac{61}{495} = .1\overline{23} = .12323\overline{23}$$

■

Consider a number of common fractions and their decimal form.

$\frac{1}{2} = .5\overline{0} = .5$

$\frac{1}{3} = .\overline{3} = .333333333\overline{3}$

$\frac{1}{4} = .25\overline{0} = .25$

$\frac{1}{5} = .2\overline{0} = .2$

$\frac{1}{6} = .1\overline{6} = .1666666666\overline{6}$

$\frac{1}{7} = .\overline{142857}$

$\frac{1}{8} = .125\overline{0} = .125$

$\frac{1}{9} = .\overline{1} = .111111111\overline{1}$

$\frac{1}{10} = .1\overline{0} = .1$

$\frac{1}{11} = .\overline{09} = .09090909\overline{09}$

$\frac{1}{12} = .08\overline{3} = .08333333\overline{3}$

$\frac{1}{13} = .\overline{076923}$

All of the fractions repeat some part of their decimal representation.

Characterization of Rational Numbers

> Every rational number has a repeating decimal representation. Irrational numbers do *not* have repeating decimal representations.

So, we have a reasonable way to differentiate between a rational and an irrational number. This should provide some feeling for the numbers of our number system. The set of real numbers emerges as a collection of subsets that have expanded to allow for more operations with the numbers.

Example 4. Arrange the following numbers on the number line from smallest to largest.

$$\frac{2}{3}, \frac{1}{\sqrt{2}}, \frac{3}{5}, \frac{\pi}{5}, 0.66$$

Solution First obtain decimal approximations. (A calculator would be helpful.)

$$\frac{2}{3} \approx 0.67 \qquad \frac{1}{\sqrt{2}} \approx \frac{1}{1.414} \approx 0.71$$

$$\frac{3}{5} \approx 0.60 \qquad \frac{\pi}{5} \approx \frac{3.1416}{5} \approx 0.63$$

Then arrange in order: 0.60, 0.63, 0.66, 0.67, 0.71, and choose a scale so that the number line will accommodate all the values as separate points.

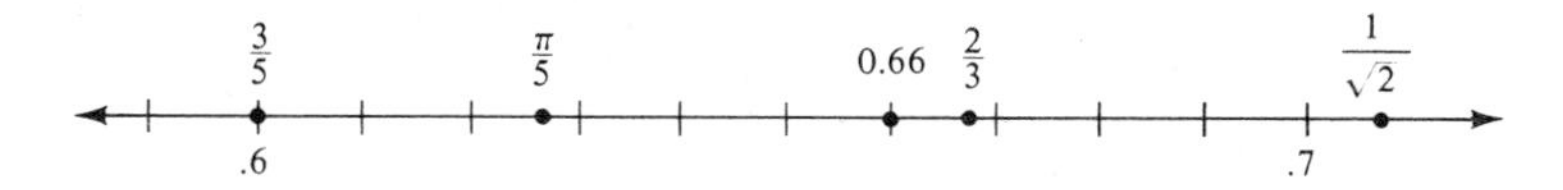

■

Potential Problems

The set of real numbers does not include the square roots of negative numbers. For example,

$$\sqrt{-4}$$

does not exist in the set of real numbers, because any number squared must be nonnegative. This statement means that

$$\sqrt{x}$$

makes sense only if x is nonnegative. To avoid one potential problem, therefore, in this book we assume that the domain of all variables is restricted so that the expressions under radical symbols represent nonnegative numbers.

Examples *Find the square roots of the following rational numbers. If any is not rational, state whether it is irrational or not real.*

5. $\sqrt{81} = 9$

6. $-\sqrt{4} = -2$

7. $\sqrt{\frac{1}{4}} = \frac{1}{2}$

8. $-\sqrt{\frac{4}{9}} = -\frac{2}{3}$

9. $\sqrt{15}$ is irrational

10. $-\sqrt{8}$ is irrational

11. $\sqrt{-9}$ is not real

12. $-\sqrt{9} = -3$

Another problem occurs when using a calculator. There is a common misconception that square roots are operation symbols only, as in

$$\sqrt{4} = 2 \text{ (4, operation square root, answer 2).}$$

But a square root is more than an operation symbol; it is also a symbol used to represent a number, as in

$$\sqrt{2}$$

Just as 2, 19, $\frac{2}{3}$, or 6.7 represent numbers, $\sqrt{2}$ repesents a number. Compare this notation with the division symbol and the fraction symbol:

$$\frac{2}{3} \text{ can mean } 2 \div 3\text{, but it also represents a single number } \frac{2}{3}$$

If you wish to know what is meant by $\sqrt{2}$, do not go to a calculator and press

[2] [√] display: 1.414213562

and then say $\sqrt{2}$ is 1.414213562. This is just as incorrect as saying

1 is exactly the same as .99.

The correct terminology is to say that these numbers are approximately equal, and use the symbol $\approx$. Thus,

$$\sqrt{2} \approx 1.414213562, \quad 1 \approx .99$$

$$\frac{1}{3} \approx .333333\text{, but} \quad \frac{1}{3} = .333\overline{333}$$

The number square root of 2 is that number that equals 2 when it is multiplied times itself. *This number* is denoted by $\sqrt{2}$. Remember, calculators can only give decimal approximations for irrational numbers.

When working problems involving radicals, approximate answers should be given only when approximate results are requested. Otherwise, exact answers are required.

Examples *Answer true or false; you may use a calculator if you wish.*

13. $\frac{1}{3} = .333333333$

14. $\frac{2}{3} \approx .666666667$

15. $\sqrt{3} = 1.732050808$

16. $\pi = 3.141592654$

13. false

14. true

15. false

16. false

17. $\pi = \frac{22}{7}$ **17.** false

18. $\pi \approx 3.1416$ **18.** true ■

Problem Set 9.1

In Problems 1–10 use the Pythagorean Theorem to determine if the numbers represent the sides of a right triangle. See Examples 1 and 2.

1. 8, 15, 17 **2.** 16, 63, 65 **3.** 20, 21, 29
4. 20, 99, 101 **5.** 12, 35, 37 **6.** 28, 45, 53
7. 33, 56, 65 **8.** 36, 77, 85 **9.** 39, 80, 89
10. 48, 55, 73

Use the number line to arrange the numbers in Problems 11–20 from smallest to largest. Approximate each before you set the scale. See Examples 3 and 4.

11. $\frac{2}{3}, \frac{3}{4}, \frac{4}{5}, \frac{5}{6}, \frac{6}{7}, \frac{7}{8}$ **12.** $\frac{1}{2}, \frac{3}{5}, \frac{5}{7}, \frac{7}{9}, \frac{9}{11}, \frac{11}{13}$

13. $\frac{1}{4}, \frac{2}{9}, \frac{1}{5}, \frac{4}{13}, \frac{3}{11}, \frac{4}{15}$ **14.** $\frac{1}{5}, \frac{2}{11}, \frac{1}{6}, \frac{3}{20}, \frac{2}{13}, \frac{3}{17}$

15. $\frac{2}{3}, \frac{33}{50}, \frac{\pi}{5}, \frac{7}{11}$ **16.** $\frac{3}{4}, \frac{31}{40}, \frac{\pi}{4}, \frac{8}{11}$

17. $\frac{34}{11}, \sqrt{10}, \pi, \frac{22}{7}$ **18.** $\frac{45}{11}, \sqrt{17}, \frac{4\pi}{3}, \frac{29}{7}$

19. $\frac{7}{4}, \frac{87}{50}, \sqrt{3}, (1.2)^3$ **20.** $\frac{56}{25}, \frac{57}{25}, \sqrt{5}, (1.5)^2$

Write out the expressions in Problems 21–40 without using a radical symbol if the number is rational. If it is not rational, state whether it is irrational or not real. See Examples 5–12.

21. $\sqrt{25}$ **22.** $\sqrt{16}$ **23.** $\sqrt{64}$ **24.** $-\sqrt{9}$ **25.** $\sqrt{-4}$

26. $\sqrt{5}$ **27.** $\sqrt{\frac{1}{9}}$ **28.** $-\sqrt{\frac{9}{4}}$ **29.** $\sqrt{\frac{1}{16}}$ **30.** $\sqrt{-\frac{4}{25}}$

31. $\sqrt{\frac{49}{64}}$ **32.** $-\sqrt{\frac{25}{81}}$ **33.** $\sqrt{144}$ **34.** $\sqrt{-196}$ **35.** $\sqrt{121}$

36. $\sqrt{0.36}$ **37.** $\sqrt{0.9}$ **38.** $\sqrt{-0.49}$ **39.** $-\sqrt{1.44}$ **40.** $-\sqrt{1.96}$

Answer true or false in Problems 41–50. See Examples 13–18.

41. $\frac{1}{9} = 0.11111111$ **42.** $\frac{1}{6} \approx 0.16666667$

43. $\frac{34}{11} \approx 3.09090909$ **44.** $\frac{29}{7} = 4.14285714$

45. $\frac{333}{106} = 3.141509$ **46.** $\frac{355}{113} \approx 3.14159$

47. $\sqrt{4} \approx 2$

48. $\sqrt{2} = 1.41421356$

49. $\pi \approx \dfrac{355}{113}$

50. $\pi = \dfrac{103993}{33102}$

9.2 Multiplication and Division of Square Roots

Multiplying and dividing expressions with square roots is quite different from that of rational numbers. An important property of square roots provides the key to these operations. Notice that

$$\sqrt{9}\sqrt{25} = 3 \cdot 5 = 15 \quad \text{and} \quad \sqrt{9 \cdot 25} = \sqrt{225} = 15$$

thus $\sqrt{9}\sqrt{25} = \sqrt{9 \cdot 25}$. Also,

$$\frac{\sqrt{9}}{\sqrt{25}} = \frac{3}{5} \quad \text{and} \quad \sqrt{\frac{9}{25}} = \frac{3}{5}$$

thus $\dfrac{\sqrt{9}}{\sqrt{25}} = \sqrt{\dfrac{9}{25}}$. These examples can be generalized as the following property.

Multiplication and Division Properties of Square Roots

For all nonnegative real numbers a and b,

MULTIPLICATION: $\sqrt{a} \cdot \sqrt{b} = \sqrt{ab}$

DIVISION: $\dfrac{\sqrt{a}}{\sqrt{b}} = \sqrt{\dfrac{a}{b}} \quad b \neq 0$

Examples *In Examples 1–4, perform the indicated operations.*

1. $\sqrt{5}\sqrt{6} = \sqrt{5 \cdot 6} = \mathbf{\sqrt{30}}$

2. $\sqrt{2}\sqrt{x} = \mathbf{\sqrt{2x}}$

3. $\dfrac{\sqrt{42}}{\sqrt{7}} = \sqrt{\dfrac{42}{7}} = \mathbf{\sqrt{6}}$

4. $\dfrac{\sqrt{125}}{\sqrt{5}} = \sqrt{\dfrac{125}{5}} = \sqrt{25} = \mathbf{5}$ ■

Compare the last two examples. In Example 3 we stopped after dividing, but in Example 4 we were not finished after dividing and obtaining $\sqrt{25}$. The reason is that $\sqrt{6}$ is simplified whereas $\sqrt{25}$ is not. An expression containing square roots is considered to be in ***simplest form*** if it meets the following three conditions.

1. ***There are no square factors in the radicand.***

If we completely factor the ***radicand*** (the number inside the radical sign) and write the factors in exponent form, there will be no factors raised to a power greater than 1 when the radical is simplified.

Examples

5. $\sqrt{2}$ — *This is the simplified form.*

6. $\sqrt{4} = \sqrt{2^2}$ — *First, write the expression in completely factored form. It is not simplified, because 2 is raised to a power greater than 1.*

   $= \mathbf{2}$ — *This is the simplified form.*

7. $\sqrt{8} = \sqrt{2^3}$ — *First, write the expression in completely factored form. It is not yet simplified, because 2 is raised to a power greater than 1.*

   $= \sqrt{2^2}\sqrt{2}$ — *To simplify, mentally use the multiplication property of square roots.*

   $= \mathbf{2\sqrt{2}}$ — *This is the simplified form.*

8. $\sqrt{50} = \sqrt{5^2 \cdot 2}$ — *First, write the expression in completely factored form. It is not yet simplified, because 5 is raised to a power greater than 1.*

   $= \sqrt{5^2}\sqrt{2}$ — *To simplify, mentally use the multiplication property of square roots.*

   $= \mathbf{5\sqrt{2}}$ — *This is the simplified form.*

9. $\sqrt{52x^2y} = \sqrt{2^2 \cdot 13 \cdot x^2 \cdot y}$ — *First, write the expression in completely factored form.*

   $= \sqrt{2^2x^2}\sqrt{13y}$ — *Associate those terms that are squared, and mentally use the multiplication property of square roots.*

   $= \mathbf{2x\sqrt{13y}}$ — *This is the simplified form.*

10. $\sqrt{8x^3} = \sqrt{2^3x^3}$ — *Factor completely.*

    $= \sqrt{2^2 \cdot 2 \cdot x^2 \cdot x}$ — *If there are exponents larger than 1, separate them into factors that are squared.*

    $= \sqrt{2^2x^2}\sqrt{2x}$

    $= \mathbf{2x\sqrt{2x}}$ ■

2. ***No radical appears in a denominator.***

Examples

11. $\dfrac{3}{\sqrt{2}}$ — *This expression is not simplified, because there is a radical in the denominator.*

    $= \dfrac{3}{\sqrt{2}} \cdot \dfrac{\sqrt{2}}{\sqrt{2}}$ — *Multiply by 1, written as a fraction, so the number that appears in the denominator is repeated.*

    $= \dfrac{3\sqrt{2}}{2}$

    $= \mathbf{\tfrac{3}{2}\sqrt{2}}$

12. $\dfrac{10x^2}{\sqrt{2x}} = \dfrac{10x^2}{\sqrt{2x}} \cdot \dfrac{\sqrt{2x}}{\sqrt{2x}}$

    $= \dfrac{10x^2\sqrt{2x}}{2x}$

    $= \mathbf{5x\sqrt{2x}}$ ■

This technique of clearing the denominator of radicals is called ***rationalizing the denominator.*** Remember the technique multiplies the numerator and denominator by a number that will give a perfect square in the radicand of the denominator. This creates a fraction in which no radical appears in the denominator.

3. No fraction appears within a radical.

Examples 13. $\sqrt{\frac{1}{2}}$ — *This expression is not simplified, because there is a fraction under the radical.*

$= \frac{\sqrt{1}}{\sqrt{2}}$ — *To simplify, use the division property of radicals, and then complete the problem by rationalizing the denominator.*

$= \frac{1}{\sqrt{2}} \cdot \frac{\sqrt{2}}{\sqrt{2}}$

$= \frac{\sqrt{2}}{2}$

$= \frac{1}{2}\sqrt{2}$

14. $\sqrt{\frac{5}{3x}} = \frac{\sqrt{5}}{\sqrt{3x}} \cdot \frac{\sqrt{3x}}{\sqrt{3x}}$

$= \frac{\sqrt{15x}}{3x}$ *This is in simplified form.* ■

Now to summarize, you should remember the three conditions that have been illustrated.

Simplest Form of Square Roots

A square root is simplified if all three of the following conditions are satisfied:

1. There are no square factors in the radicand.
2. No radical appears in a denominator.
3. No fraction appears within a radical.

Problem Set 9.2

Perform the indicated operations in Problems 1–30 and simplify. All variables represent positive real numbers. See Examples 1–4.

1. $\sqrt{2}\sqrt{5}$
2. $\sqrt{10}\sqrt{7}$
3. $\sqrt{3}\sqrt{5}$
4. $\sqrt{3}\sqrt{7a}$
5. $\sqrt{2}\sqrt{3b}$
6. $\sqrt{5}\sqrt{2c}$
7. $\sqrt{2}\sqrt{xy}$
8. $\sqrt{3}\sqrt{yz}$
9. $\sqrt{5}\sqrt{xz}$
10. $\sqrt{5a}\sqrt{3b}$
11. $\sqrt{2m}\sqrt{7n}$
12. $\sqrt{5r}\sqrt{7s}$
13. $\frac{\sqrt{18}}{\sqrt{6}}$
14. $\frac{\sqrt{24}}{\sqrt{8}}$
15. $\frac{\sqrt{28}}{\sqrt{7}}$
16. $\frac{\sqrt{30}}{\sqrt{5}}$
17. $\frac{\sqrt{18}}{\sqrt{2}}$
18. $\frac{\sqrt{42}}{\sqrt{6}}$
19. $\frac{\sqrt{2a}}{\sqrt{a}}$
20. $\frac{\sqrt{3b}}{\sqrt{b}}$

21. $\frac{5\sqrt{c}}{\sqrt{5}}$ **22.** $\frac{\sqrt{12m}}{\sqrt{3m}}$ **23.** $\frac{\sqrt{18n}}{\sqrt{2n}}$ **24.** $\frac{\sqrt{15mn}}{\sqrt{5m}}$

25. $\frac{\sqrt{6}\sqrt{2a}}{\sqrt{3a}}$ **26** $\frac{\sqrt{5}\sqrt{10b}}{\sqrt{2b}}$ **27.** $\frac{\sqrt{14}\sqrt{6c}}{\sqrt{7c}}$ **28.** $\frac{\sqrt{6xy}\sqrt{5z}}{\sqrt{10y}}$

29. $\frac{\sqrt{21a}\sqrt{2bc}}{\sqrt{6ab}}$ **30.** $\frac{\sqrt{3m}\sqrt{22n}}{\sqrt{6n}}$

Simplify Problems 31–70. All variables represent positive real numbers. See Examples 5–12.

31. $\sqrt{27}$ **32.** $\sqrt{12}$ **33.** $\sqrt{18}$ **34.** $-\sqrt{8}$

35. $-\sqrt{50}$ **36.** $-\sqrt{49}$ **37.** $\frac{1}{\sqrt{2}}$ **38.** $\frac{1}{\sqrt{3}}$

39. $-\frac{1}{\sqrt{5}}$ **40.** $-\sqrt{\frac{2}{3}}$ **41.** $\sqrt{\frac{3}{5}}$ **42.** $\sqrt{\frac{5}{2}}$

43. $\sqrt{48}$ **44.** $\sqrt{75}$ **45.** $-\sqrt{75}$ **46.** $\sqrt{125}$

47. $-\sqrt{150}$ **48.** $\sqrt{128}$ **49.** $\sqrt{3}\sqrt{6}$ **50.** $-\sqrt{5}\sqrt{10}$

51. $\sqrt{2}\sqrt{14}$ **52.** $\sqrt{15}\sqrt{45}$ **53.** $\sqrt{12}\sqrt{48}$ **54.** $\sqrt{20}\sqrt{45}$

55. $\frac{3}{2\sqrt{5}}$ **56.** $\frac{7}{3\sqrt{3}}$ **57.** $\frac{2}{5\sqrt{6}}$ **58.** $\frac{14\sqrt{6}}{2\sqrt{3}}$

59. $\frac{10\sqrt{10}}{5\sqrt{5}}$ **60.** $\frac{26\sqrt{14}}{2\sqrt{7}}$ **61.** $\sqrt{x^3}$ **62.** $\sqrt{x^4}$

63. $\sqrt{x^5}$ **64.** $\sqrt{12m^5}$ **65.** $-\sqrt{9n^3}$ **66.** $\sqrt{18p^4}$

67. $2\sqrt{\frac{2x}{25}}$ **68.** $\sqrt{\frac{3x}{7y}}$ **69.** $3\sqrt{\frac{6}{5y}}$ **70.** $\frac{x}{\sqrt{y}}$

The following is necessary in Problems 71 and 72.

The *escape velocity* v at the surface of a planet with radius r and acceleration due to gravity g is given by

$$v = \sqrt{2gr}$$

71. The gravitational constant g is $\frac{1}{165}$ miles per second^2 on earth, whose radius is approximately 3960 miles. What is the escape velocity of a projectile from earth?

72. The planet Mars has a radius of approximately 2106 miles and gravitational constant of $\frac{1}{432}$ miles per second^2. What is the escape velocity from the surface of Mars?

9.3 Addition and Subtraction of Square Roots

We have now developed all of the mechanics necessary for the addition and subtraction of expressions involving square roots. The procedure is to find similar terms. For example, the following expressions can be simplified further, as shown.

$$5x + 3x = (5 + 3)x = 8x \qquad 5\sqrt{2} + 3\sqrt{2} = (5 + 3)\sqrt{2} = 8\sqrt{2}$$

But neither $5x + 3y$ nor $5\sqrt{2} + 3\sqrt{3}$ can be simplified further.

Examples *Simplify.*

1. $4\sqrt{5} + 9\sqrt{5} = (4 + 9)\sqrt{5} = \mathbf{13\sqrt{5}}$

2. $2\sqrt{x} - 5\sqrt{x} = (2 - 5)\sqrt{x} = \mathbf{-3\sqrt{x}}$

3. $\sqrt{8} + \sqrt{18} = \sqrt{2^2 \cdot 2} + \sqrt{3^2 \cdot 2} = 2\sqrt{2} + 3\sqrt{2} = \mathbf{5\sqrt{2}}$

4. $\sqrt{12} - \sqrt{32} = \sqrt{2^2 \cdot 3} - \sqrt{2^4 \cdot 2} = 2\sqrt{3} - 2^2\sqrt{2} = \mathbf{2\sqrt{3} - 4\sqrt{2}}$

5. $2\sqrt{75} + \sqrt{12} - \dfrac{3}{\sqrt{3}} = 2\sqrt{5^2 \cdot 3} + \sqrt{2^2 \cdot 3} - \dfrac{3}{\sqrt{3}}\dfrac{\sqrt{3}}{\sqrt{3}}$ *Simplify each radical first.*

$= 2 \cdot 5\sqrt{3} + 2\sqrt{3} - \dfrac{3\sqrt{3}}{3}$ *Do this mentally.*

$= 10\sqrt{3} + 2\sqrt{3} - \sqrt{3}$ *Combine similar radicals.*

$= \mathbf{11\sqrt{3}}$ ■

Such expressions can be simplified by adding or subtracting *if* there are similar radical terms. To recognize these terms, each term is simplified before attempting to add or subtract. Notice how several operations are performed in the following examples.

Examples *Simplify.*

6. $4(\sqrt{2} - \sqrt{8}) = 4\sqrt{2} - 4\sqrt{8} = 4\sqrt{2} - 4 \cdot 2\sqrt{2} = 4\sqrt{2} - 8\sqrt{2} = \mathbf{-4\sqrt{2}}$

7. $\sqrt{3x}(\sqrt{x} + \sqrt{6}) = \sqrt{3x^2} + \sqrt{18x} = \mathbf{x\sqrt{3} + 3\sqrt{2x}}$

8. $\dfrac{5 + \sqrt{75}}{5} = \dfrac{5 + 5\sqrt{3}}{5}$ *Simplify.*

$= \dfrac{5(1 + \sqrt{3})}{5}$ *Factor.*

$= \mathbf{1 + \sqrt{3}}$ *Reduce.*

9. $\dfrac{\sqrt{2}(\sqrt{6} + \sqrt{5})}{\sqrt{3}} = \dfrac{\sqrt{3}\sqrt{2}(\sqrt{6} + \sqrt{5})}{\sqrt{3}\sqrt{3}}$ *Rationalize the denominator.*

$= \dfrac{\sqrt{6}(\sqrt{6} + \sqrt{5})}{3}$ *Multiply.*

$= \mathbf{\dfrac{6 + \sqrt{30}}{3}}$ *Apply distributive property.* ■

Recall the product of binomials in Chapter 4. The product sums involving radicals can be treated with a similar pattern. For example,

$$(x + 2)(x + 1) = x \cdot x + (x + 2x) + 2 \cdot 1 = x^2 + 3x + 2$$

$$(\sqrt{3} + 2)(\sqrt{3} + 1) = \sqrt{3}\sqrt{3} + (\sqrt{3} + 2\sqrt{3}) + 2 \cdot 1 = 3 + 3\sqrt{3} + 2 = 5 + 3\sqrt{3}$$

Examples *Simplify.*

10. $(\sqrt{2}-3)(\sqrt{2}+5) = \sqrt{2}\sqrt{2} + (5\sqrt{2} - 3\sqrt{2}) + (-3)(5)$
$= 2 + 2\sqrt{2} - 15$
$= -13 + 2\sqrt{2}$

11. $(3-2\sqrt{x})(2+\sqrt{x}) = 3\cdot 2 + (3\sqrt{x} - 4\sqrt{x}) - 2\sqrt{x}\sqrt{x}$
$= 6 - \sqrt{x} - 2x$

12. $(\sqrt{3}-1)(\sqrt{3}+1) = \sqrt{3}\sqrt{3} + (\sqrt{3} - \sqrt{3}) + (-1)(1)$
$= 3 + 0 - 1$
$= 2$ ■

Problem Set 9.3

Simplify the expressions by combining similar radical terms in Problems 1–20. See Examples 1–5.

1. $\sqrt{5} + 3\sqrt{5}$
2. $2\sqrt{7} - \sqrt{7}$
3. $\sqrt{8} - \sqrt{2}$
4. $\sqrt{3} + \sqrt{27}$
5. $\sqrt{27} + 2\sqrt{12}$
6. $\sqrt{24} + \sqrt{30}$
7. $2\sqrt{2} - 5\sqrt{2} + \sqrt{2}$
8. $\sqrt{3} - 2\sqrt{3} + 7\sqrt{3}$
9. $5\sqrt{17} - 3\sqrt{17} + \sqrt{17}$
10. $\sqrt{19} + 3\sqrt{19} - 4\sqrt{19}$
11. $\sqrt{8} + \sqrt{50} + \sqrt{18}$
12. $\sqrt{27} + \sqrt{75} - \sqrt{12}$
13. $\sqrt{48} + 2\sqrt{147} - \sqrt{108}$
14. $\sqrt{50} + 2\sqrt{98} - \sqrt{72}$
15. $\sqrt{20} + \sqrt{80} - \sqrt{125}$
16. $\sqrt{28} - \sqrt{63} + \sqrt{175}$
17. $\dfrac{6}{\sqrt{3}} - \sqrt{75} + 2\sqrt{3}$
18. $3\sqrt{32} - \dfrac{4}{\sqrt{2}} + \sqrt{50}$
19. $\dfrac{13}{\sqrt{13}} + 2\sqrt{52} + 8\sqrt{13}$
20. $2\sqrt{44} + \dfrac{33}{\sqrt{11}} + \sqrt{176}$

Perform the indicated operations in Problems 21–60 and simplify. See Examples 6–9.

21. $\sqrt{2} + \sqrt{6} + \sqrt{8}$
22. $\sqrt{3} + \sqrt{12} + \sqrt{18}$
23. $\sqrt{3}(\sqrt{3} - \sqrt{12} + \sqrt{48})$
24. $\sqrt{2}(\sqrt{2} + \sqrt{18} - \sqrt{72})$
25. $\sqrt{5}(\sqrt{10} + \sqrt{15} + \sqrt{20})$
26. $\sqrt{7}(\sqrt{14} - \sqrt{21} + \sqrt{28})$
27. $3(\sqrt{3} - \sqrt{27})$
28. $5(\sqrt{18} - \sqrt{2})$
29. $\sqrt{405} - \sqrt{320} + \sqrt{125}$
30. $\sqrt{252} - \sqrt{343} + \sqrt{175}$
31. $\dfrac{\sqrt{12} - 6}{2}$
32. $\dfrac{\sqrt{18} - 6}{3}$
33. $\dfrac{3\sqrt{2} + 9}{6}$
34. $\dfrac{2\sqrt{3} + 4}{6}$
35. $\dfrac{2\sqrt{2} + \sqrt{4} + \sqrt{8}}{4}$
36. $\dfrac{3\sqrt{3} + \sqrt{9} + \sqrt{27}}{9}$
37. $\dfrac{14 - \sqrt{98}}{21}$
38. $\dfrac{12 - \sqrt{96}}{30}$
39. $\dfrac{72 + \sqrt{252}}{54}$
40. $\dfrac{72 - \sqrt{192}}{56}$

41. $\dfrac{\sqrt{8} + \sqrt{12}}{6}$

42. $\dfrac{\sqrt{18} + \sqrt{27}}{6}$

43. $\dfrac{\sqrt{8} + \sqrt{12}}{\sqrt{6}}$

44. $\dfrac{\sqrt{18} + \sqrt{27}}{\sqrt{6}}$

45. $\dfrac{8 + \sqrt{12}}{\sqrt{6}}$

46. $\dfrac{18 + \sqrt{27}}{\sqrt{6}}$

47. $\dfrac{2\sqrt{2} + 5\sqrt{2} + \sqrt{6}}{\sqrt{2}}$

48. $\dfrac{7\sqrt{3} - 2\sqrt{3} - \sqrt{6}}{\sqrt{3}}$

49. $\dfrac{\sqrt{2} - \sqrt{18} + 2\sqrt{3}}{\sqrt{3}}$

50. $\dfrac{\sqrt{8} + \sqrt{50} + \sqrt{18}}{\sqrt{2}}$

51. $\dfrac{\sqrt{10} + \sqrt{15} + \sqrt{20}}{\sqrt{5}}$

52. $\dfrac{\sqrt{14} - \sqrt{21} + \sqrt{28}}{\sqrt{7}}$

53. $\dfrac{3(\sqrt{5} + 3\sqrt{5})}{2}$

54. $\dfrac{2(2\sqrt{7} - 5\sqrt{7})}{3}$

55. $\dfrac{(\sqrt{8} + \sqrt{50} + \sqrt{18})\sqrt{5}}{5}$

56. $\dfrac{\sqrt{7}(\sqrt{27} + \sqrt{75} - \sqrt{12})}{3}$

57. $\dfrac{\sqrt{2}(\sqrt{98} - 3\sqrt{2} + \sqrt{8})}{3}$

58. $\dfrac{(\sqrt{75} + 5\sqrt{3} - \sqrt{27})\sqrt{3}}{7}$

59. $\dfrac{\sqrt{24} + \sqrt{54} + \sqrt{50}}{\sqrt{2}}$

60. $\dfrac{\sqrt{60} + \sqrt{15} - \sqrt{135}}{\sqrt{3}}$

Simplify the expressions in Problems 61–70. See Examples 10–12.

61. $(2 + \sqrt{3})(1 - \sqrt{3})$
62. $(1 + \sqrt{5})(3 - \sqrt{5})$
63. $(5 - \sqrt{7})(3 + \sqrt{7})$
64. $(7 - \sqrt{2})(3 + \sqrt{2})$
65. $(1 + 2\sqrt{11})(3 - \sqrt{11})$
66. $(5 - 3\sqrt{13})(1 + \sqrt{13})$
67. $(4 - 3\sqrt{2})(3 + 2\sqrt{2})$
68. $(5 + 2\sqrt{3})(2 - 5\sqrt{3})$
69. $(2 + \sqrt{5})^2$
70. $(3 - \sqrt{7})^2$

The following is necessary for Problems 71 and 72.

The distance that can be seen (in miles) to the horizon is approximated by

$$d \approx \sqrt{\frac{3h}{2}}$$

where h is the height (in feet) of the observer.

71. Standing on a cliff above the ocean, you see a ship on the horizon. If your line of sight is 54 feet above the ocean, how far out to sea is the vessel?
72. If you climb the mast of a ship to the height of 24 feet above the water, how far can you see to the horizon?

9.4 Simplifying Square Root Expressions

Multiplying radical sums like $(\sqrt{2} + 3)(\sqrt{2} + 5)$ uses the pattern of the binomial product $(x + 3)(x + 5)$.

$$(x + 3)(x + 5) = x^2 + 8x + 15$$
$$(\sqrt{2} + 3)(\sqrt{2} + 5) = (\sqrt{2})^2 + 8(\sqrt{2}) + 15$$
$$= 2 + 8\sqrt{2} + 15$$
$$= 17 + 8\sqrt{2}$$

Thus it is natural to consider the special products: perfect square and difference of squares.

$$\mathbf{(x + a)^2 = x^2 + 2ax + a^2} \qquad \textit{Perfect Square}$$
$$(x + \sqrt{2})^2 = x^2 + 2\sqrt{2}x + (\sqrt{2})^2$$
$$= x^2 + 2\sqrt{2}x + 2$$
$$(\sqrt{2} + \sqrt{3})^2 = (\sqrt{2})^2 + 2\sqrt{2}\sqrt{3} + (\sqrt{3})^2$$
$$= 2 + 2\sqrt{6} + 3$$
$$= 5 + 2\sqrt{6}$$

$$\mathbf{(x + a)(x - a) = x^2 - a^2} \qquad \textit{Difference of Squares}$$
$$(2 + \sqrt{3})(2 - \sqrt{3}) = (2)^2 - (\sqrt{3})^2$$
$$= 4 - 3$$
$$= 1$$

Notice in this example that the product of two irrational numbers can be a rational number. The difference of two squares proves to be a useful concept in dealing with radicals. The difference of two squares is the product of a sum and a difference of two numbers.

Difference of Squares

$$(x + a)(x - a) = x^2 - a^2$$

The quantities $x + a$ and $x - a$ are called ***conjugates.*** In the preceding instance, $2 + \sqrt{3}$ and $2 - \sqrt{3}$ are conjugates. We may also say that $2 - \sqrt{3}$ is the conjugate of $2 + \sqrt{3}$.

Examples *Simplify.*

1. $(5 + \sqrt{19})(5 - \sqrt{19}) = (5)^2 - (\sqrt{19})^2$
   $= 25 - 19$
   $= \mathbf{6}$

2. $(1 + 2\sqrt{x})(1 - 2\sqrt{x}) = (1)^2 - (2\sqrt{x})^2$
   $= \mathbf{1 - 4x}$

3. $(a - \sqrt{5})(a + \sqrt{5}) = \mathbf{a^2 - 5}$

4. $(\sqrt{2} + \sqrt{5})(\sqrt{2} - \sqrt{5}) = (\sqrt{2})^2 - (\sqrt{5})^2$
   $= 2 - 5$
   $= \mathbf{-3}$

■

Since the product of two conjugates produces an expression free of square roots, the idea is useful in simplifying certain square root expressions. Recall *rationalizing denominators* in the last section. For instance,

$$\frac{3\sqrt{2}}{\sqrt{3x}} = \frac{3\sqrt{2}}{\sqrt{3x}} \cdot \frac{\sqrt{3x}}{\sqrt{3x}} = \frac{3\sqrt{6x}}{3x} = \frac{\sqrt{6x}}{x}$$

To simplify square root expressions with two terms in the denominator, multiply by a fraction (equal to 1) that is represented as the conjugate of the denominator.

$$\frac{3\sqrt{2}}{3+\sqrt{x}} \cdot \frac{\mathbf{3-\sqrt{x}}}{\mathbf{3-\sqrt{x}}} = \frac{9\sqrt{2}-3\sqrt{2x}}{9-x}$$

Examples *Simplify the expressions in Examples 5–7.*

5. $$\frac{1}{\sqrt{2}+1} = \frac{1}{\sqrt{2}+1} \cdot \frac{\mathbf{\sqrt{2}-1}}{\mathbf{\sqrt{2}-1}} = \frac{\sqrt{2}-1}{2-1} = \mathbf{\sqrt{2}-1}$$

6. $$\frac{x}{\sqrt{5}-\sqrt{3}} = \frac{x}{\sqrt{5}-\sqrt{3}} \cdot \frac{\mathbf{\sqrt{5}+\sqrt{3}}}{\mathbf{\sqrt{5}+\sqrt{3}}} = \frac{x(\sqrt{5}+\sqrt{3})}{5-3} = \frac{\mathbf{x\sqrt{5}+x\sqrt{3}}}{\mathbf{2}}$$

7. $$\frac{1+\sqrt{x}}{1-\sqrt{x}} = \frac{1+\sqrt{x}}{1-\sqrt{x}} \cdot \frac{\mathbf{1+\sqrt{x}}}{\mathbf{1+\sqrt{x}}} = \frac{(1+\sqrt{x})(1+\sqrt{x})}{(1-\sqrt{x})(1+\sqrt{x})} = \frac{\mathbf{1+2\sqrt{x}+x}}{\mathbf{1-x}}$$

■

Problem Set 9.4

In Problems 1–10 find the product of the number and its conjugate. See Examples 1–4.

1. $2-\sqrt{5}$ **2.** $5+\sqrt{2}$ **3.** $\sqrt{3}+7$ **4.** $\sqrt{7}-3$

5. $4-\sqrt{15}$ **6.** $5+\sqrt{21}$ **7.** $2-5\sqrt{3}$ **8.** $3+2\sqrt{5}$

9. $\sqrt{2}+\sqrt{3}$ **10.** $\sqrt{7}-\sqrt{3}$

Simplify the expressions in Problems 11–30. Assume that all the variables represent positive real numbers. See Examples 1–4.

11. $(\sqrt{7} - 1)(\sqrt{7} + 1)$
12. $(\sqrt{2} + 3)(\sqrt{2} - 3)$
13. $(\sqrt{5} - 2)(\sqrt{5} + 2)$
14. $(3 + \sqrt{5})(3 - \sqrt{5})$
15. $(\sqrt{7} + 2)(\sqrt{7} - 2)$
16. $(\sqrt{6} + 4)(\sqrt{6} - 4)$
17. $(2\sqrt{3} + 3)(2\sqrt{3} - 3)$
18. $(5 - 3\sqrt{2})(5 + 3\sqrt{2})$
19. $(5\sqrt{2} - 7)(5\sqrt{2} + 7)$
20. $(2 + \sqrt{a})(2 - \sqrt{a})$
21. $(5 - \sqrt{b})(5 + \sqrt{b})$
22. $(\sqrt{c} - 3)(\sqrt{c} + 3)$
23. $(x + \sqrt{y})(x - \sqrt{y})$
24. $(1 + 2\sqrt{m})(1 - 2\sqrt{m})$
25. $(n - \sqrt{6})(n + \sqrt{6})$
26. $(2\sqrt{3} + \sqrt{11})(2\sqrt{3} - \sqrt{11})$
27. $(\sqrt{15} - 3\sqrt{2})(\sqrt{15} + 3\sqrt{2})$
28. $(2\sqrt{5} - \sqrt{19})(2\sqrt{5} + \sqrt{19})$
29. $\sqrt{6}(2 - 4\sqrt{6})(12 + \sqrt{6})$
30. $\sqrt{7}(3 + \sqrt{7})(7 - 3\sqrt{7})$

Simplify the expressions in Problems 31–60. All variables represent positive reals. See Examples 5–7.

31. $\frac{2}{3 - \sqrt{5}}$
32. $\frac{3}{4 + \sqrt{7}}$
33. $\frac{5}{3 + \sqrt{6}}$
34. $\frac{12}{\sqrt{17} - 3}$
35. $\frac{18}{5 - \sqrt{19}}$
36. $\frac{10}{\sqrt{11} + 4}$
37. $\frac{\sqrt{2}}{\sqrt{2} - 3}$
38. $\frac{\sqrt{7}}{\sqrt{7} + 1}$
39. $\frac{\sqrt{5}}{\sqrt{5} + 2}$
40. $\frac{3\sqrt{5}}{3 + \sqrt{5}}$
41. $\frac{2\sqrt{7}}{\sqrt{7} - 2}$
42. $\frac{4\sqrt{6}}{\sqrt{6} + 4}$
43. $\frac{5 - \sqrt{3}}{1 + \sqrt{3}}$
44. $\frac{1 - \sqrt{3}}{2 - \sqrt{3}}$
45. $\frac{2 + \sqrt{3}}{\sqrt{3} - 1}$
46. $\frac{1 + \sqrt{2}}{1 - \sqrt{2}}$
47. $\frac{\sqrt{2} - 3}{\sqrt{2} - 1}$
48. $\frac{1 - \sqrt{2}}{3 + \sqrt{2}}$
49. $\frac{1 + \sqrt{t}}{1 - \sqrt{t}}$
50. $\frac{\sqrt{r} - 1}{\sqrt{r} + 1}$
51. $\frac{\sqrt{s} + 1}{\sqrt{s} - 1}$
52. $\frac{3}{\sqrt{5} + \sqrt{3}}$
53. $\frac{5}{\sqrt{3} - \sqrt{2}}$
54. $\frac{-7}{\sqrt{5} + \sqrt{2}}$

The following is needed for Problems 55–58.

The pulse rate P of an adult (in beats per minute) is approximated by

$$P = \frac{600}{\sqrt{h}}$$

where h is the height of the person (in inches).

Find the expected pulse rate (to the nearest integer) for people with each of the following heights.

55. 64 inches
56. 81 in.
57. 68 in.
58. 72″

The following is needed for Problems 59–61.

The evaporation e (in inches per day) of a large body of water on a particular day is approximated by

$$e = \frac{w}{20\sqrt{a}}$$

where a is the surface area (in square miles) and w is the wind speed (in miles per hour) of the air over the water.

Approximate the evaporation (to the nearest hundredth of an inch) if the wind speed is 10 mph.

59. The largest natural lake in New York is Lake Oneida, which has a surface area of 57 square miles.
60. Lake Winnipesaukee is New Hampshire's largest natural lake. It has a surface area of 70 square miles.
61. Lake Tahoe on the California-Nevada border has a surface area of 192 square miles.

9.5 Equations with Square Roots

When solving certain types of problems, we sometimes will find it necessary to square both sides of an equation. And, at those times, you may run into an unexpected difficulty, as illustrated by the following example.

Solve: $x - 2 = \sqrt{x}$

Square both sides to eliminate the radical.

$$\begin{aligned}(x-2)^2 &= x \\ x^2 - 4x + 4 &= x \\ x^2 - 5x + 4 &= 0 \\ (x-4)(x-1) &= 0 \\ x &= 4, 1\end{aligned}$$

But now check these values:

$$\begin{aligned}x = 4; \quad & 4 - 2 \stackrel{?}{=} \sqrt{4} \\ & 2 = 2, \text{ which checks} \\ x = 1: \quad & 1 - 2 \stackrel{?}{=} \sqrt{1} \\ & -1 \neq 1, \text{ which does not check}\end{aligned}$$

Notice from this example that if both sides of an equation are squared, the solution of the resulting equation does not necessarily solve the original equation. However, read the following property of equality.

Squaring Property of Equality

> If both sides of an equation are squared, the solution set of the resulting equation will contain all of the solutions of the original equation.

This statement means that we *must* check all of the solutions in the original equation. Those values that do not satisfy the original equation are called ***extraneous roots*** and are eliminated from the solution set. Thus, the solution set of the previous example is $x = 4$.

The general procedure for solving a simple radical equation is to isolate the radical and then square both sides. Simplify and solve for the variable, but be sure to check for extraneous roots.

Example 1. $\sqrt{x} = 5$

Solution First, square both sides.

$$(\sqrt{x})^2 = 5^2$$
$$x = 25$$

Check in the original equation.

$$\sqrt{25} \stackrel{?}{=} 5 \quad \checkmark$$

The solution is $\boldsymbol{x = 25}$. ■

Example 2. $\sqrt{x} = -4$

Solution First square.

$$(\sqrt{x})^2 = (-4)^2$$
$$x = 16$$

Check: $\sqrt{16} \stackrel{?}{=} -4$
$$4 \neq -4$$

No solution, with extraneous root $x = 16$. ■

Example 3. $\sqrt{1 - x} = 2$

Solution

$$(\sqrt{1 - x})^2 = (2)^2$$
$$1 - x = 4$$
$$-x = 3$$
$$x = -3$$

Check: $\sqrt{1 - (-3)} \stackrel{?}{=} 2,\ \sqrt{4} = 2 \quad \checkmark$

The solution is $\boldsymbol{x = -3}$. ■

Example 4. $\sqrt{2x - 7} = \sqrt{3}$

Solution

$$(\sqrt{2x - 7})^2 = (\sqrt{3})^2$$
$$2x - 7 = 3$$
$$2x = 10$$
$$x = 5$$

Check: $\sqrt{2 \cdot 5 - 7} \stackrel{?}{=} \sqrt{3},\ \sqrt{3} = \sqrt{3} \quad \checkmark$

The solution is $\boldsymbol{x = 5}$. ■

Example 5. $\sqrt{x - 1} = \sqrt{11 - 3x}$

Solution

$$(\sqrt{x - 1})^2 = (\sqrt{11 - 3x})^2$$
$$x - 1 = 11 - 3x$$
$$4x = 12$$
$$x = 3$$

Check: $\sqrt{3 - 1} \stackrel{?}{=} \sqrt{11 - 9},\ \sqrt{2} = \sqrt{2} \quad \checkmark$

The solution is $\boldsymbol{x = 3}$. ■

Example 6. $\sqrt{3-x} = x - 3$

Solution

$$(\sqrt{3-x})^2 = (x-3)^2$$
$$3 - x = x^2 - 6x + 9$$
$$0 = x^2 - 5x + 6$$
$$0 = (x-2)(x-3)$$
$$x = 2, 3$$

Check: $x = 2$: $\sqrt{3-2} \stackrel{?}{=} 2 - 3$, $1 \neq -1$, does not check
$x = 3$: $\sqrt{3-3} \stackrel{?}{=} 3 - 3$, $0 = 0$ ✓

The solution is $\boldsymbol{x = 3}$**,** with extraneous root $x = 2$. ■

Example 7. $\sqrt{x-1} + 3 = x$

Solution Isolate the radical before squaring.

$$\sqrt{x-1} = x - 3$$
$$x - 1 = x^2 - 6x + 9$$
$$0 = x^2 - 7x + 10$$
$$(x-2)(x-5) = 0$$
$$x = 2, 5$$

Check: $x = 2$: $\sqrt{1} + 3 \stackrel{?}{=} 2$, $4 \neq 2$
$x = 5$: $\sqrt{4} + 3 \stackrel{?}{=} 5$, $5 = 5$ ✓

The solution is $\boldsymbol{x = 5}$**,** with extraneous root $x = 2$. ■

Example 8. A certain number is equal to the square root of the sum of the number and 6. Find the number.

Solution

$$\text{NUMBER} = \sqrt{\text{NUMBER} + 6}$$

Let $n = \text{NUMBER}$, and

$$n = \sqrt{n+6}$$
$$n^2 = n + 6$$
$$n^2 - n - 6 = 0$$
$$(n+2)(n-3) = 0$$
$$n = -2, 3$$

The root $n = 3$ checks and $n = -2$ does not.
The certain number is 3. ■

Problem Set 9.5.

Solve and note the extraneous roots, if any, in Problems 1–30. See Examples 1–5.

1. $\sqrt{x} = 3$
2. $\sqrt{y} = 5$
3. $\sqrt{z} = 4$
4. $\sqrt{a} = -2$
5. $\sqrt{b} = -3$
6. $\sqrt{c} = -1$
7. $\sqrt{x+3} = 5$
8. $\sqrt{1-x} = -2$

9. $\sqrt{x-1}=7$
10. $\sqrt{3x+4}=5$
11. $\sqrt{2x-1}=7$
12. $\sqrt{5-4x}=3$
13. $2\sqrt{x}-1=5$
14. $7=3\sqrt{x}+1$
15. $7-5\sqrt{x}=3$
16. $\sqrt{4+3x}=x$
17. $\sqrt{x+6}=x$
18. $\sqrt{4x+12}=x$
19. $\sqrt{x-2}=\sqrt{3}$
20. $\sqrt{x+4}=\sqrt{7}$
21. $\sqrt{5-x}=\sqrt{2}$
22. $\sqrt{x+3}=2\sqrt{x}$
23. $\sqrt{5+7x}=3\sqrt{x}$
24. $\sqrt{2x}=\sqrt{x+4}$
25. $\sqrt{3x-1}=\sqrt{7-x}$
26. $\sqrt{1-4x}=\sqrt{2x+13}$
27. $\sqrt{2x-5}=\sqrt{5x+4}$
28. $\sqrt{6x+7}=\sqrt{2x-5}$
29. $\sqrt{13-2x}=\sqrt{3x-2}$
30. $\sqrt{5-2x}=\sqrt{3x-10}$

Solve Problems 31–44. Note extraneous roots, if any. See Examples 6 and 7.

31. $\sqrt{5-x}=x-5$
32. $\sqrt{x+3}=3+x$
33. $\sqrt{x-2}+x=2$
34. $x=1+\sqrt{1-x}$
35. $\sqrt{2x+3}=x-6$
36. $\sqrt{3x+1}=x-1$
37. $\sqrt{5x-4}=x-2$
38. $\sqrt{7x-3}=x+1$
39. $2\sqrt{3x+4}-x=3$
40. $x=3\sqrt{2x+6}-7$
41. $2=x+\sqrt{6-5x}$
42. $\sqrt{10-3x}+2=2x$
43. $\sqrt{3x}+2=2x-1$
44. $\sqrt{2x}+5=x+1$

For Problems 45–50, design an equation that can be used to answer each question, solve the equation, and state an answer. See Example 8.

45. The square root of the sum of a number and five is three. Find the number.
46. The square root of the difference of a number less two is three. Find the number.
47. A number is equal to the square root of the product of six and that number. Find the number.
48. A number is equal to twice the square root of the product of three and that number. Find the number.
49. Two less than a certain number is equal to the square root of the sum of four and that number. Find the number.
50. Five less than a certain number is equal to the square root of the difference of the number less three. Find the number.

9.6 Review

This chapter introduced you to the simplification and manipulation of irrational numbers. The discussion was limited to square roots only, and you will have a great deal of practice with other roots in your subsequent courses. It is important to *think* of irrational numbers as *numbers,* not as "decimals that go on." For example, $\sqrt{2}$, π, and $3+\sqrt{5}$ are all numbers in their simplest form, and you need not say "$\sqrt{2}$ is about 1.4142" any more than you need to say "2 is about 1.9989."

In the next chapter you will need irrational numbers to solve quadratic equations that are not factorable.

Self-Test

Simplify the expressions in Problems 1–7 and express your answer in simplest form. All variables represent positive real numbers.

[9.1] **1.** **a.** $\sqrt{36}$ **b.** $-\sqrt{4}$ **c.** $\sqrt{\frac{1}{25}}$ **d.** $\sqrt{\frac{4}{9}}$

[9.2] **2.** **a.** $\sqrt{27}$ **b.** $\sqrt{a^3}$ **c.** $\sqrt{18x^2}$ **d.** $\sqrt{32x^7}$

[9.2] **3.** **a.** $\frac{1}{\sqrt{8}}$ **b.** $\frac{2}{\sqrt{x}}$ **c.** $\frac{6}{2\sqrt{3}}$ **d.** $\sqrt{\frac{3x}{5}}$

[9.3] **4.** **a.** $\sqrt{8} + \sqrt{18}$ **b.** $\sqrt{2} - \sqrt{12} + \sqrt{32}$
c. $4\sqrt{3} - 5\sqrt{18}$ **d.** $\sqrt{8} + \frac{1}{\sqrt{2}}$

[9.3] **5.** **a.** $\frac{4 + \sqrt{8}}{6}$ **b.** $\frac{12 - \sqrt{32}}{16}$
c. $\frac{\sqrt{2} + \sqrt{3}}{\sqrt{3}}$ **d.** $\frac{3 + \sqrt{3}}{\sqrt{3}}$

[9.4] **6.** **a.** $(\sqrt{2} - 1)(\sqrt{2} + 1)$ **b.** $(\sqrt{a} + 1)^2$
c. $\sqrt{2}(\sqrt{8} + \sqrt{6})$ **d.** The product of $\sqrt{3} + 1$ and its conjugate.

[9.4] **7.** **a.** $\frac{3}{2 - \sqrt{3}}$ **b.** $\frac{2}{3 + \sqrt{5}}$
c. $\frac{5\sqrt{2}}{\sqrt{6} - 4}$ **d.** $\frac{2 - \sqrt{3}}{1 + \sqrt{3}}$

[9.5] **8.** *Solve and note the extraneous roots, if any.*
a. $\sqrt{x} = 4$ **b.** $\sqrt{x} = -25$
c. $\sqrt{3x + 4} = x$ **d.** $\sqrt{2 - x} - 3x = 8$

[9.2, 9.3, 9.4] **9.** *Answer the following. Approximate, if necessary, to the nearest tenth.*
a. The time t in seconds required to learn a list of n symbols is given by

$$t = k\sqrt{n^3}$$

where k is the difficulty factor. Find the time for a list of 10 symbols if $k = 0.2$
b. The original speed s (in mph) of a car can be determined by the length D (in ft) of the skid marks by

$$s = \sqrt{30Df}$$

where f is the coefficient of friction for the particular tire and surface. Find the speed of a car that leaves 150 ft of skid marks on a surface with $f = 0.45$

[9.5] **10.** Design an equation to answer the question, solve the equation, and state the answer. One more than twice a certain number is equal to the square root of the sum of the number and 2. Find the number(s).

Review Problems

Simplify the expressions in Problems 1–87 and express your answer in simplest form. All variables represent positive real numbers.

1. $\sqrt{9}$ **2.** $-\sqrt{25}$ **3.** $\sqrt{81}$

4. $\sqrt{\frac{1}{49}}$ **5.** $\sqrt{\frac{1}{25}}$ **6.** $-\sqrt{\frac{16}{81}}$

7. $-\sqrt{0.25}$ **8.** $\sqrt{1.21}$ **9.** $\sqrt{2.25}$

10. $\sqrt{8}$ **11.** $-\sqrt{125}$ **12.** $\sqrt{98}$

13. $\sqrt{2}\sqrt{6a}$ **14.** $\sqrt{45b^3}$ **15.** $\sqrt{3a}\sqrt{6a}$

16. $\frac{1}{\sqrt{7}}$ **17.** $-\frac{1}{\sqrt{12}}$ **18.** $\frac{1}{\sqrt{5}}$

19. $-\sqrt{\frac{3}{5}}$ **20.** $\sqrt{\frac{4}{7}}$ **21.** $\sqrt{\frac{3}{2}}$

22. $\sqrt{\frac{2x}{3}}$ **23.** $-\sqrt{\frac{2x}{5y}}$ **24.** $\sqrt{\frac{5z}{2}}$

25. $\sqrt{54}$ **26.** $\sqrt{56}$ **27.** $-\sqrt{50}$

28. $2\sqrt{3} + \sqrt{3}$ **29.** $\sqrt{5} - 2\sqrt{5}$ **30.** $3\sqrt{7} + 4\sqrt{7}$

31. $\sqrt{12} - \sqrt{3} + \sqrt{27}$ **32.** $\sqrt{8} - \sqrt{18} - \sqrt{2}$ **33.** $\sqrt{20} + \sqrt{25} - \sqrt{80}$

34. $2\sqrt{3} + \frac{1}{\sqrt{3}}$ **35.** $\frac{1}{\sqrt{2}} - 3\sqrt{8}$ **36.** $\sqrt{6} + \sqrt{\frac{2}{3}}$

37. $\frac{5}{\sqrt{5}} + \sqrt{75}$ **38.** $\sqrt{\frac{3}{5}} - \sqrt{60}$ **39.** $\frac{5}{\sqrt{11}} + \frac{6}{\sqrt{11}}$

40. $\sqrt{2} + \sqrt{6} + \sqrt{8}$ **41.** $\sqrt{3} + \sqrt{6} + \sqrt{9}$ **42.** $\sqrt{3} + \sqrt{9} + \sqrt{12}$

43. $\sqrt{2}(\sqrt{2} + \sqrt{6} + \sqrt{8})$ **44.** $\sqrt{3}(\sqrt{3} + \sqrt{6} + \sqrt{9})$ **45.** $\sqrt{3}(\sqrt{3} + \sqrt{9} + \sqrt{12})$

46. $\frac{\sqrt{27} + 6}{3}$ **47.** $\frac{4 - \sqrt{12}}{2}$ **48.** $\frac{9 + 2\sqrt{18}}{3}$

49. $\frac{\sqrt{27} + 6}{\sqrt{3}}$ **50.** $\frac{4 - \sqrt{12}}{\sqrt{2}}$ **51.** $\frac{9 + 2\sqrt{18}}{\sqrt{3}}$

52. $\frac{\sqrt{8} + \sqrt{18}}{\sqrt{2}}$ **53.** $\frac{\sqrt{27} + \sqrt{12}}{\sqrt{3}}$ **54.** $\frac{\sqrt{20} - \sqrt{45}}{\sqrt{5}}$

55. $\frac{2(\sqrt{3} + 5\sqrt{3})}{3}$ **56.** $\frac{3(\sqrt{2} + 5\sqrt{2})}{6}$ **57.** $\frac{5(2\sqrt{5} + 5\sqrt{5})}{7}$

58. $(1 + \sqrt{3})(1 - \sqrt{3})$ **59.** $(2 - \sqrt{2})(2 + \sqrt{2})$ **60.** $(2 - \sqrt{5})(3 + \sqrt{5})$

61. $(2 + \sqrt{3})(1 - \sqrt{3})$ **62.** $(3 - \sqrt{2})(2 + \sqrt{2})$ **63.** $(5 - \sqrt{5})(3 + \sqrt{5})$

64. $(2\sqrt{3} - \sqrt{5})(2\sqrt{3} + \sqrt{5})$ **65.** $(3\sqrt{2} - \sqrt{7})(3\sqrt{2} + \sqrt{7})$ **66.** $(5\sqrt{3} + \sqrt{7})(5\sqrt{3} - \sqrt{7})$

67. $(5 + \sqrt{x})(5 - \sqrt{x})$ **68.** $(7 - \sqrt{y})(7 + \sqrt{y})$ **69.** $(3 + 2\sqrt{z})(3 - 2\sqrt{z})$

70. $\frac{2}{1 + \sqrt{3}}$ **71.** $\frac{3}{2 - \sqrt{2}}$ **72.** $\frac{5}{3 - \sqrt{5}}$

73. $\frac{5}{3 - \sqrt{2}}$ **74.** $\frac{2}{5 - 3\sqrt{3}}$ **75.** $\frac{5}{5 + 2\sqrt{5}}$

76. $\dfrac{\sqrt{5}}{4+\sqrt{5}}$ **77.** $\dfrac{\sqrt{3}}{2\sqrt{3}-5}$ **78.** $\dfrac{\sqrt{7}}{3-\sqrt{7}}$

79. $\dfrac{5+\sqrt{3}}{1-\sqrt{3}}$ **80.** $\dfrac{1+\sqrt{3}}{2+\sqrt{3}}$ **81.** $\dfrac{2-\sqrt{3}}{1-\sqrt{3}}$

82. $\dfrac{1-\sqrt{a}}{1+\sqrt{a}}$ **83.** $\dfrac{\sqrt{b}+1}{\sqrt{b}-1}$ **84.** $\dfrac{\sqrt{c}-2}{\sqrt{c}+2}$

85. $\dfrac{3-\sqrt{5}}{2+\sqrt{5}}$ **86.** $\dfrac{5-\sqrt{7}}{3+\sqrt{7}}$ **87.** $\dfrac{\sqrt{11}+3}{\sqrt{11}-4}$

Solve and note the extraneous roots, if any, in Problems 88–100.

88. $\sqrt{x}=5$ **89.** $-3=\sqrt{y}$ **90.** $\sqrt{z}=7$

91. $\sqrt{2x-1}=5$ **92.** $7=\sqrt{1-3x}$ **93.** $\sqrt{5x-1}=-7$

94. $x-2=\sqrt{2x-4}$ **95.** $x+4=\sqrt{2x+23}$ **96.** $\sqrt{7-x}=x-1$

97. $\sqrt{-(x+2)}=x+2$ **98.** $\sqrt{12x}-x=3$ **99.** $x+2=\sqrt{4x+5}$

100. $x+\sqrt{3x+15}=1$

10

Quadratic Equations

*<u>Why</u> and <u>how</u> are words
so important that
they cannot be used too often.*

Napoleon Bonaparte

*Imagination is more important
than knowledge.*

Albert Einstein

10.1 Square Root Property

Recall that a quadratic equation has the form $ax^2 + bx + c = 0$, where $a \neq 0$. In Section 5.5 we solved such equations by factoring. Here is an example.

$$
\begin{aligned}
x^2 &= 25 \\
x^2 - 25 &= 0 \\
(x-5)(x+5) &= 0 \\
x = 5 \quad \text{or} \quad x &= -5
\end{aligned}
$$

However, as explained in Chapter 5, not all quadratics are factorable. In this chapter we will develop more general methods for solving these equations. Consider

$$x^2 = 25$$

It is tempting to say "take the square root of both sides" so that

$$x = 5$$

but this is only a partial solution, since $x = -5$ also satisfies the equation. Therefore, we use the following property of square root for solving quadratic equations.

Property of Square Roots

If $x^2 = a$, then $x = \sqrt{a}$ or $x = -\sqrt{a}$ or both.

The expression $x = \sqrt{a}$ or $x = -\sqrt{a}$ is sometimes written as $x = \pm\sqrt{a}$. The symbol $\pm$ is commonly used to mean $+$ or $-$.

Solve the following equations.

Example 1. $x^2 = 49$

Solution

$$
\begin{aligned}
x &= \pm\sqrt{49} \\
x = 7 \quad &\text{or} \quad x = -7
\end{aligned}
$$

■

Example 2. $x^2 = 7$

Solution

$$
\begin{aligned}
x &= \pm\sqrt{7} \\
x = \sqrt{7} \quad &\text{or} \quad x = -\sqrt{7}
\end{aligned}
$$

■

Example 3. $(x+1)^2 = 25$

Solution

$$
\begin{aligned}
(x+1) &= \pm\sqrt{25} \\
x + 1 &= \pm 5 \\
x &= -1 \pm 5 \\
x = 4 \quad &\text{or} \quad x = -6
\end{aligned}
$$

■

Sometimes one or more of the roots may be excluded because of a restriction on the problem.

Example 4. One square is two units longer on each side than a smaller square. The larger square has an area of 36 square units. What is the length of a side of the smaller square?

Solution

$$(\text{LENGTH OF LARGER SIDE})^2 = 36$$
$$[(\text{LENGTH OF SHORTER SIDE}) + 2]^2 = 36$$

Let $s =$ the LENGTH OF THE SHORTER SIDE.

$$(s + 2)^2 = 36$$
$$(s + 2) = \pm\sqrt{36}$$
$$s + 2 = \pm 6$$
$$s = -2 \pm 6$$
$$s = 4 \quad \text{or} \quad s = -8$$

The value of -8 is excluded, because length is nonnegative.

The side of the smaller square is four units. ■

The equation in Example 4 could have been solved by factoring.

$$(s + 2)^2 = 36$$
$$s^2 + 4s + 4 = 36$$
$$s^2 + 4s - 32 = 0$$
$$(s - 4)(s + 8) = 0$$
$$s - 4 = 0 \quad \text{or} \quad s + 8 = 0$$
$$s = 4 \quad \text{or} \quad s = -8$$

However, note that the equation in the next example cannot be solved by factoring.

Example 5. A number is three more than a second number whose square is seven. Find the number.

Solution

$$(\text{SECOND NUMBER})^2 = 7$$
$$[(\text{FIRST NUMBER}) - 3]^2 = 7$$

Let $n =$ the FIRST NUMBER.

$$(n - 3)^2 = 7$$
$$(n - 3) = \pm\sqrt{7}$$
$$n - 3 = \pm\sqrt{7}$$
$$n = 3 \pm \sqrt{7}$$

The number is either $3 + \sqrt{7}$ or $3 - \sqrt{7}$. ■

Example 5, like Example 4, can be written as a quadratic equation.

$$(n - 3)^2 = 7$$
$$n^2 - 6n + 9 = 7$$
$$n^2 - 6n + 2 = 0$$

But this equation does not factor with integers and hence must be solved by employing the property of square root.

Example 6. $(w-1)^2 = 4$

$(w-1) = \pm\sqrt{4}$ *Use the square root property.*

$w - 1 = \pm 2$ *Remember $\pm$.*

$w = 1 \pm 2$ *Solve for w.*

$w = 1 + 2$ or $w = 1 - 2$

$w = 3$ or $w = -1$ ■

Example 7. $(y+2)^2 = 3$

$(y+2) = \pm\sqrt{3}$

$y + 2 = \pm 3$

$y = -2 \pm \sqrt{3}$

$y = -2 + \sqrt{3}$ or $y = -2 - \sqrt{3}$ ■

Example 8. $(2z-5)^2 = 7$

$(2z-5) = \pm\sqrt{7}$

$2z - 5 = \pm\sqrt{7}$

$2z = 5 \pm \sqrt{7}$

$$z = \frac{5 \pm \sqrt{7}}{2}$$

$$z = \frac{5 + \sqrt{7}}{2} \quad \text{or} \quad z = \frac{5 - \sqrt{7}}{2}$$ ■

Problem Set 10.1

Solve each equation in Problems 1–30 by the square-root method. See Examples 1–3.

1. $x^2 = 36$ **2.** $y^2 = 64$ **3.** $z^2 = 121$
4. $x^2 = 5$ **5.** $y^2 = 10$ **6.** $z^2 = 12$
7. $x^2 - 9 = 0$ **8.** $y^2 - 1 = 0$ **9.** $z^2 - 4 = 0$
10. $4x^2 = 25$ **11.** $9y^2 = 100$ **12.** $25z^2 = 4$
13. $9x^2 - 16 = 0$ **14.** $16y^2 - 25 = 0$ **15.** $4z^2 - 9 = 0$
16. $(x-1)^2 = 25$ **17.** $(y-2)^2 = 9$ **18.** $(z-3)^2 = 4$
19. $(x+3)^2 = 16$ **20.** $(y+2)^2 = 9$ **21.** $(z+4)^2 = 4$
22. $(2x-5)^2 = 81$ **23.** $(5y-2)^2 = 64$ **24.** $(3y-2)^2 = 49$
25. $(2x-1)^2 = 1$ **26.** $(3y-1)^2 = 1$ **27.** $(5z-1)^2 = 1$
28. $(7x+3)^2 = 9$ **29.** $(5y-2)^2 = 4$ **30.** $(3z+5)^2 = 25$

Solve each of Problems 31–40, and state an answer. See Examples 4 and 5.

31. One square is 5 ft longer on each side than a smaller square. If the larger square has an area of 169 sq ft, what is the length of the sides of the smaller square?

32. One square is 3 m longer on each side than a smaller square. If the larger square has an area of 196 sq m, what is the length of the sides of the smaller square?

33. The area of one square is 36 sq cm greater than another square. If the length of the side of the smaller square is four-fifths that of the larger square, what is the length of the large square?

34. The area of one square is 81 sq in. greater than another square. If the length of the side of the smaller square is four-fifths that of the larger square, what is the length of the side of the larger square?
35. A number is 5 less than a second number whose square is 3. Find the first number.
36. A number is seven more than another number whose square is two. What is the first number?
37. If two more than a given number is squared, then the result is five. Find the number.
38. If three less than a number is squared, then the result is seven. What is the number?
39. If two less than three times a certain number is squared, the result is forty-nine. Find the number.
40. If five less than twice a given number is squared, the result is eighty-one. Find the number.

Solve each equation in Problems 41–46 by the square root method. If you have a calculator, estimate the roots to the nearest thousandth. See Examples 6–8.

41. $(x-2)^2 = 5$ **42.** $(y-5)^2 = 2$ **43.** $(2x-1)^2 = 8$
44. $(3y-1)^2 = 27$ **45.** $(5x-3)^2 = 6$ **46.** $(7y-5)^2 = 3$

10.2 Completing the Square

In the previous section certain equations were solved by using the square root property. But what about quadratics that do not appear as perfect squares? The work of this section is to rewrite them in the form of a perfect square and then apply the square root method to solve them. First, consider what makes a perfect square and how the coefficients are related.

$$(x-a)^2 = x^2 - 2ax + a^2$$
$$(x+a)^2 = x^2 + 2ax + a^2$$

Notice that in both cases if you take half of the coefficient of x and square it, you obtain the constant term. That is, $\frac{1}{2}(\pm 2a) = \pm a$ and $(\pm a)^2 = a^2$. Consider a particular example.

$x^2 + 6x + ?$
$x^2 + 6x + (\frac{1}{2}\cdot 6)^2$ *Take half of 6 and square.*
$x^2 + 6x + 3^2$
$x^2 + 6x + 9 = (x+3)^2$ *Write in factored form as a perfect square.*

In this operation, the necessary constant term is added to create a complete or perfect square. This process is called ***completing the square.***

Examples *Complete the square for each expression.*

Solutions

1. $x^2 + 10x + ?$

$x^2 + 10x + (\frac{1}{2}\cdot \mathbf{10})^2$ *Halve the middle coefficient and square it.*
$x^2 + 10x + 5^2$
$x^2 + 10x + 25 = (x+5)^2$

2. $x^2 - 24x + ?$ $\quad x^2 - 24x + (\frac{1}{2} \cdot \mathbf{24})^2$
$x^2 - 24x + 12^2$
$x^2 - 24x + 144 = (x - 12)^2$ *Watch the sign of the middle term.*

3. $x^2 + 3x + ?$ $\quad x^2 + 3x + (\frac{1}{2} \cdot \mathbf{3})^2$
$x^2 + 3x + (\frac{3}{2})^2$
$x^2 + 3x + \frac{9}{4} = (x + \frac{3}{2})^2$ ■

Now this process will be used to solve equations. Be careful that when you add a constant to complete the square, that it is added to *both* sides of the equation. The method is illustrated by the following examples.

Solve the following equations by completing the squares.

Example 4. $x^2 + 8x = 9$

Solution

$x^2 + 8x + ? = 9 + ?$

$x^2 + 8x + (\frac{1}{2} \cdot \mathbf{8})^2 = 9 + (\frac{1}{2} \cdot \mathbf{8})^2$ *Complete the square.*

$x^2 + 8x + 16 = 9 + 16$ *Simplify.*

$(x + 4)^2 = 25$ *Write in factored form.*

$x + 4 = \pm\sqrt{25}$ *Square root property.*

$x + 4 = \pm 5$

$x = -4 \pm 5$ *Solve for x.*

$x = -4 + 5$ or $x = -4 - 5$

$x = 1$ or $x = -9$ *State the solutions.* ■

Example 5. $x^2 - 4x + 1 = 0$

Solution

$x^2 - 4x + \mathbf{?} = -1 + \mathbf{?}$

$x^2 - 4x + \mathbf{2^2} = -1 + \mathbf{2^2}$

$(x - 2)^2 = 3$

$x - 2 = \pm\sqrt{3}$

$x = 2 \pm \sqrt{3}$

$x = 2 + \sqrt{3}$ or $x = 2 - \sqrt{3}$ ■

Irrational solutions are frequently approximated in decimal form. The solutions in Example 5 could be written as

$x = 2 \pm \sqrt{3}$

$\approx 2 \pm 1.732$

$x \approx 2 + 1.732$ or $x \approx 2 - 1.732$

$x \approx 3.732$ or $x \approx 0.268$

In each of the previous examples, the coefficient of x^2 was 1. If it is not 1, divide by the coefficient of x^2 before completing the square. This procedure is illustrated in the following example.

Example 6. Solve for x and approximate irrational solutions to the nearest tenth.

Solution

$$3x^2 + 10x + 6 = 0$$

$$x^2 + \frac{10}{3}x + 2 = 0$$

Remember to divide by the coefficient of x^2 before completing the square.

$$x^2 + \frac{10}{3}x = -2$$

$$x^2 + \frac{10}{3}x + \left(\frac{5}{3}\right)^2 = -2 + \left(\frac{5}{3}\right)^2$$

Complete the square.

$$\left(x + \frac{5}{3}\right)^2 = -2 + \left(\frac{5}{3}\right)^2$$

Factor and simplify.

$$= -2 + \frac{25}{9}$$

$$= -2 + \frac{25}{9}$$

$$= \frac{-18 + 25}{9}$$

$$= \frac{7}{9}$$

$$\left(x + \frac{5}{3}\right) = \pm\sqrt{\frac{7}{9}}$$

Apply the square root method.

$$x + \frac{5}{3} = \pm\frac{\sqrt{7}}{3}$$

$$x = -\frac{5}{3} \pm \frac{\sqrt{7}}{3}$$

Solve for x.

$$= \frac{-5 \pm \sqrt{7}}{3}$$

$$x = \frac{-5 + \sqrt{7}}{3} \quad \text{or} \quad x = \frac{-5 - \sqrt{7}}{3}$$

$$x \approx \frac{-5 + 2.65}{3} \quad \text{or} \quad x \approx \frac{-5 - 2.65}{3}$$

Approximate, using $\sqrt{7} \approx 2.65$.

$$\approx \frac{-2.35}{3} \qquad \approx \frac{-7.65}{3}$$

$$x \approx -0.8 \quad \text{or} \quad x \approx -2.6$$

■

Completing the Square and Areas

Most of the concepts that are developed in this algebra course also have a geometric approach. Completing the square is no exception and is, in fact, more graphic than most. Notice in the following illustrations that $x^2 + 6x$ can be interpreted as the area of the rectangle with sides x and $x + 6$.

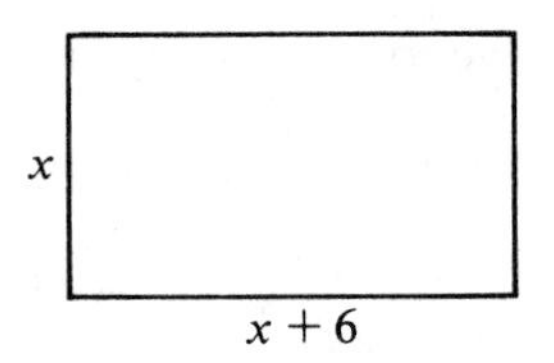

The side with length $x + 6$ can be divided into two parts, x and 6, and the 6 can be further divided into two equal lengths of 3 each.

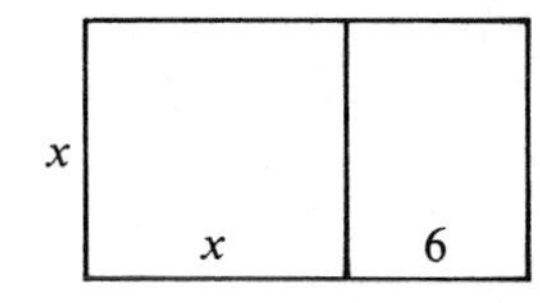

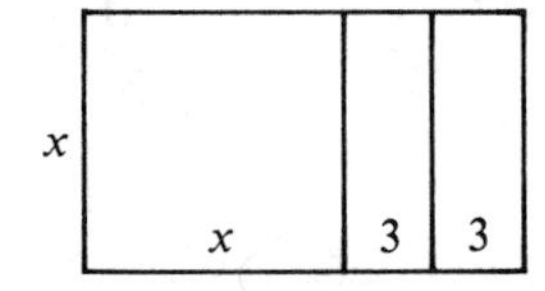

One of the long narrow rectangles of length 3 can be moved so that there are two lengths of $x + 3$. But doing this leaves a small square missing from a complete perfect square. If we add this little square of side 3, we add 3^2 or 9 to $x^2 + 6x$ and obtain a square with area $(x + 3)^2$ or $x^2 + 6x + 9$. These pictures show just what you do to complete the square to solve a quadratic.

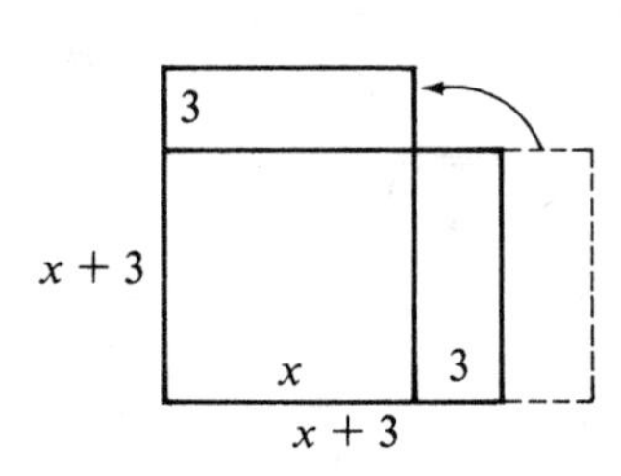

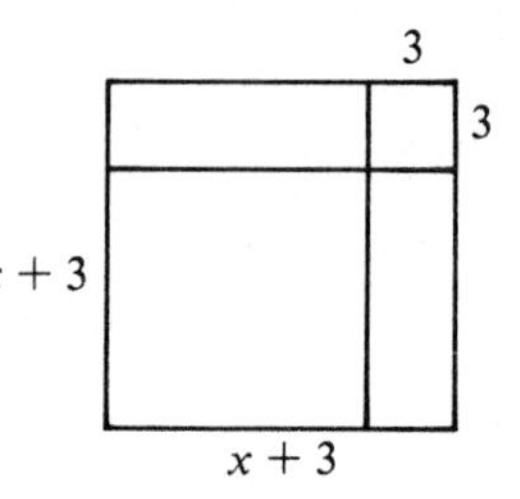

Problem Set 10.2

In Problems 1–20, give the number that is necessary to complete the square and write the perfect square in factored form. See Examples 1–3.

1. $x^2 + 8x$ **2.** $y^2 + 16y$ **3.** $z^2 + 2z$ **4.** $x^2 - 4x$
5. $y^2 - 10y$ **6.** $z^2 - 6z$ **7.** $x^2 + 24x$ **8.** $y^2 - 32y$
9. $z^2 + 28z$ **10.** $x^2 - 80x$ **11.** $y^2 + 50y$ **12.** $z^2 - 60z$
13. $x^2 + 3x$ **14.** $y^2 - 5y$ **15.** $z^2 + 7z$ **16.** $x^2 + x$
17. $y^2 - y$ **18.** $z^2 - 9z$ **19.** $x^2 - 11x$ **20.** $y^2 + 13y$

Solve each equation in Problems 21–40 by completing the square. See Examples 4 and 5.

21. $x^2 + 8x + 12 = 0$ **22.** $z^2 + 2z - 3 = 0$
23. $x^2 - 4x - 5 = 0$ **24.** $z^2 - 6z + 8$
25. $x^2 - 24x - 180 = 0$ **26.** $y^2 - 32y + 240 = 0$
27. $x^2 + 3x + 2 = 0$ **28.** $z^2 + 7z + 10 = 0$
29. $y^2 - y = 2$ **30.** $z^2 = 9z - 20$
31. $x^2 - 2x = 2$ **32.** $y^2 + 2y = 1$
33. $x^2 + 1 = 4x$ **34.** $z^2 + 4 = 6z$
35. $2y^2 + y = 6$ **36.** $3z^2 + z = 2$

37. $2x^2 - 3x + 1 = 0$ **38.** $2x^2 - 3x - 1 = 0$
39. $4x^2 + 16x + 3 = 0$ **40.** $4x^2 + 12x = 1$

Solve each equation in Problems 41–50 by completing the square. Approximate each solution to the nearest thousandth. See Example 6.

41. $9x^2 = 8x - 1$ **42.** $9y^2 = 18y - 2$
43. $4h^2 - 21 = 5h$ **44.** $7k - 10 = k^2$
45. $3m^2 - 10m + 7 = 0$ **46.** $9n^2 - 42n + 40 = 0$
47. $x^2 - 3x - 2 = 0$ **48.** $2y^2 - 10y + 3 = 0$
49. $x^2 + 2\sqrt{5}x + 5 = 0$ **50.** $y^2 - 2\sqrt{7}y + 7 = 0$

Illustrate geometrically the completion of the square in Problems 51–54. See example and discussion in section.

51. $y^2 + 4y$ **52.** $y^2 + 2y$ **53.** $x^2 + 2ax$ **54.** $x^2 + 6ax$

10.3 The Quadratic Formula

The purpose of the past two sections has been the solution of *any* quadratic, whether factorable or nonfactorable. Completing the square gives that ability, so consider *any* quadratic $ax^2 + bx + c = 0$. That is, if a, b, and c are any real numbers, $ax^2 + bx + c = 0$, $a \neq 0$, represents any quadratic, and if we solve for x we will have a formula that can be applied to any set of coefficients. We can proceed to complete the square and solve for x with this general equation. (Compare the steps here with the numerical Example 6 of the previous section.)

$$ax^2 + bx + c = 0 \qquad a \neq 0$$

$$x^2 + \frac{b}{a}x + \frac{c}{a} = 0$$

Divide by the leading coefficient, a.

$$x^2 + \frac{b}{a}x = -\frac{c}{a}$$

Isolate the x-terms.

$$x^2 + \frac{b}{a}x + \left(\frac{1}{2}\cdot\frac{b}{a}\right)^2 = \left(\frac{1}{2}\cdot\frac{b}{a}\right)^2 - \frac{c}{a}$$

Complete the square.

$$\left(x + \frac{b}{2a}\right)^2 = \frac{b^2}{4a^2} - \frac{c}{a}$$

Rewrite the factored form.

$$= \frac{b^2}{4a^2} - \frac{4ac}{4aa}$$

Add fractions, and simplify.

$$= \frac{b^2 - 4ac}{4a^2}$$

$$\left(x + \frac{b}{2a}\right) = \pm\sqrt{\frac{b^2 - 4ac}{4a^2}}$$

Apply the square root property.

$$x + \frac{b}{2a} = \frac{\pm\sqrt{b^2 - 4ac}}{2a}$$

Solve for x, and simplify.

$$x = \frac{-b \pm \sqrt{b^2 - 4ac}}{2a}$$

This result is called the ***Quadratic Formula,*** which can be used to solve any quadratic equation with real coefficients. You need only identify the coefficients a, b, and c and then substitute the values into the formula.

The Quadratic Formula

If $ax^2 + bx + c = 0$, $a \neq 0$, then

$$x = \frac{-b \pm \sqrt{b^2 - 4ac}}{2a}$$

In Examples 1–2, solve each equation, using the Quadratic Formula.

Example 1. $x^2 + 6x = 7$

First write the equation in the form $ax^2 + bx + c = 0$, and identify a, b, and c.

Solution

$\mathbf{1x^2 + 6x - 7 = 0}$

$\mathbf{a = 1 \quad b = 6 \quad c = -7}$

$$\mathbf{x = \frac{-b \pm \sqrt{b^2 - 4ac}}{2a}}$$

Recall the formula, and

$$= \frac{-6 \pm \sqrt{6^2 - 4(1)(-7)}}{2(1)}$$

substitute for a, b, and c.

$$= \frac{-6 \pm \sqrt{36 + 28}}{2}$$

Simplify.

$$= \frac{-6 \pm \sqrt{64}}{2}$$

$$= \frac{-6 \pm 8}{2}$$

$$= -3 \pm 4$$

$x = -3 + 4$ or $x = -3 - 4$

State both solutions.

$\mathbf{x = 1}$ **or** $\mathbf{x = -7}$ ■

Example 2. $x^2 - 4x + 2 = 0$

Solution

$\mathbf{1x^2 + (-4)x + 2 = 0}$

$\mathbf{a = 1 \quad b = -4 \quad c = 2}$

$$\mathbf{x = \frac{-b \pm \sqrt{b^2 - 4ac}}{2a}}$$

$$= \frac{-(-4) \pm \sqrt{(-4)^2 - 4(1)(2)}}{2(1)}$$

$$= \frac{4 \pm \sqrt{16 - 8}}{2}$$

$$= \frac{4 \pm \sqrt{8}}{2}$$

$$= \frac{4 \pm 2\sqrt{2}}{2}$$

$$= 2 \pm \sqrt{2}$$

$\mathbf{x = 2 + \sqrt{2}}$ **or** $\mathbf{x = 2 - \sqrt{2}}$ ■

In Examples 3–4, solve and approximate irrational roots to the nearest thousandth.

Example 3. $2x^2 = 2x + 1$

Solution

$$\mathbf{2x^2 + (-2)x + (-1) = 0}$$

$$\boldsymbol{a = 2 \quad b = -2 \quad c = -1}$$

$$\boldsymbol{x = \frac{-b \pm \sqrt{b^2 - 4ac}}{2a}}$$

$$= \frac{-(-2) \pm \sqrt{(-2)^2 - 4(2)(-1)}}{2(2)}$$

$$= \frac{2 \pm \sqrt{4 + 8}}{4}$$

$$= \frac{2 \pm \sqrt{12}}{4}$$

$$= \frac{2 + 2\sqrt{3}}{4}$$

$$= \frac{1 \pm \sqrt{3}}{2}$$

$$x = \frac{1 + \sqrt{3}}{2} \quad \text{or} \quad x = \frac{1 - \sqrt{3}}{2}$$

$$\boldsymbol{x \approx 1.366 \quad \text{or} \quad x \approx -0.366}$$

■

Example 4. $3x^2 - 4x - 1 = 0$ *Here $a = 3$, $b = -4$, and $c = -1$.*

Solution

$$x = \frac{4 \pm \sqrt{16 - 4(3)(-1)}}{2(3)}$$

Substitute into the formula.

$$= \frac{4 \pm \sqrt{28}}{6}$$

Many of the mental steps shown in the other examples are omitted.

$$= \frac{4 \pm 2\sqrt{7}}{6}$$

$$= \frac{2 \pm \sqrt{7}}{3}$$

$$x = \frac{2 + \sqrt{7}}{3} \quad \text{or} \quad x = \frac{2 - \sqrt{7}}{3}$$

$$\boldsymbol{x \approx 1.549 \quad \text{or} \quad x \approx -0.215}$$

■

Problem Set 10.3

In Problems 1–10, specify the constants a, b, and c for each equation when written in the form $ax^2 + bx + c = 0$. See Examples 1–3.

1. $x^2 + 2x - 15 = 0$
2. $3x^2 - 4x + 1 = 0$
3. $6x^2 - 7x - 3 = 0$
4. $x^2 = 2x + 2$
5. $x^2 + 4 = 6x$
6. $x^2 + 8x + 1 = 0$
7. $4x^2 - 4x = 2$
8. $9x^2 = 6x + 2$
9. $9x^2 + 24x + 1 = 0$
10. $6x + 7 = 9x^2$

Solve Problems 11–31 using the Quadratic Formula. See Examples 1–4.

11. $x^2 + 5x - 14 = 0$
12. $x^2 + 2x - 15 = 0$
13. $x^2 + 6x - 16 = 0$
14. $2x^2 - 3x + 1 = 0$
15. $3x^2 - 4x + 1 = 0$
16. $4x^2 + 3x - 1 = 0$
17. $6x^2 + 7x + 2 = 0$
18. $6x^2 - 7x - 3 = 0$
19. $6x^2 + 5x + 1 = 0$
20. $x^2 = 2x - 1$
21. $x^2 = 4x - 4$
22. $x^2 = 2x + 14$
23. $x^2 + 1 = 4x$
24. $x^2 + 4 = 6x$
25. $x^2 + 23 = 10x$
26. $x^2 + 6x - 2 = 0$
27. $x^2 + 8x + 1 = 0$
28. $x^2 + 4x - 9 = 0$
29. $4x^2 - 4x = 1$
30. $4x^2 - 4x = 2$
31. $4x^2 - 4x = 4$

Solve each equation in Problems 32–40 and approximate irrational solutions to the nearest thousandth. See Examples 3–4.

32. $9x^2 = 6x + 1$
33. $9x^2 = 6x + 2$
34. $9x^2 = 6x + 4$
35. $2x^2 + 7 = 10x$
36. $9x^2 + 24x + 1 = 0$
37. $9x^2 + 1 = 4x$
38. $1 = 2x^2 + 3x$
39. $6x + 7 = 9x^2$
40. $6x^2 + x + 1 = 0$
41. Reconsider the equation:

$$ax^2 + bx + c = 0, \qquad a \neq 0$$

Subtract c from each side.

$$ax^2 + bx = -c$$

Multiply by $4a$.

$$4a^2x^2 + 4abx = -4ac$$

Now, *complete the square* and continue to derive the quadratic form. Does this alternative derivation offer any advantages over the standard derivation shown in this section? Comment.

10.4 Solution of Quadratic Equations

Have you ever watched an experienced mechanic reach instinctively for the correct wrench from a jumble of various tools? A lot of work has resulted in an intimate familiarity with the use of each tool. You are in a similar position with equations. You now have several ways to examine and solve a quadratic equation. Before you pick up a particular wrench, are you certain that it is the best one for the situation? Let us stand back and look at the methods applicable to a quadratic equation and see if we can develop some instinct for the best method to apply in various cases.

The simplest quadratic has the form $ax^2 = b$ or $x^2 = \frac{b}{a}$, *where* a and b are constants. The square of the variable is equal to some number. This particular equation is probably easiest to solve by taking the square root of each side of the equation. However, we could also factor or use the quadratic formula. Which of these methods seems most efficient?

Example **1.** $x^2 = 4$

Square Root	**Factoring**	**Formula**
$x^2 = 4$	$x^2 = 4$	$x^2 = 4$
$x = \pm\sqrt{4}$	$x^2 - 4 = 0$	$x^2 + 0x - 4 = 0$
$x = \pm 2$	$(x - 2)(x + 2) = 0$	$x = \dfrac{0 \pm \sqrt{0 + 16}}{2} = \pm\dfrac{4}{2}$
		$= \pm 2$
$x = 2$ or $x = -2$	$x = 2$ or $x = -2$	$x = 2$ or $x = -2$ ■

The formula method does look overpowering for this equation. Consider another, more involved equation and compare the methods.

Example **2.** $x^2 = 6x$

Square Root	**Factoring**	**Formula**
$x^2 = 6x$	$x^2 = 6x$	$x^2 = 6x$
$x^2 - 6x = 0$	$x^2 - 6x = 0$	$x^2 - 6x + 0 = 0$
$x^2 - 6x + 9 = 9$	$x(x - 6) = 0$	$x = \dfrac{6 \pm \sqrt{36 - 0}}{2}$
$(x - 3)^2 = 9$	$x = 0$ or $x = 6$	$x = \dfrac{6 \pm 6}{2} = 3 \pm 3$
$x - 3 = \pm 3$		$x = 6$ or $x = 0$
$x = 3 \pm 3$		
$x = 6$ or $x = 0$		■

To apply the square root method in Example 2, you first had to complete the square. Again, the formula was not really needed.

Example **3.** $x^2 - 8x = 20$

Square Root	**Factoring**	**Formula**
$x^2 - 8x = 20$	$x^2 - 8x = 20$	$x^2 - 8x = 20$
$x^2 - 8x + 16 = 20 + 16$	$x^2 - 8x - 20 = 0$	$x^2 - 8x - 20 = 0$
$(x - 4)^2 = 36$	$(x + 2)(x - 10) = 0$	
$(x - 4) = \pm\sqrt{36}$	$x = -2$ or $x = 10$	$x = \dfrac{8 \pm \sqrt{64 + 80}}{2}$
$x - 4 = \pm 6$		$x = \dfrac{8 \pm \sqrt{144}}{2} = \dfrac{8 \pm 12}{2}$
$x = 4 \pm 6$		$= 4 \pm 6$
$x = 10$ or $x = -2$		$x = 10$ or $x = -2$ ■

Completing the square is becoming more involved as the equations become more complex. Factoring looks like the best method for Example 3.

Example 4. $(2x - 3)^2 = 25$

Square Root

$$(2x-3)^2 = 25$$
$$(2x-3) = \pm\sqrt{25}$$
$$2x - 3 = \pm 5$$
$$2x = 3 \pm 5$$
$$x = \frac{3 \pm 5}{2}$$
$$x = 4 \quad \text{or} \quad x = -1$$

Factoring

$$(2x-3)^2 = 25$$
$$4x^2 - 12x + 9 = 25$$
$$4x^2 - 12x - 16 = 0$$
$$x^2 - 3x - 4 = 0$$
$$(x+1)(x-4) = 0$$
$$x = -1 \quad \text{or} \quad x = 4$$

Formula

$$(2x-3)^2 = 25$$
$$4x^2 - 12x + 9 = 25$$
$$4x^2 - 12x - 16 = 0$$
$$x^2 - 3x - 4 = 0$$
$$x = \frac{3 \pm \sqrt{9+16}}{2}$$
$$x = \frac{3 \pm 5}{2}$$
$$x = 4 \quad \text{or} \quad x = -1$$ ■

In Example 4, the members are perfect squares, so the square root method applies nicely. Factoring works, but takes several steps before you can actually factor. The formula seems a little more involved than the other methods in this example.

Example 5. $x^2 - 6x - 1 = 0$

Square Root

$$x^2 - 6x - 1 = 0$$
$$x^2 - 6x = 1$$
$$x^2 - 6x + 9 = 1 + 9$$
$$(x-3)^2 = 10$$
$$(x-3) = \pm\sqrt{10}$$
$$x - 3 = \pm\sqrt{10}$$
$$x = 3 \pm \sqrt{10}$$
$$x = 3 + \sqrt{10} \quad \text{or}$$
$$x = 3 - \sqrt{10}$$

Factoring

$$x^2 - 6x - 1 = 0$$

(does not factor over the integers)

Formula

$$x^2 - 6x - 1 = 0$$
$$x = \frac{6 \pm \sqrt{36+4}}{2}$$
$$x = \frac{6 \pm \sqrt{40}}{2} = \frac{6 \pm 2\sqrt{10}}{2}$$
$$x = 3 \pm \sqrt{10}$$
$$x = 3 + \sqrt{10} \quad \text{or}$$
$$x = 3 - \sqrt{10}$$ ■

The equation in Example 5 does not factor, so that method is inoperative. The square root method and the formula can always be used. The formula seems more efficient here, because the equation is in the form to which the formula can be directly applied.

Example 6. $2x^2 - 3x - 1 = 0$

Square Root

$$2x^2 - 3x - 1 = 0$$
$$2x^2 - 3x = 1$$
$$x^2 - \frac{3}{2}x = \frac{1}{2}$$

Factoring

$$2x^2 - 3x - 1 = 0$$

(does not factor)

Formula

$$2x^2 - 3x - 1 = 0$$
$$x = \frac{3 \pm \sqrt{9+8}}{4}$$
$$x = \frac{3 \pm \sqrt{17}}{4}$$

$$x^2 - \frac{3}{2}x + \frac{9}{16} = \frac{9}{16} + \frac{1}{2}$$

$$\sqrt{\left(x - \frac{3}{4}\right)^2} = \pm\sqrt{\frac{17}{16}}$$

$$x - \frac{3}{4} = \pm\frac{\sqrt{17}}{4}$$

$$x = \frac{3 \pm \sqrt{17}}{4}$$

■

Finally, factoring will not work in Example 6. Completing the square becomes more difficult, so the quadratic formula is the most efficient method. It may be helpful to summarize the results of these examples.

Solution of Quadratic Equations

> To solve a quadratic equation efficiently:
>
> 1. If both sides are *perfect squares,* use the square root method.
> 2. If both sides are not perfect squares, use the form $ax^2 + bx + c = 0$ and solve by factoring, *if the equation is factorable.*
> 3. If it is not factorable, use the *quadratic formula*
>
> $$x = \frac{-b \pm \sqrt{b^2 - 4ac}}{2a}$$

In the problem set we present a wide variety of quadratics. Select the most efficient method of solution after considering the alternatives. The word problems in the problem set are geometric, and both the Pythagorean Theorem and the computation of area provide quadratic expressions. The concluding example illustrates the solution of such a geometric problem.

Example 7. The length of a certain rectangle is 1 less than twice the width. The diagonal of the rectangle is 1 more than twice the width. What are the dimensions of the figure?

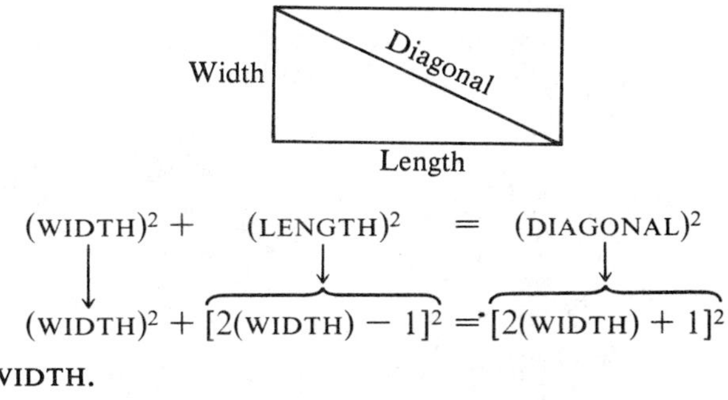

Solution Let w = WIDTH.

$$w^2 + (2w - 1)^2 = (2w + 1)^2$$
$$w^2 + 4w^2 - 4w + 1 = 4w^2 + 4w + 1$$

$$w^2 - 8w = 0 \quad \text{See Example 2.}$$
$$w(w - 8) = 0$$
$$w = 0 \quad \text{or} \quad w = 8 \quad \text{Reject } w = 0, \text{ since the width would be zero.}$$
$$2w - 1 = 15$$

The dimensions of the rectangle are 8 units by 15 units. ■

Problem Set 10.4

Solve each of the quadratic equations in Problems 1–30 by the most efficient method. See Examples 1–6.

1. $x^2 = 9$
2. $x^2 = 36$
3. $x^2 = 5$
4. $x^2 - 24 = 0$
5. $x^2 - 48 = 0$
6. $x^2 - 16 = 0$
7. $x^2 = 9x$
8. $x^2 = 6x$
9. $5x^2 = 3x$
10. $2x^2 + 3x = 0$
11. $3x^2 + 4x = 0$
12. $3x^2 - 2x = 0$
13. $x^2 + 8x = 20$
14. $x^2 + 9x = 36$
15. $x^2 + 7x + 2 = 0$
16. $6x^2 - 19x + 15 = 0$
17. $9x^2 + 9x - 28 = 0$
18. $7x^2 + 5 = 2x$
19. $(x - 5)^2 = 16$
20. $(x + 7)^2 = 9$
21. $(x + 3)^2 = 16$
22. $(2x + 1)^2 = 49$
23. $(3x - 2)^2 = 64$
24. $(2x - 5)^2 = 81$
25. $x^2 = 2x + 6$
26. $x^2 = 4x + 4$
27. $x^2 = 8x - 3$
28. $9x^2 + 12x = 1$
29. $5x^2 = 8x - 1$
30. $4x^2 + 1 = 6x$

For Problems 31–40, design an equation that can be used to answer each question, solve the equation, and state an answer.

31. The length of a rectangle is 7 cm more than twice is width. If the area is 114 sq cm, what are the dimensions of the figure?

32. One leg of a right triangle is 7 in. longer than the other. The hypotenuse is 3 in. longer than the longer leg. What are the dimensions of the triangle to the nearest inch?

33. A side of one square is 8 mm larger than the side of a second square. The sum of their areas is 424 sq mm. Find the side of the larger square.

34. A polygon of n sides has $\frac{1}{2}n(n - 3)$ diagonals. How many sides has a polygon with 135 diagonals?

35. The perimeter of a rectangle is 34 cm, and its diagonal is 13 cm. Find the dimensions of the figure.

36. One side of a rectangle is 60 ft. The diagonal is 50 ft longer than the other side. Find the dimensions of the figure.

37. One side of a triangle is three times the length of the other. If the hypotenuse is 30 cm, what are the lengths of the sides?

38. The length of a rectangle is one in. less than twice the width. What are the dimensions of the figure if the diagonal is 17 in.?

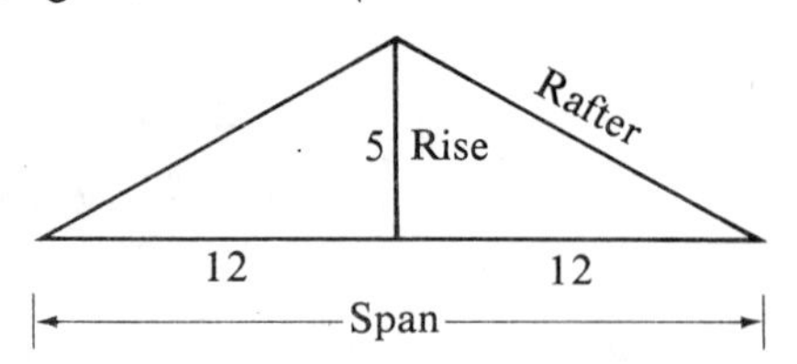

39. The pitch of a roof is the ratio of the rise to the half-span. The rafter length is the hypotenuse of a right triangle formed as shown in the figure. A common pitch is 5 : 12. For this pitch, what is the vertical rise for a rafter length of 20 ft?

40. If the pitch of a roof is 7 : 24, what is the span for a rafter length of 20 ft?

10.5 Graph of the Equation $y = ax^2 + bx + c$

When cannons were first introduced in the 13th century, their primary use was to demoralize the enemy. It was much later that they were employed for strategic purposes. In fact, cannons existed nearly three centuries before enough was known about the behavior of a projectile to use them with accuracy. The cannonball does not travel in a straight line, it was discovered, because of an unseen force called gravity. The path described by a projectile is called a ***parabola.*** Any projectile—a ball, an arrow, a bullet, a rock from a slingshot, and even water from the nozzle of a hose or sprinkler—travels a parabolic path. Consider Figure 10.1. It is a scale drawing (a graph) of the path of a cannonball fired in a particular way.

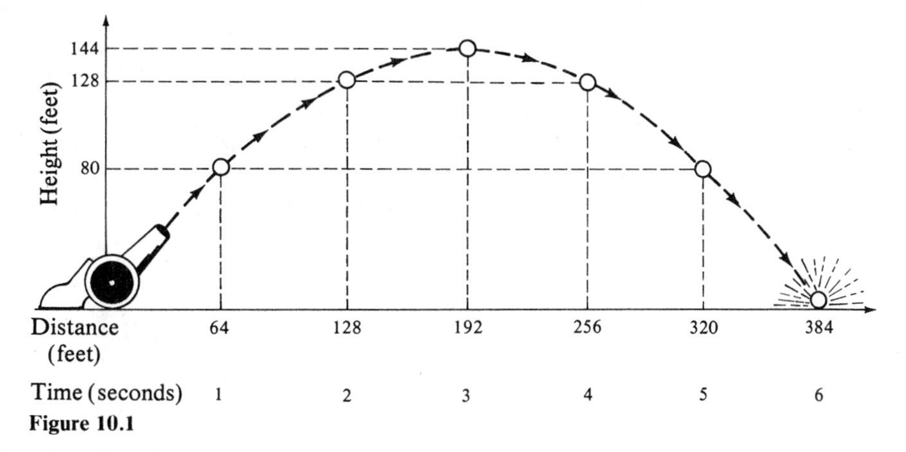

Figure 10.1

Note that the parabolic curve has a maximum height and is symmetric about a vertical line through that height. In other words, the ascent and descent paths are symmetric. Perhaps this curve looks familiar. If you have ever used a flashlight to illuminate your way in the dark, the shape of the lighted area was probably parabolic. This smooth curve has many interesting properties and applications. Most interesting, at the moment, is the fact that the parabola is the graph of a quadratic equation. Every equation of the form $y = ax^2 + bx + c$ has a parabola as its graph. Consider an example with $a = 1$, $b = -6$, and $c = 4$.

$$y = x^2 - 6x + 4$$

First compile the values for x and y,

x	y	$y = x^2 - 6x + 4$
0	4	$y = \mathbf{0}^2 - 6 \cdot \mathbf{0} + 4 = 4$
1	-1	$y = \mathbf{1}^2 - 6 \cdot \mathbf{1} + 4 = -1$
2	-4	$y = \mathbf{2}^2 - 6 \cdot \mathbf{2} + 4 = -4$
3	-5	$y = \mathbf{3}^2 - 6 \cdot \mathbf{3} + 4 = -5$
4	-4	$y = (?)^2 - 6(?) + 4 = ?$
5	-1	$y = ?$
6	4	

Now plot these points, as shown in Figure 10.2, and then sketch the parabola through these points, as in Figure 10.3. The graph displays the same shape as the

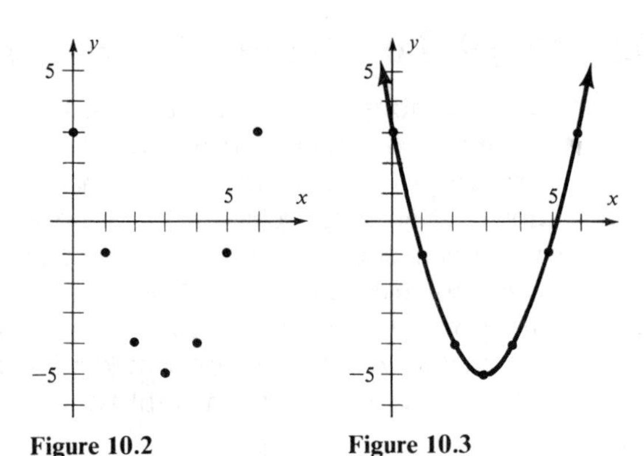

Figure 10.2 Figure 10.3

cannonball but opens upward instead of downward. If the constant a is positive, the parabola opens up; if a is negative, the parabola opens down. The equation of the path of the cannonball shown in Figure 10.1 is $y = -16x^2 + 96x$, where $a = -16$; a is negative, so the parabola opens down. The constant a also governs the shape of the parabola—that is, whether it is narrow, wide, steep, or flat. To illustrate, let us compile the values (see Table 10.1) and then sketch $y = ax^2$ for $a = \frac{1}{2}$, 1, 2, and 4 (see Figure 10.4).

Table 10.1
Values for $y = ax^2$

$a = \frac{1}{2}$

x	$\frac{1}{2}x^2$
-4	8
-3	$4\frac{1}{2}$
-2	2
-1	$\frac{1}{2}$
0	0
1	$\frac{1}{2}$
2	2
3	$4\frac{1}{2}$
4	8

$a = 1$

x	x^2
-4	16
-3	9
-2	4
-1	1
0	0
1	1
2	4
3	9
4	16

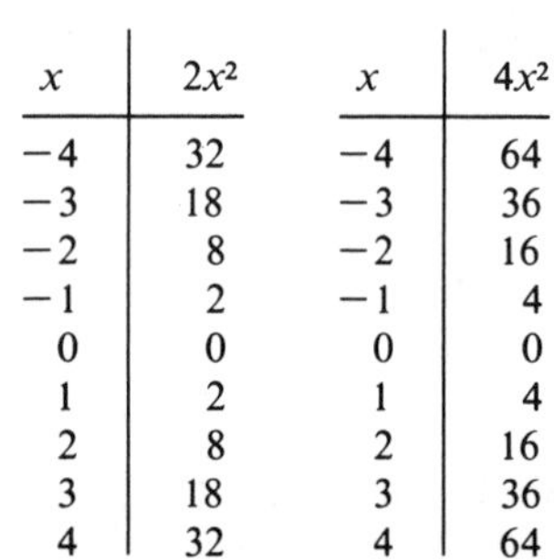

$a = 2$

x	$2x^2$
-4	32
-3	18
-2	8
-1	2
0	0
1	2
2	8
3	18
4	32

$a = 4$

x	$4x^2$
-4	64
-3	36
-2	16
-1	4
0	0
1	4
2	16
3	36
4	64

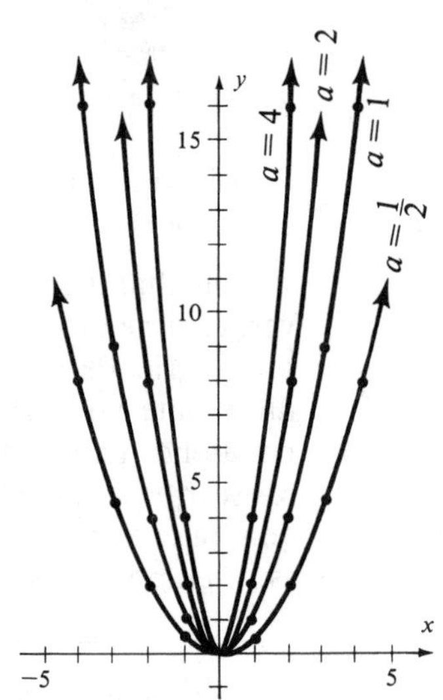

As the absolute value of the constant $|a|$ increases, the parabola is narrower, and, as $|a|$ decreases, the curve is wider. The constant a seems to control the shape of the parabola just as the constant m, in the linear equation $y = mx + b$, controls the slope of the line. Wider and narrower are, of course, relative terms. In this case, we take $y = x^2$ as the norm. Thus, we are saying that, as $|a|$ becomes greater than 1, the parabola is steeper than $y = x^2$.

In Examples 1–2, determine how the parabola opens from the constant a, and verify by plotting points and sketching the curve.

Example 1. $y = -2x^2 + 16x - 24$

Solution $a < 0$ (opens downward)
$|a| > 1$ (opens narrow)

x	y
0	−24
1	−10
2	0
3	6
4	8
5	6
6	0

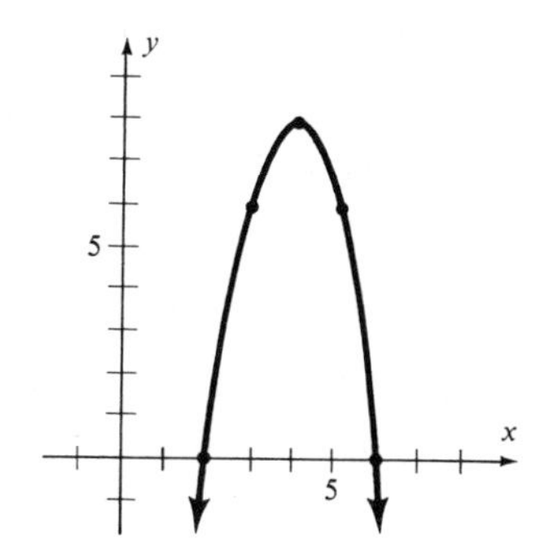

■

Example 2. $y = \frac{1}{4}x^2 + x$

Solution $a > 0$ (opens upward)
$|a| < 1$ (opens wide)

x	y
−6	3
−5	$1\frac{1}{4}$
−4	0
−3	$-\frac{3}{4}$
−2	−1
−1	$-\frac{3}{4}$
0	0
1	$1\frac{1}{4}$
2	3

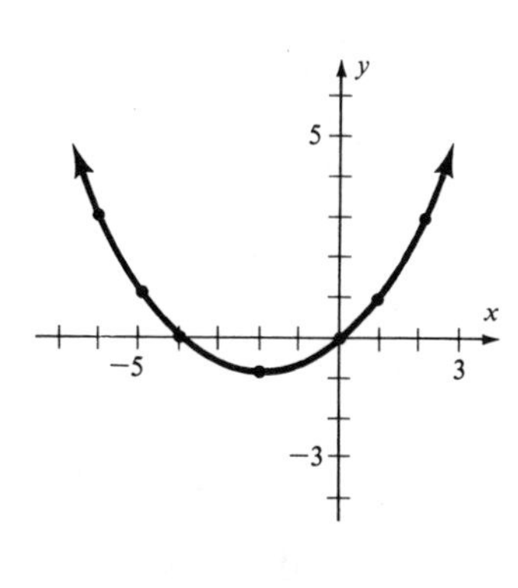

■

We have seen that the equation $y = ax^2 + bx + c$ has a parabola as its graph. Further, we can see that the shape of the curve depends on the constant a. But what about its position? We positioned a straight line by its y-intercept. To find the y-intercept of a parabola, let $x = 0$.

$$\begin{aligned} y &= ax^2 + bx + c \\ &= a \cdot 0^2 + b \cdot 0 + c \\ &= c \end{aligned}$$

Thus, $(0, c)$ is the **y-intercept** and can be determined by inspection, because c is the constant term in the equation. To find the x-intercept, let $y = 0$.

$$\begin{aligned} 0 &= ax^2 + bx + c \\ x &= \frac{-b \pm \sqrt{b^2 - 4ac}}{2a} \end{aligned}$$

Thus, the **x-intercepts** are the roots of a quadratic equation. These roots are the points on the x-axis that are also on the parabola.

Example 3. Find the intercepts of the parabola $y = 3x^2 - x - 10$.

Solution $(0, -10)$ is the y-intercept (by inspection). To find the x-intercepts, let $y = 0$ and solve.

$$0 = 3x^2 - x - 10$$
$$0 = (3x + 5)(x - 2)$$
$$3x + 5 = 0 \qquad x - 2 = 0$$
$$3x = -5 \qquad x = 2$$
$$x = -\frac{5}{3}$$

$(0, -10)$, $(-\frac{5}{3}, 0)$, and $(2, 0)$ are the intercepts. ■

Example 4. Find the intercepts of $y = x^2 - 2x - 2$.

Solution $(0, -2)$ is the y-intercept. Let $y = 0$.

$$0 = x^2 - 2x - 2$$

$$x = \frac{2 \pm \sqrt{4 + 8}}{2} = \frac{2 \pm \sqrt{12}}{2} = \frac{2 \pm 2\sqrt{3}}{2}$$

$$x = 1 \pm \sqrt{3}$$

$(0, -2)$, $(1 + \sqrt{3}, 0)$, and $(1 - \sqrt{3}, 0)$ are the intercepts. ■

Example 5. Determine how the parabola $y = -2x^2 + 3x - 5$ opens, find its intercepts. Make a rough sketch.

Solution $a = -2 < 0$ *The parabola opens downward.*
(0, −5) is the y-intercept.
There are no x-intercepts.

x	y
1	−4
2	−7
3	−14

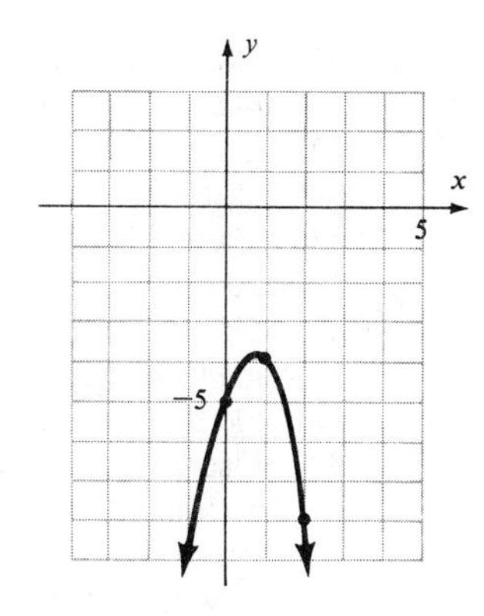

■

Problem Set 10.5

Compile a table of values, and sketch the graph of each curve in Problems 1–10. See Examples 1 and 2.

1. $y = 2x^2$
2. $y = -3x^2$
3. $y = -\frac{1}{2}x^2$
4. $y = \frac{1}{3}x^2$
5. $y = x^2 - 4$
6. $y = 9 - x^2$
7. $y = -3x^2 + 4$
8. $y = 2x^2 - 3$
9. $3y = x^2 + x + 6$
10. $5y = 2x^2 - 15$

In Problems 11–20 find the x- and y-intercepts of the curve. See Examples 3 and 4.

11. $y = x^2 + 8x - 33$
12. $y = x^2 - 11x - 26$
13. $y = 2x^2 + 5x - 1$
14. $y = 2x^2 - 9x - 5$
15. $y = x^2 - 3$
16. $y = x^2 - 5$
17. $y = x^2 + 4x - 1$
18. $y = x^2 - 6x + 7$
19. $y = 4x^2 - 8x - 3$
20. $y = 9x^2 - 6x - 4$

Determine how the parabolas in Problems 21–30 open, find the y-intercepts and sketch. See Example 5.

21. $y = 6x^2 - x - 15$
22. $y = 4x^2 - 12x + 9$
23. $y = \frac{1}{2}x^2 - x - 2$
24. $y = -\frac{1}{3}x^2 + x - \frac{1}{2}$
25. $y = -\frac{1}{5}x^2 + \frac{2}{5}x - 1$
26. $y = \frac{1}{4}x^2 + x + \frac{5}{4}$
27. $y = 3x^2 - 24x + 40$
28. $y = 3x^2 - 30x + 69$
29. $y = -\frac{1}{4}x^2 + 2x$
30. $y = -\frac{1}{2}x^2 + 4x - 3$

If an object is thrown vertically upward with an initial velocity of v feet per second, it is h feet high after t seconds. The height is given by $h = vt - 16t^2$. Problems 31–38 refer to this formula.

31. Sketch the graph for the height of a cannonball fired upward with an initial velocity of 128 fps. (fps means feet per second)
32. Use the graph in Problem 31 to determine when the cannonball reaches its maximum altitude.
33. Sketch the graph for the height of a baseball hurled upward with an initial velocity of 200 fps.
34. Use the graph in Problem 33 to determine when the ball reaches 600 ft in height. A given height is reached twice—going up and coming down.
35. Sketch the graph for the path of a projectile fired with an initial velocity of 1600 fps.
36. Use the graph in Problem 35 to determine how long it will take the projectile to reach an altitude of approximately 7000 ft.
37. An arrow is shot upward with an initial velocity of 160 fps. Use the graph to estimate how long it will take to reach 320 ft.
38. A rock is thrown upward at 200 fps. Use the graph to estimate how long it takes to reach 200 ft.

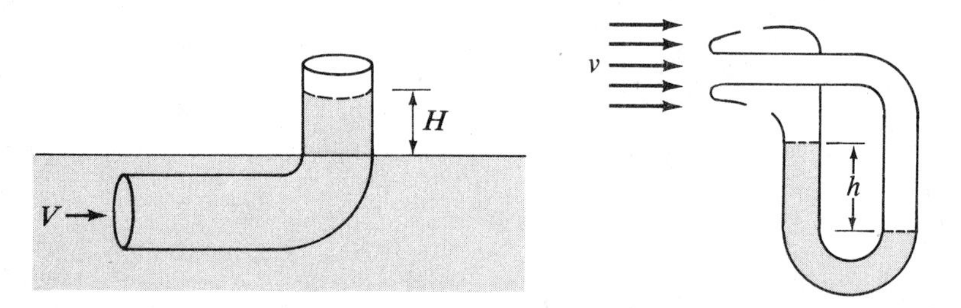

39. Torricelli's Law in physics may be used to measure the speed of moving water. The law states that if an L-shaped tube is used as shown, the height H of the column of water pushed up by the stream is related to the speed or velocity V by the quadratic equation $V^2 = 2gH$. The gravitational constant g is approximately 32 ft per second per second. The device could be used as a simple speedometer for a boat traveling in still water. Use the graph of this relationship to determine how the tube should be marked to be used as a speedometer.
40. Similar to the tube based on Torricelli's Law is a tube used to measure the flow of a gas. This one, shown in the diagram, is based on Bernoulli's principles and is called a Pitot tube. It may also be used as a speedometer indicating airspeed. The air rushes in and forces a column of liquid down, causing a difference h in the heights of the columns. The velocity v of the gas is given by the equation $v^2 = 2grh$, where g is the gravitational constant (approximately 32 feet per second per second) and r is the ratio of the densities of the fluid and the gas. Use the graph of the relationship for $r = \frac{5}{8}$ to determine how the tube might be marked to be used as a speedometer.

10.6 Review

In this, the last chapter, you see factoring (completing the square), radicals (square root method), and graphing (parabolas) brought together to treat the quadratic equation. You should now be able to find the real solutions for any linear or quadratic equation and graph any linear or quadratic function. The differences between the various methods of solution of quadratics should be clear and you should be able to choose the most efficient method for the given equation.

Self-Test

[10.1] **1.** Solve each equation by the square root method.
a. $x^2 = 169$ **b.** $3y^2 = 81$
c. $(x - 3)^2 = 121$ **d.** $(2x + 1)^2 = 25$

[10.2] **2.** Supply the missing number so as to complete the square.
a. $x^2 - 10x + ?$ **b.** $x^2 + 4x + ?$
c. $y^2 + 11y + ?$ **d.** $x^2 + x + ?$

[10.2] **3.** Solve each equation by completing the square.
a. $x^2 - 2x - 3 = 0$ **b.** $3y^2 + 2y - 1 = 0$
c. $x^2 - 4x + 1 = 0$ **d.** $9y^2 - 6y - 4 = 0$

[10.3] **4.** Solve each equation using the quadratic formula.
a. $x^2 - 2x - 3 = 0$ **b.** $3y^2 + 2y - 1 = 0$
c. $x^2 - 4x + 1 = 0$ **d.** $9y^2 - 6y - 4 = 0$

[10.4] **5.** Solve the equations by the most efficient method; estimate the roots to the nearest hundredth.
a. $5x^2 = 1$ **b.** $21y^2 - 29y = 10$
c. $x^2 = 4x + 3$ **d.** $3y^2 + 4y - 1 = 0$

[10.4] **6.** A length of a large field is 4 m more than four times its width. What are the dimensions of the field if the field is 41 m diagonally corner to corner?

[10.5] **7.** Find the x- and y-intercepts.
a. $y = x^2 - 5x - 24$ **b.** $y = 6x^2 + 11x - 10$
c. $y = x^2 - 4x - 1$ **d.** $y = x^2 - 3x - 2$

[10.5] **8.** Sketch the graph of each curve.
a. $y = 2x^2$ **b.** $y = -x^2 - 8x - 13$
c. $y = \frac{2}{3}x^2 - 4x$ **d.** $y = -\frac{2}{9}x^2 + 4x - 16$

[10.5] *If an object is thrown vertically upward with an initial velocity of v feet per second, it is h feet high after t seconds. The height is given by* $h = vt - 16t^2$.

9. Sketch the graph for the path of a cannonball fired with an initial velocity of 1280 fps.
10. Use the graph to determine how long it will take the cannonball to first reach an altitude of approximately 10,000 ft.

Review Problems

Solve each of the equations in Problems 1–15 by the square-root method.

1. $x^2 = 49$ **2.** $y^2 = 144$ **3.** $z^2 = 196$
4. $x^2 = 7$ **5.** $y^2 = 18$ **6.** $z^2 = 2$
7. $9x^2 = 4$ **8.** $16y^2 = 9$ **9.** $25z^2 = 64$
10. $(x + 3)^2 = 25$ **11.** $(y - 2)^2 = 4$ **12.** $(z - 5)^2 = 3$
13. $(2x - 3)^2 = 4$ **14.** $(3y - 4)^2 = 2$ **15.** $(4z + 1)^2 = 3$

Solve each of the equations in Problems 16–30 by completing the square.

16. $x^2 - 3x - 4 = 0$
17. $y^2 + 4x - 5 = 0$
18. $z^2 - 7z + 6 = 0$
19. $x^2 = 7x - 12$
20. $y^2 + 2y = 63$
21. $z^2 = 24 - 10z$
22. $x^2 - 4x + 1 = 0$
23. $y^2 - 6y + 4 = 0$
24. $z^2 - 4z + 2 = 0$
25. $2x^2 + 5x = 3$
26. $4y^2 = 5y + 6$
27. $3z^2 = 4z - 1$
28. $9x^2 + 1 = x$
29. $2y^2 = y - 1$
30. $2z^2 = 1 - z$

Solve each of the equations in Problems 31–45 using the quadratic formula.

31. $x^2 + 3x - 10 = 0$
32. $y^2 - y - 12 = 0$
33. $z^2 + 2z - 15 = 0$
34. $x^2 - 2x - 1 = 0$
35. $y^2 - 2y - 2 = 0$
36. $z^2 - 4z - 1 = 0$
37. $9x^2 = 9x - 2$
38. $4y^2 = 8y + 3$
39. $6z^2 = z + 1$
40. $2x + 1 = 2x^2$
41. $4y^2 + 8y + 1 = 0$
42. $6z = 2z^2 + 3$
43. $9x^2 = 24x + 1$
44. $9y^2 + 24y + 1 = 0$
45. $9z^2 + 24y = 1$

Solve the equations in Problems 46–75 by the most efficient method; estimate irrational roots to the nearest hundredth.

46. $x^2 = 4$
47. $y^2 = 8$
48. $z^2 = 16$
49. $x^2 - 32 = 0$
50. $y^2 - 64$
51. $z^2 - 96 = 0$
52. $x^2 = 8x$
53. $2y^2 - 3y = 0$
54. $3z^2 + 5z = 0$
55. $x^2 + 5x - 14 = 0$
56. $y^2 + 2y = 15$
57. $z^2 + 8z = 9$
58. $x^2 = 2x + 4$
59. $y^2 + 4y = 3$
60. $z^2 = 6z - 6$
61. $(x - 3)^2 = 25$
62. $(y + 2)^2 = 16$
63. $(z - 5)^2 = 1$
64. $6x^2 + 5x = 6$
65. $10y^2 = 1 - 3y$
66. $15z^2 = 4z + 4$
67. $(2x - 3)^2 = 5$
68. $(3y + 1)^2 = 2$
69. $(5z - 3)^2 = 7$
70. $x^2 - 3x + 1 = 0$
71. $9y^2 = 1 - 6y$
72. $30z = 25z^2 + 2$
73. $9x^2 + 12x = 7$
74. $5y^2 = 8y + 1$
75. $5z + 3 = z^2$

Find the x- and y-intercepts of each curve in Problems 76–85. Give all values in decimal form, estimating irrational values to the nearest hundredth.

76. $y = x^2 - 2$
77. $y = x^2 - 9$
78. $y = 2x^2 - 6x$
79. $y = x^2 - 4x - 21$
80. $y = x^2 - 4x - 20$
81. $y = x^2 + 7x - 19$
82. $y = x^2 + 6x + 2$
83. $y = x^2 - 6x - 2$
84. $y = 6x^2 + 11x - 7$
85. $y = 14x^2 + 13x - 12$

Sketch the graph of each curve given in Problems 86–100.

86. $y = x^2 - 6x + 7$
87. $y = x^2 + 4x$
88. $y = -x^2 + 6x - 7$
89. $y = 4x - x^2$
90. $y = 2x^2 - 4x + 2$
91. $y = 2x^2 - 8$
92. $2y = x^2 + 6x + 15$
93. $-2y = x^2 + 6x + 3$
94. $3y = -2x^2 + 4x + 16$
95. $3y = 2x^2 + 16x + 14$
96. $2y = x^2 - 4$
97. $y = \frac{3}{4}x^2 - 6x + 9$
98. $y = -\frac{5}{2}x^2 + 30x - 70$
99. $y = -\frac{2}{3}x^2 + 4x$
100. $y = \frac{2}{5}x^2 - 4x + 4$

Appendix

Answers to Selected Problems
Glossary
Index

Experience is the name
everyone gives to
their mistakes.

Oscar Wilde

Answers to Selected Problems

Problem Set 1.1, Pages 8–9

1. a. constant **b.** constant **c.** constant **d.** constant **e.** constant **f.** constant **g.** constant **h.** variable **i.** variable **3.** counting or natural **5.** 3.1416 (or 3.14 or 3.14159265358979323846 if you like) **7.** sum, terms **9.** product, factors **11.** $<$ or $\leq$ **13.** $>$ or $\geq$ **15.** $\leq$ or $\geq$ **17.** $<$ or $\leq$ **19.** $>$ or $\geq$ **21.** $30 = 5 \cdot 6$, correct **23.** $54 = 8 \cdot 7$, not correct, $\frac{54}{8} \neq 7$ **25.** $52 = 13 \cdot 4$, correct **27.** $128 = 4 \cdot 32$, correct **29.** $5 = 0 \cdot 0$, not correct, $\frac{5}{0} \neq 0$ **31.** $9 = 4 + 5$ **33.** $17 = 9 + 8$ **35.** $7 = 9 - 2$ **37.** $7 = \frac{63}{9}$ **39.** $63 > 5 \cdot 12$ In 41–60, n is used as the variable, although your answer may contain a different choice. **41.** $n + 1$ **43.** $3n$ **45.** $5 - n$ **47.** $7 + n$ or $n + 7$ **49.** $n \cdot 9$ **51.** $1 - n$ **53.** $8 + n$ or $n + 8$ **55.** $n + 10$ or $10 + n$ **57.** $3 + 2n$ or $2n + 3$ **59.** $n + m$

Problem Set 1.2, Pages 12–13

1. numerical expression **3.** neither (no operation) **5.** numerical expression **7.** numerical expression **9.** variable expression **11.** variable expression **13.** numerical expression **15.** answers will vary **17.** 23 **19.** 93 **21.** 6 **23.** 15 **25.** 2 **27.** 64 **29.** 20 **31.** 12 **33.** 6 **35.** 22 **37.** 16 **39.** 21 **41.** $3 + 2 \cdot 1 = 5$ **43.** $3 + \frac{2}{1} = 5$ **45.** $8(9 + 10) = 152$ **47.** $n + 3$ **49.** $10n + 3$

Problem Set 1.3, Pages 16–17

1. 2^3 **3.** 2^4 **5.** 2^5 **7.** $2^2 \cdot 3^2$ **9.** $2^3 \cdot 3^2$ **11.** 5^4 **13.** $2 \cdot 3^2 \cdot 5^2 \cdot 7$ **15.** $2 \cdot 3^4 \cdot 5 \cdot 11$ **17.** $2^2 \cdot 5^2 \cdot 7^2$ **19.** $2^6 \cdot 5^6$ **21.** $2 \cdot 2 \cdot 2 \cdot 2 \cdot 2 = 32$ **23.** $2 \cdot 2 \cdot 3 \cdot 3 \cdot 3 = 108$ **25.** $xxxx$ **27.** $tttttt$ **29.** $2xx$ **31.** $8aa$ or $2 \cdot 2 \cdot 2aa$ **33.** $5xx$ **35.** $10aabb$ or $2 \cdot 5aabb$ **37.** $3 \cdot 3 \cdot 3 \cdot 2 \cdot 2 \cdot 2 = 216$ **39.** $4 \cdot 4 \cdot 3 \cdot 3 = 144$ **41.** 1 **43.** 230 **45.** 215 **47.** 6 **49.** 42 **51.** 15 **53.** 148 **55.** a^6 **57.** a^5 **59.** x^7y^5 **61.** x^2y^4z **63.** 32 **65.** $25x^5$ **67.** $a^{10}b^5$ **69.** $(x + y)^5$ **71.** $3^2 + 5^2 = 9 + 25 = 34$ **73.** $3^2 + 2^3 = 9 + 8 = 17$ **75.** $n^2 + 5$ **77.** $7^2 + n^2 = n^2 + 49$ **79.** $n^3 - m^3$

Problem Set 1.4, Pages 20–21

1. 19 **3.** 19 **5.** 7 **7.** 12 **9.** 49 **11.** 1 **13.** 23 **15.** 1 **17.** 33 **19.** 100
21. 2, 4, 6 **23.** 350 ft **25.** 30π or 94.25 in. **27.** $1650.00 **29.** $7.60 **31.** $A = bh$
33. $A = \frac{1}{2}pq$ **35.** $V = s^3$ **37.** $V = \pi r^2 h$ **39.** $V = \frac{4}{3}\pi r^3$

Problem Set 1.5, Pages 25–26

1. commutative **3.** associative **5.** commutative **7.** commutative **9.** commutative
11. associative **13.** associative **15.** commutative **17.** associative **19.** commutative
21. sum **23.** sum **25.** sum **27.** product **29.** product **31.** sum **33.** sum
35. sum **37.** sum **39.** difference **41.** $ab + ac$ **43.** $xy + xz$ **45.** $7y + 28$
47. $6x + 8$ **49.** $15z + 20$ **51.** $y^2 + yz$ **53.** $2x^2 + 2xy + 2xz$ **55.** $5x^4 + 10x^3 + 35x^2$
57. $a(x + w)$ **59.** $c(x + y)$ **61.** $uv(w + s)$ **63.** $3(x + y)$ **65.** $3(3x + 2y)$ **67.** $7(2y + 1)$
69. $x(x + 5)$ **71.** $n + 1 = 1 + n$ **73.** $3(n + 4) = 16$ **75.** $5(n + 1) = 5n + 5$
77. $n(7 + n) = 0$ **79.** $n(9 + n) = 9n + n^2$

1.6 Review Self-Test, Pages 26–27

1. a. $>$ or $\geq$ **b.** $\leq$ or $\geq$ **c.** $>$ or $\geq$ **d.** $<$ or $\leq$ **2. a.** $45 = 9 \cdot 5$, correct **b.** $35 = 6 \cdot 7$, not correct $\frac{35}{6} \neq 7$ **c.** $54 = 8 \cdot 7$, not correct $\frac{54}{8} \neq 7$ **d.** $5 = 0 \cdot 0$, not correct $\frac{5}{0} \neq 0$
3. a. $12 + 3 = 15$ **b.** $5 = 9 - 4$ **c.** $8 = \frac{40}{5}$ **d.** $6 \cdot 8 < 7^2$ **4. a.** $2n + 5$
b. $n + (n + 1)$ **c.** $2(n - 5)$ **d.** $2n - 5$ **5. a.** 11 **b.** 13 **c.** 1 **d.** 86
6. a. $7.13 \cdot 100{,}000{,}000 = 713{,}000{,}000$ **b.** exponent **c.** factor **d.** base **7. a.** $2^3 \cdot 3^2 \cdot 5$
b. $2s^3t^2$ **c.** $5^5x^2y^3$ **d.** $3^3a^3b^3$ or $27a^3b^3$ **8. a.** $18 > 14$ or $18 \geq 14$ **b.** $25 > 13$ or $25 \geq 13$
c. $35 > 23$ or $35 \geq 23$ **d.** $48 \geq 48$ or $48 \leq 48$ **9. a.** $5x + 5y$ **b.** $6a + 9b$ **c.** $6x^3y^2 + 2x^2y^3$
d. $3a^2bc + c$ **10. a.** $st(u + v)$ **b.** $5(x + 3y)$ **c.** $2mn(1 + 3m)$ **d.** $x^2y^2(1 + xy)$

Review Problems, Pages 27–28

1. 23 **3.** 1 **5.** 5 **7.** 11 **9.** 6 **11.** 13 **13.** 6 **15.** 20 **17.** 20 **19.** 17
21. 30 **23.** 24 **25.** 1 **27.** 20 **29.** 100 **31.** d^4 **33.** k^6 **35.** n^8 **37.** $3x^6$
39. $2z^4$ **41.** b^6c^9 **43.** $x^4y^2z^5$ **45.** xy^3z^4 **47.** 1 **49.** 10 **51.** 7 **53.** 12 **55.** 2
57. 7 **59.** 26 **61.** 1 **63.** 3 **65.** 4 **67.** 24 **69.** 45 **71.** 2 **73.** 49 **75.** 1
77. $wx + wy$ **79.** $3s + 6$ **81.** $7m + 21$ **83.** $15x + 6$ **85.** $s^2 + 2s$ **87.** $m^2 + 3m$
89. $b(x + z)$ **91.** $xy(z + w)$ **93.** $mn(p + q)$ **95.** $2(x + 3y)$ **97.** $7(x + 1)$ **99.** $7(7x + 1)$

Problem Set 2.1, Pages 33–34

1. a. $+6$ **b.** a loss of six dollars **c.** -6 **3. a.** -7 **b.** seven steps to the right **c.** $+7$
5. a. $+2$ **b.** 2° below zero **c.** -2 **7. a.** $+12$ **b.** 12° latitude South **c.** -12
9. a. $+10$ **b.** a 10-yard loss **c.** -10 **11. a.** $+6200$ **b.** an ocean floor 6200 ft deep
c. -6200 **13. a.** $+500$ **b.** a $500 penalty **c.** -500 **15. a.** $+5$ **b.** a five dollar decrease in price **c.** -5 **17. a.** $+70$ **b.** 70° longitude West **c.** -70 **19. a.** -50 **b.** 50° latitude North **c.** $+50$ **21.** minus **23.** opposite **25.** opposite **27.** negative
29. opposite **31.** $(-5) + (+4) + (-3) = -4$ **33.** $(-5) + (+8) + (-3) = 0$
35. $(+2) + (-5) + (+6) = +3$ **37.** $(-10) + (+12) + (-20) = -18$
39. $(-20) + (-12) + (+15) = -17$ **41.** $+1$ **43.** $+3$ **45.** -5 **47.** -8 **49.** $+8$
51. 4, 4 **53.** 5, 7 **55.** -10, 11 **57.** 4, 13 **59.** 4, 4 **61.** No, explanations vary

Problem Set 2.2, Pages 37–38

1. -3 **3.** -4 **5.** -4 **7.** -3 **9.** 3 **11.** 11 **13.** 5 **15.** 4 **17.** 5 **19.** 5
21. 39 **23.** 71 **25.** 60 **27.** -40 **29.** -20 **31.** -27 **33.** -10 **35.** 2 **37.** 3

39. -3 **41.** 3 **43.** 6 **45.** -5 **47.** 4 **49.** 7 **51.** 0 **53.** -2 **55.** 0
57. -6 **59.** -12 **61.** -13 **63.** 0 **65.** 6 **67.** 6 **69.** 1 **71.** -6 **73.** -12
75. -6 **77.** -1 **79.** -3 **81.** -28 **83.** -3 **85.** -7 **87.** 2 **89.** 0

Problem Set 2.3, Pages 40–41

1. -2 **3.** -10 **5.** -7 **7.** -15 **9.** -7 **11.** 16 **13.** -2 **15.** 3 **17.** 0
19. 15 **21.** -1 **23.** 5 **25.** 6 **27.** -5 **29.** -7 **31.** 5 **33.** -3 **35.** -3
37. -9 **39.** -5 **41.** 8 **43.** 9 **45.** -4 **47.** -14 **49.** 16 **51.** -4 **53.** 26
55. 104 **57.** -49 **59.** 31 **61.** $(5-7)+[(-8)-(-6)]=-4$
63. $[(-6)+3]-[(-4)-5]=6$ **65.** $(-2)+[(-2)-8]=-12$
67. $[(-3)+(-6)]-(-6)=-3$ **69.** $(-3)-[1+(-3)]=-1$

Problem Set 2.4, Pages 46–47

1. -48 **3.** -15 **5.** -81 **7.** 35 **9.** 0 **11.** -40 **13.** -64 **15.** -35 **17.** -42
19. -72 **21.** 10 **23.** -54 **25.** 8 **27.** 6 **29.** -84 **31.** 6 **33.** -4 **35.** 20
37. 5 **39.** -4 **41.** 12 **43.** 7 **45.** -4 **47.** 16 **49.** -32 **51.** -1 **53.** 43
55. 49 **57.** -17 **59.** 7 **61.** 4 **63.** -2 **65.** -3
67. $\dfrac{2+(-5)+4+(-7)+(-4)}{5}=-2$ **69.** $\dfrac{(-3)+(-6)+5+8+0+(-2)+5}{7}=1$ **71.** 9
73. 1 **75.** -7 **77.** -12 **79.** -1 **81.** 8 **83.** 30 **85.** 2 **87.** 1 **89.** -3

Problem Set 2.5, Pages 49–50

1. a. 7, 6, 1, 5 **b.** 4, 3, 2, 1 **3.** $3x$, $(5\cdot 2)x$, and $11x$ are similar; $4xy$ and $7xy$ are similar; $2x^2y$ and $6x^2y$ are similar. **5. a.** $21ab$ **b.** $21ab$ **7. a.** $-36st$ **b.** $-36st$ **9. a.** $12xyz$
b. $-12xyz$ **11. a.** $36xy^2$ **b.** $12xy^2$ **13. a.** $16m^3$ **b.** $-32n^4$ **15. a.** $20y^6$ **b.** $20y^6$
17. a. $-21a^3b^3$ **b.** $-49b^6$ **19. a.** $-63m^7$ **b.** $441m^8$ **21.** $7x$ **23.** $15z$ **25.** w
27. $4a$ **29.** $-10c$ **31.** $-3k$ **33.** $-3m$ **35.** $-5p$ **37.** $-7y$ **39.** $62a$ **41.** $13x$
43. $9a$ **45.** $2c+3d$ **47.** $-4x$ **49.** $2a$ **51.** $6w$ **53.** $-3y$ **55.** $-8x$ **57.** 0
59. 0 **61.** $9x-24$ **63.** $7x-10$ **65.** $6x^2-6x+10$ **67.** $7x^3-x^2y+8x^2-3x$
69. $8r^3-9r^2s+5rs^2-3$

2.6 Review Self-Test, Page 51

1. a. -12 **b.** $+50$ **c.** -17 **d.** -17 **e.** 17 **2.** -2 **3. a.** -2 **b.** -8 **c.** 1
d. -2 **e.** 8 **4. a.** 12 **b.** 4 **c.** -7 **d.** -6 **e.** $-5y$
5. a. $[(-11)+9]-[12+(-8)]=-6$ **b.** $\dfrac{9+(-5)+(-11)+3+(-6)}{5}=-2$
c. $n+(n-8)=2n-8$ **6. a.** -35 **b.** 12 **c.** 24 **d.** 14 **e.** 1 **7. a.** -1 **b.** 0
c. 0 **d.** 1 **e.** 30 **8. a.** $1, -2, 7$ **b.** $3, 2, 1$ **9. a.** $-35ab$ **b.** $24x^4$ **c.** $16y^4$
d. $24y^7$ **e.** $2b^8$ **10. a.** $2x+3y$ **b.** $-8y$ **c.** $2z$ **d.** $6a+14$ **e.** a^3+b^2

Review Problems, Pages 51–53

1. -31 **3.** -7 **5.** -40 **7.** 0 **9.** -2 **11.** 24 **13.** -34 **15.** -11 **17.** 11
19. 39 **21.** -48 **23.** 98 **25.** 54 **27.** -156 **29.** -36 **31.** 64 **33.** -1
35. -1 **37.** -31 **39.** 23 **41.** -5 **43.** -3 **45.** -13 **47.** -1 **49.** 2 **51.** 2
53. -3 **55.** 1 **57.** -2 **59.** -1 **61.** 0 **63.** 4 **65.** -1 **67.** -1 **69.** -3
71. -13 **73.** -7 **75.** 11 **77.** $-12y^2$ **79.** $-a^5$ **81.** c^4 **83.** $30a^2b^2$ **85.** $-14x^3y^2$
87. $6x$ **89.** $-4y$ **91.** 0 **93.** $-3z-2$ **95.** $3y-6x$ **97.** $-2x-3z$ **99.** $2y+6$

Problem Set 3.1, Page 57

1. no **3.** yes **5.** no **7.** yes **9.** yes **11.** yes **13.** yes **15.** yes **17.** yes
19. no **21.** no **23.** no Answers to Problems 25–50 may vary. **25.** 3 + NUMBER = 11;

$3 + N = 11$ **27.** $-2 + \text{NUMBER} = -6; -2 + N = -6$ **29.** $4(\text{NUMBER}) = 72; 4N = 72$
31. $\text{NUMBER} - 6 = 15;$ $N - 6 = 15$ **33.** $\frac{\text{NUMBER}}{7} = 12;$ $\frac{N}{7} = 12$ **35.** $5(\text{NUMBER}) - 3 = 12;$ $5N - 3 = 12$ **37.** $\text{NUMBER} + 12 = 4(\text{ANOTHER NUMBER}); N + 12 = 4M$
39. $\frac{3(\text{NUMBER})}{2} = 5(\text{NUMBER}); \frac{3N}{2} = 5N$ **41.** $\frac{\text{NUMBER} + 7}{2} = 11(\text{NUMBER}); \frac{N+7}{2} = 11N$
43. $2(\text{DAUGHTER'S AGE}) = \text{SON'S AGE}; 2D = S$ **45.** $\text{WIDTH} + 10 = 2(\text{LENGTH}); W + 10 = 2L$
47. $\text{CIRCUMFERENCE} = 3(\text{RADIUS}) + .7; C = 3R + .7$ **49.** $\text{DOSAGE} = \frac{(\text{CHILD'S AGE}) + 2}{50(\text{CHILD'S AGE})}$

Problem Set 3.2, Page 61

1. 6 **3.** 16 **5.** 17 **7.** -18 **9.** -34 **11.** -43 **13.** -18 **15.** -4 **17.** 15 **19.** -14 **21.** -6 **23.** -112 **25.** -6 **27.** -7 **29.** -9 **31.** 9 **33.** 10 **35.** 2 **37.** 3 **39.** 4 **41.** 4 **43.** 3 **45.** -8 **47.** 0 **49.** 17

Problem Set 3.3, Pages 64–65

1. x **3.** m **5.** y **7.** z **9.** t **11.** 5 **13.** 6 **15.** 13 **17.** 12 **19.** 2 **21.** 10 **23.** 12 **25.** 52 **27.** 112 **29.** 12 **31.** -6 **33.** -96 **35.** -60 **37.** -44 **39.** -30 **41.** -64 **43.** 1 **45.** 2 **47.** $-\frac{1}{6}$ **49.** 3 **51.** 2 **53.** 21 **55.** -20 **57.** 12 **59.** 4

Problem Set 3.4, Page 67

1. 0 **3.** 20 **5.** 7 **7.** 2 **9.** 6 **11.** 81 **13.** 13 **15.** -39 **17.** 62 **19.** 41 **21.** 14 **23.** 8 **25.** 2 **27.** 9 **29.** 13 **31.** -9 **33.** 19 **35.** 2 **37.** 2 **39.** -3 **41.** 5 **43.** 13 **45.** all values **47.** 7 **49.** 0 **51.** -12 **53.** -4 **55.** 0 **57.** no roots **59.** 8 **61.** 0

Problem Set 3.5, Page 70

1. $B = c - a$ **3.** $Y = -x + 7$ **5.** $Y = 3x - 5$ **7.** $P = m - rt$ **9.** $D = rt$ **11.** $R = \frac{d}{t}$
13. $R = \frac{c}{2\pi}$ **15.** $x = 2y - 2a$ **17.** $y = 6x + 5$ **19.** $y = -3x + 4$ **21.** $y = -2x - 4$
23. $y = -3x - 8$ **25.** $y = 4x + 5$ **27.** $y = x - 5$ **29.** $y = -2x + 4$ **31.** $y = 2x - 3$
33. $y = 2x - 7$ **35.** $y = 2x - 4$ **37.** $r = \frac{C}{2\pi}$ **39.** $R = \frac{k}{I}$ **41.** $F = \frac{W}{d}$ **43.** $l = \frac{k}{f}$
45. $s = \frac{D}{4\pi b^2}$ **47.** **a.** varies **b.** extra material needed

$$\begin{aligned} &= \text{NEW CIRCUMFERENCE} - \text{OLD CIRCUMFERENCE} \\ &= \pi(d + 12) - \pi d \\ &= \pi d + 12\pi - \pi d \\ &= 12\pi \end{aligned}$$

The amount needed is about 38 ft.

Problem Set 3.6, Pages 74–76

1. $2(\text{NUMBER}) + 7 = 17$
$$2N + 7 = 17$$
$$2N = 10$$
$$N = 5$$
The number is 5.

3. $2(\text{NUMBER}) + 15 = 7$
$$2N + 15 = 7$$
$$2N = -8$$
$$N = -4$$
The number is -4.

5. $5(\text{NUMBER}) - (-10) = -30$
$$5N + 10 = -30$$
$$5N = -40$$
$$N = -8$$
The number is -8.

7. $$\begin{aligned} \text{NUMBER} + 24 &= 4(\text{NUMBER}) \\ N + 24 &= 4N \\ 24 &= 3N \\ 8 &= N \end{aligned}$$
The number is 8.

9. $$\begin{aligned} 2(\text{NUMBER}) - 12 &= 4(\text{NUMBER}) \\ 2N - 12 &= 4N \\ -12 &= 2N \\ -6 &= N \end{aligned}$$
The number is -6.

11. $$\begin{aligned} \text{NUMBER} + \text{NEXT NUMBER} &= 117 \\ N + (N + 1) &= 117 \\ 2N + 1 &= 117 \\ 2N &= 116 \\ N &= 58 \end{aligned}$$
The numbers are 58 and 59.

13. $$\begin{aligned} \text{NUMBER} + \text{2nd NUMBER} + \text{3rd NUMBER} &= 54 \\ \text{NUMBER} + (\text{NUMBER} + 1) + (\text{NUMBER} + 2) &= 54 \\ N + (N + 1) + (N + 2) &= 54 \\ 3N + 3 &= 54 \\ 3N &= 51 \\ N &= 17 \end{aligned}$$
The numbers are 17, 18, 19.

15. $$\begin{aligned} \text{NUMBER} + \text{2nd NUMBER} + \text{3rd NUMBER} + \text{4th NUMBER} &= 74 \\ \text{NUMBER} + (\text{NUMBER} + 1) + (\text{NUMBER} + 2) + (\text{NUMBER} + 3) &= 74 \\ N + (N + 1) + (N + 2) + (N + 3) &= 74 \\ 4N + 6 &= 74 \\ 4N &= 68 \\ N &= 17 \end{aligned}$$
The numbers are 17, 18, 19, and 20.

17. $$\begin{aligned} \text{ODD NUMBER} + \text{NEXT ODD NUMBER} &= 48 \\ D + (D + 2) &= 48 \\ 2D + 2 &= 48 \\ 2D &= 46 \\ D &= 23 \end{aligned}$$
The smaller odd number is 23.

19. $$\begin{aligned} \text{EVEN INTEGER} + \text{NEXT EVEN INTEGER} &= 2(\text{EVEN INTEGER}) + 2 \\ \text{EVEN INTEGER} + (\text{EVEN INTEGER} + 2) &= 2(\text{EVEN INTEGER}) + 2 \\ E + (E + 2) &= 2E + 2 \\ 2E + 2 &= 2E + 2 \\ 2 &= 2 \end{aligned}$$
Any two even integers will satsify the conditions of the problem.

21. $$\begin{aligned} 2\text{WIDTH} + 2\text{LENGTH} &= \text{PERIMETER} \\ 2(75) + 2L &= 750 \\ 150 + 2L &= 750 \\ 2L &= 600 \\ L &= 300 \end{aligned}$$
The length is 300 m.

23. $$\begin{aligned} 3x + 2x + 4x &= 117 \\ 9x &= 117 \\ x &= 13 \end{aligned}$$
The sides are 26, 39, and 52.

25. $$\begin{aligned} \text{LENGTH} &= \text{WIDTH} + 2 \\ &= 48 + 2 \\ &= 50 \end{aligned} \qquad \begin{aligned} \text{AREA} &= (\text{LENGTH})(\text{WIDTH}) \\ &= 48(50) \\ &= 2400 \end{aligned}$$
The area is 2400 sq m.

27. $$\begin{aligned} \begin{pmatrix} \text{VALUE OF} \\ \text{TYPE I CABINET} \end{pmatrix} + \begin{pmatrix} \text{VALUE OF} \\ \text{TYPE II CABINET} \end{pmatrix} &= 2075 \\ \begin{pmatrix} \text{VALUE OF} \\ \text{TYPE I CABINET} \end{pmatrix} + 4\begin{pmatrix} \text{VALUE OF} \\ \text{TYPE II CABINET} \end{pmatrix} &= 2075 \\ V + 4V &= 2075 \\ 5V &= 2075 \\ V &= 415 \\ 4V &= 1660 \end{aligned}$$
The value of the cabinets are \$415 and \$1660.

29. DEGREES OF SMALLEST ANGLE + DEGREES OF LARGEST ANGLE + DEGREES OF THIRD ANGLE = 180

$$\text{DEGREES OF SMALLEST ANGLE} + 2\left(\text{DEGREES OF SMALLEST ANGLE}\right) + \left[2\left(\text{DEGREES OF SMALLEST ANGLE}\right) - 20\right] = 180$$

$$A + 2A + [2A - 20] = 180$$
$$5A - 20 = 180$$
$$5A = 200$$
$$A = 40$$

The angles are 40°, 80° and 60°.

31. $\text{IQ} = \frac{18}{15}(100)$
$= 120$
The IQ is 120.

33. $132 = \frac{\text{MA}}{9}(100)$
$1188 = \text{MA}(100)$
$11.88 = \text{MA}$
A mental age of 11.88.

Problem Set 3.7, Page 80

1. 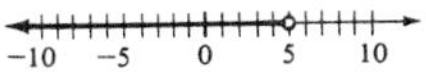 −10 −5 0 5 10

3. −3; −10 −5 0 5 10

5. 4; −10 −5 0 5 10

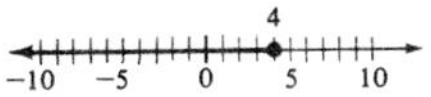

7. 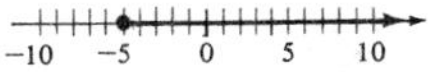 −10 −5 0 5 10

9. 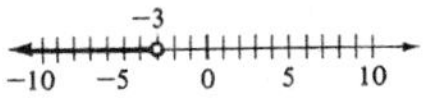 −3; −10 −5 0 5 10

11. 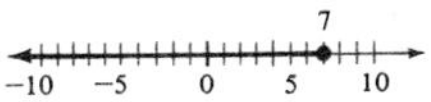 7; −10 −5 0 5 10

13. 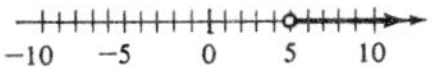 −10 −5 0 5 10

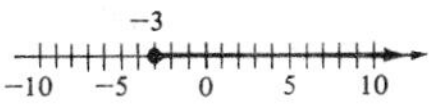

15. 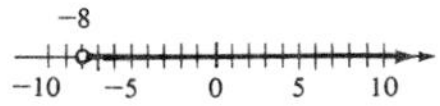 −8; −10 −5 0 5 10

17. 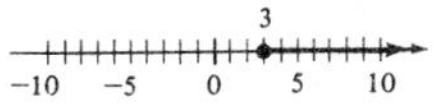 3; −10 −5 0 5 10

19. 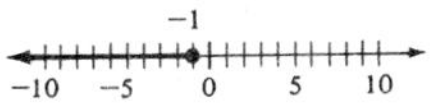 −1; −10 −5 0 5 10

21. 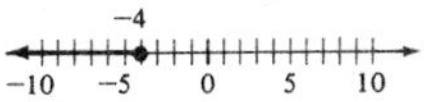 −4; −10 −5 0 5 10

23. 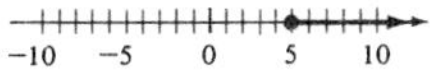 −10 −5 0 5 10

25. 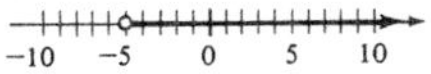 −10 −5 0 5 10

27. 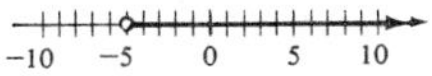 −10 −5 0 5 10

29. 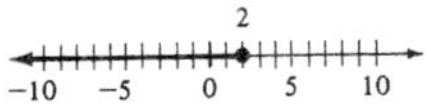 2; −10 −5 0 5 10

31. $7(\text{NUMBER}) + 35 > 0$
$$7N + 35 > 0$$
$$7N > -35$$
$$N > -5$$
Any number greater than −5.

33. PERIMETER OF SQUARE > PERIMETER OF TRIANGLE

$$4\left(\text{SIDE OF SQUARE}\right) > 3\left(\text{SIDE OF TRIANGLE}\right)$$
$$4\left(\text{SIDE OF SQUARE}\right) > 3\left(\text{SIDE OF SQUARE} + 2\right)$$
$$4S > 3(S + 2)$$
$$4S > 3S + 6$$
$$S > 6$$

The length of the side of the square must be greater than 6.

3.8 Review Self-Test, Pages 80–81

1. a. no **b.** yes **c.** no **d.** no **2. a.** −17 **b.** 0 **c.** 19 **d.** 18 **3. a.** −5

b. -12 **c.** 2 **d.** 48 **4. a.** -2 **b.** 2 **c.** 1 **d.** 15 **5. a.** 3 **b.** 2 **c.** no roots **d.** 8 **6. a.** $p \geq 2$; **b.** $y \leq 0$; **c.** $x < 2$; **d.** $y > 6$; **7. a.** $b = \frac{A}{h}$ **b.** $y = 3x + 2$ **c.** $y = -3x - 6$ **d.** $n = \frac{A - b - 2p}{p}$

8. 2(NUMBER) + 6 = NUMBER

$$2N + 6 = N$$
$$N + 6 = 0$$
$$N = -6$$

The number is -6.

9. 1st ODD INTEGER + 2nd ODD INTEGER + 3rd ODD INTEGER = 2(1st ODD INTEGER) − 8

Let x = 1st ODD INTEGER

$x + 2$ = 2nd ODD INTEGER

$x + 4$ = 3rd ODD INTEGER

$$x + (x + 2) + (x + 4) = 2x - 8$$
$$3x + 6 = 2x - 8$$
$$x = -14$$
$$x + 2 = -12$$
$$x + 4 = -10$$

The middle integer is -12.

10. 2LENGTH + 2WIDTH = PERIMETER

2(2WIDTH) + 2WIDTH = 24

$$4W + 2W = 24$$
$$-6W = 24$$
$$W = 4$$
$$2W = 8$$

The drawing is 4 in. by 8 in., and the room is 16 ft by 32 ft.

Review Problems, Pages 81–82

1. 13 **3.** 16 **5.** 1 **7.** -6 **9.** 9 **11.** -4 **13.** 6 **15.** 9 **17.** 3 **19.** 4 **21.** -5 **23.** 3 **25.** 24 **27.** 80 **29.** 2 **31.** 5 **33.** -2 **35.** -8 **37.** 1 **39.** 0 **41.** 8 **43.** 8 **45.** 1 **47.** 8 **49.** 3 **51.** 3 **53.** 3 **55.** 7 **57.** 4 **59.** 5 **61.** -4 **63.** 2 **65.** 1 **67.** -2 **69.** 4 **71.** $x \geq 3$ **73.** $y < 0$ **75.** $y \leq 0$ **77.** $p \leq -2$ **79.** $z > 6$ **81.** $z < 2$ **83.** $y = -3x + 5$ **85.** $y = 5x + 3$ **87.** $y = 6x - 3$ **89.** $y = -\frac{3}{2}x + 5$ **91.** $y = \frac{3}{2}x - 9$ **93.** $y = \frac{2}{3}x - 7$ **95.** $y = \frac{3}{2}x + 4$ **97.** $y = \frac{1}{3}x + 3$ **99.** $y = \frac{3}{2}x - \frac{17}{2}$

Problem Set 4.1, Pages 87–88

		Degree	Numerical Coefficient			Degree	Numerical Coefficient
1.	**a.**	4	3	**3.**	**a.**	1	8
	b.	2	5		**b.**	3	7
	c.	5	8		**c.**	1	−4
	d.	1	123		**d.**	0	10
	e.	0	46		**e.**	3	−5

5. a. binomial; 3 **b.** trinomial; 1 **c.** monomial; 3 **7.** $7a - 3b$ **9.** $3m - 4n$ **11.** $3s + t$
13. $6x + y$ **15.** $3x^2 - 2$ **17.** $-2r - 3s$ **19.** $x + 1$ **21.** $-y - 7$ **23.** $z - 8$
25. $2x^2 - 4x + 3$ **27.** $-2w^2 + 8w$ **29.** $-x + 3y$ **31.** $9b - 36$ **33.** $-h - 5$
35. $-11x - 22$ **37.** $8x - 6$ **39.** $6x - 3$ **41.** $-8m + 40$ **43.** $-2s + 5$ **45.** $10x - 38$
47. $2x - 4$ **49.** $3x - 5y - 1$ **51.** $x^2 - 5x - 4$ **53.** $x^2 - x + 4$

Problem Set 4.2, Pages 91–92

1. 2^{10} **3.** 10^7 **5.** 3^{11} **7.** 5^9 **9.** x^{10} **11.** x^3y^5 **13.** $6a^2b^4$ **15.** $3m^5n^4$
17. $3ab - 3b^2$ **19.** $e^2 + 5e - 3$ **21.** $-h^2 - h + 6$ **23.** $x^2 - 5x - 3$ **25.** $m^2 + 3m - 15$
27. $h^2 - 2h + 1$ **29.** $m^3 - n^2$ **31.** $3r^3s^3 - 6r^4s^3 + 3r^4s^2$ **33.** $4t^3u^5 - 8t^4u^5 - 4t^2u^4$ **35.** u^4v^6
37. $5x^5y^6$ **39.** $72x^5y^5$ **41.** $18m^4n^6$ **43.** x^6y^8 **45.** v^7v^6 **47.** -648 **49.** 64
51. 64 **53.** $x^2 + x - 1$ **55.** $x^3 + x^2 - 1$ **57.** x^4y^4 **59.** $-x^{18}y^9$

Problem Set 4.3, Pages 95–96

1. $x^2 + 3x + 2$ **3.** $x^2 + 12x + 35$ **5.** $x^2 - 9x + 20$ **7.** $x^2 - 49$ **9.** $x^2 - 5x - 84$
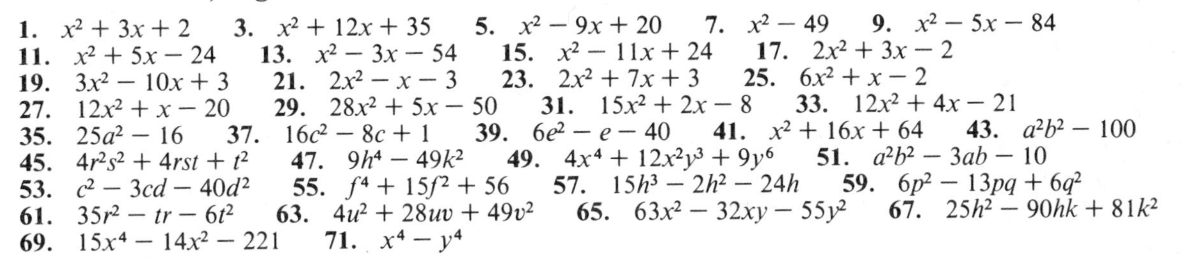
11. $x^2 + 5x - 24$ **13.** $x^2 - 3x - 54$ **15.** $x^2 - 11x + 24$ **17.** $2x^2 + 3x - 2$
19. $3x^2 - 10x + 3$ **21.** $2x^2 - x - 3$ **23.** $2x^2 + 7x + 3$ **25.** $6x^2 + x - 2$
27. $12x^2 + x - 20$ **29.** $28x^2 + 5x - 50$ **31.** $15x^2 + 2x - 8$ **33.** $12x^2 + 4x - 21$
35. $25a^2 - 16$ **37.** $16c^2 - 8c + 1$ **39.** $6e^2 - e - 40$ **41.** $x^2 + 16x + 64$ **43.** $a^2b^2 - 100$
45. $4r^2s^2 + 4rst + t^2$ **47.** $9h^4 - 49k^2$ **49.** $4x^4 + 12x^2y^3 + 9y^6$ **51.** $a^2b^2 - 3ab - 10$
53. $c^2 - 3cd - 40d^2$ **55.** $f^4 + 15f^2 + 56$ **57.** $15h^3 - 2h^2 - 24h$ **59.** $6p^2 - 13pq + 6q^2$
61. $35r^2 - tr - 6t^2$ **63.** $4u^2 + 28uv + 49v^2$ **65.** $63x^2 - 32xy - 55y^2$ **67.** $25h^2 - 90hk + 81k^2$
69. $15x^4 - 14x^2 - 221$ **71.** $x^4 - y^4$

Problem Set 4.4, Page 97

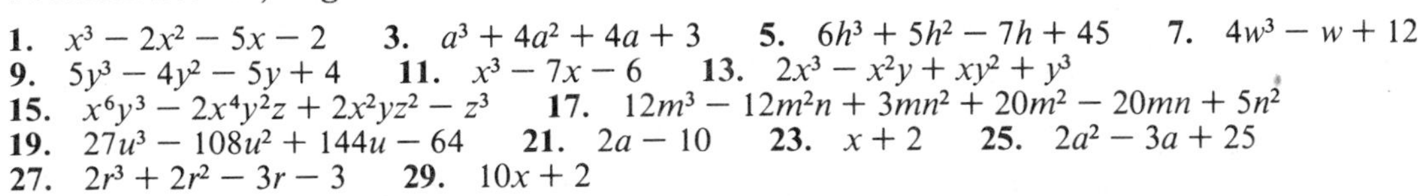
1. $x^3 - 2x^2 - 5x - 2$ **3.** $a^3 + 4a^2 + 4a + 3$ **5.** $6h^3 + 5h^2 - 7h + 45$ **7.** $4w^3 - w + 12$
9. $5y^3 - 4y^2 - 5y + 4$ **11.** $x^3 - 7x - 6$ **13.** $2x^3 - x^2y + xy^2 + y^3$
15. $x^6y^3 - 2x^4y^2z + 2x^2yz^2 - z^3$ **17.** $12m^3 - 12m^2n + 3mn^2 + 20m^2 - 20mn + 5n^2$
19. $27u^3 - 108u^2 + 144u - 64$ **21.** $2a - 10$ **23.** $x + 2$ **25.** $2a^2 - 3a + 25$
27. $2r^3 + 2r^2 - 3r - 3$ **29.** $10x + 2$

Problem Set 4.5, Pages 101–102

1. $5x^2 - 2x + 3$ **3.** $3x^3 - 2x^2 - x$ **5.** $5x^2 - 3x + 2$ **7.** $6x^2 - 3$ **9.** $2x - 4$ **11.** $4x + 7$
13. $3a^3 + 5a^2 + 3a + 4$ **15.** $u^2 - 3u + 9$ **17.** $v^2 + 2v + 4$ **19.** yes **21.** yes **23.** yes
25. $2z^2 - 7z + 4 + \dfrac{3}{z + 3}$ **27.** $x^3 + 2x^2 - x + 3 + \dfrac{-1}{x + 1}$ **29.** $y^3 - 3y^2 - 3 + \dfrac{-1}{y - 1}$
31. $7x + 5$ **33.** 40

4.6 Review Self-Test, Page 102

1. a. monomial; degree, 7 **b.** binomial; degree, 3 **c.** trinomial; degree, 4 **d.** monomial; degree, 0
e. binomial; degree, 3 **2. a.** $4h - 4k$ **b.** $x - 3$ **c.** $2m - 2n$ **d.** $-2a + 6b - 2$

e. $6x - 6$ **3. a.** 2^{10} **b.** $3b^7$ **c.** 2^8 **d.** $(-1)^{10}$ **e.** -2^6 **4. a.** $5a^2 - 5ab$ **b.** $h^2 - 8h - 5$ **c.** $m^5 - 2m^3n + mn$ **d.** $2x^3y^2 + 4x^4y^2 - 2x^3y^3$ **e.** $4x^2 - 6x + 18$ **5. a.** $x^2 + 7x + 12$ **b.** $x^2 + 4x - 5$ **c.** $x^2 - 9x + 14$ **d.** $2x^2 - 13x + 15$ **e.** $15x^2 - x - 6$ **6. a.** $x^2 + 14x + 49$ **b.** $x^2 - 49$ **c.** $4x^2 - 9$ **d.** $4x^2 - 12x + 9$ **e.** $x^2 - 4xy + 4y^2$ **7. a.** $x^3 + 2x^2 + 2x + 1$ **b.** $x^3 - 1$ **c.** $6y^3 - 17y^2 + 9y - 10$ **8. a.** $8z^3 - 36z^2 + 54z - 27$ **b.** $27x^3 - 27x^2 + 9x - 1$ **c.** $x^4 - 5x^2 + 5x - 6$ **9. a.** $x^2 - 8x + 12$ **b.** $3x^2 - x - 3$ **10. a.** $y^2 - 4y + 16$ **b.** $y^2 + 4y + 16$

Review Problems, Pages 102–103

1. $x - 4$ **3.** $7p + 5$ **5.** $2x + y + 1$ **7.** $-2x + 6y$ **9.** $12x - 3$ **11.** $7t + 1$ **13.** $y + 5$ **15.** $-2a + 3b$ **17.** $7a - 8c$ **19.** $5s - 2t - 2$ **21.** $-2x^2 - 2x - 8$ **23.** $6x^2 + 2x + 3$ **25.** $2x^2 + 2xy$ **27.** $3y^2 + 3yz$ **29.** $6x^3 - 12x^2 + 18x$ **31.** $3a^2 + 13a - 10$ **33.** $x^2 + 3x + 15$ **35.** $x^5 - 2x^4 - x^2 + x$ **37.** $x^4 + 2x^3 + x^2 + 4x + 4$ **39.** $3x^4y^2 + 3x^3y^3 + 3x^3y$ **41.** $12x^6y^4 + 4x^6y^2 + 6x^5y$ **43.** $-hk - 7k^2$ **45.** $4s^2 + 22s$ **47.** $a^2 + 7a + 10$ **49.** $c^2 - 4c + 3$ **51.** $m^2 - 3m - 10$ **53.** $6x^2 + 7x - 5$ **55.** $6x^2 - 17x - 10$ **57.** $6x^2 + 7x - 20$ **59.** $12x^2 - 119x - 10$ **61.** $t^2 - 64$ **63.** $y^2 - 25$ **65.** $36z^2 - 1$ **67.** $9t^2 - 25$ **69.** $x^2 + 6x + 9$ **71.** $y^2 - 10y + 25$ **73.** $9x^2 - 12x + 4$ **75.** $x^2 + 2xy + y^2$ **77.** $4x^2 + 12x + 9$ **79.** $x^2 - 6xy + 9y^2$ **81.** $x^2 - x + 4$ **83.** $3x^2 - 13x$ **85.** $6x^2 + 17x - 6$ **87.** $x^3 + 4x^2 + 3x$ **89.** $k^3 + 3k^2 + 3k + 1$ **91.** $z^3 - 9z^2 + 27z - 27$ **93.** $p^3 + 2p^2 - p - 2$ **95.** $x^4 + 4x^3 + x^2 + 2x + 4$ **97.** $z^2 - 2z + 4$ **99.** $z - 5$

Problem Set 5.1, Pages 107–109

1. 2^5 **3.** $2^3 \cdot 5$ **5.** $2^2 \cdot 3^2$ **7.** $2^3 \cdot 3^2 \cdot 5^3$ **9.** $2^3 \cdot 7^2$ **11.** $3(a^2 + 5b^2)$ **13.** $4(s^2 + 3t^2)$

15. $f(14 - 15f)$ **17.** $xy(3x + 5y)$ **19.** $mn(4m + 5n)$ **21.** $mn(n^2 - m)$ **23.** $\pi r^2\left(\frac{4}{3}r - 1\right)$

25. $2\pi r\left(1 - \frac{1}{3}r\right)$ **27.** $2(4y^2 - 9y + 3)$ **29.** $y^2(3y^2 - 7y - 4)$ **31.** $2(4x^2 + 6y^2 + 3z^2)$

33. $3rs(s^2 + 3r - 2s)$ **35.** $5wx(3x^2 - 5wx + 7w^2)$ **37.** $\pi r\left(\frac{1}{3}r^2 + r + 2\right)$ **39.** $\pi r\left(r + \frac{4}{3}r^2 - 2\right)$

41. $4b^2(\pi - 1)$ **43.** $5s^2$ **45.** $3\pi t^2$ **47.** $b^2\left(6 - \frac{\pi}{2}\right)$

Problem Set 5.2, Pages 113–114

1. $(x + 1)(x + 2)$ **3.** $(x - 1)(x + 2)$ **5.** $(x - 3)(x - 1)$ **7.** $(y - 2)(y - 3)$ **9.** $(y - 3)(y + 2)$ **11.** $(y + 4)(y + 3)$ **13.** $(z - 4)(z + 2)$ **15.** $(z - 8)(z + 1)$ **17.** $(z - 9)(z - 1)$ **19.** $(a + 7)(a - 2)$ **21.** $(a - 6)(a + 5)$ **23.** $(b + 6)(b - 3)$ **25.** $(c - 13)(c - 2)$ **27.** not factorable **29.** $(d - 13)(d + 11)$ **31.** $r^2 - 2r - 24 = (r - 6)(r + 4)$, so the dimensions are $(r - 6)$ by $(r + 4)$. **33.** $t^2 - 9t + 14 = (t - 7)(t - 2)$, so the orchard has $(t - 7)$ rows with $(t - 2)$ trees per row or $(t - 2)$ rows with $(t - 7)$ trees per row. **35.** **37.**

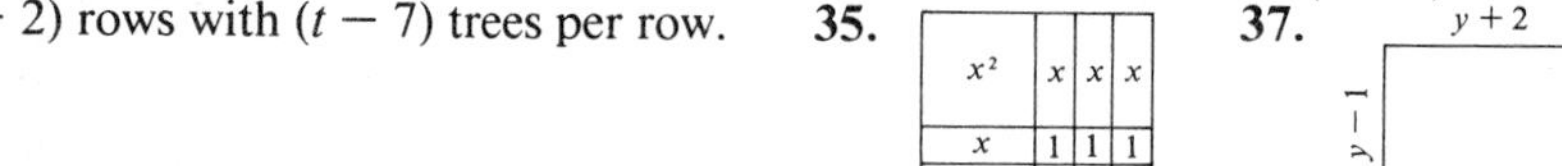

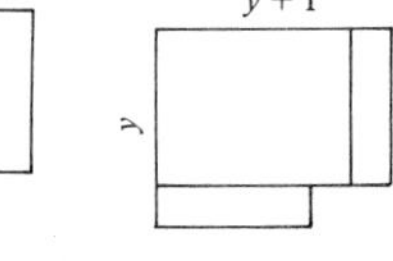

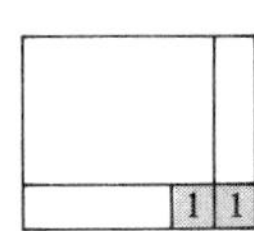

39. $x^2 + 4x + 3 = (x + 3)(x + 1)$

41. $x^2 + 6x + 8 = (x + 4)(x + 2)$

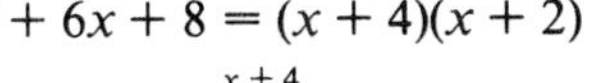

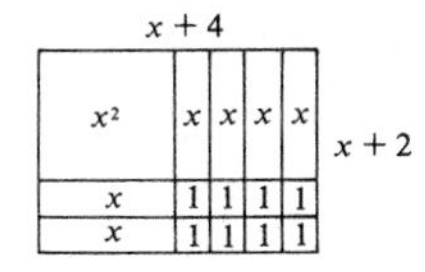

43. The following pieces cannot be arranged to form a rectangle.

Problem Set 5.3, Page 117

1. $(2y - 1)(y - 3)$ **3.** $(2u + 1)(u + 4)$ **5.** $(5v + 2)(v - 1)$ **7.** $(3w - 2)(w + 1)$ **9.** $(7x - 6)(x + 1)$ **11.** $3(z - 5)(2z + 1)$ **13.** $u(5u - 3)(u + 2)$ **15.** $3w^2(w + 4)(w - 3)$ **17.** $x(y - 9)(y - 8)$ **19.** $(3z + 4)(2z - 1)$ **21.** $(5t - 3)(3t + 2)$ **23.** $2w(2v + 3)(2v + 1)$ **25.** $(3x - 2)(4x - 1)$ **27.** $(3h + 5)(3h - 2)$ **29.** $x^2(3y - 1)(3y + 2)$ **31.** $2r + 1$ by $3r - 2$ **33.** Time is $4s - 3$.

Problem Set 5.4, Page 121

1. $(x - 8)(x + 8)$ **3.** $(x - 3)(x + 3)$ **5.** $(z - 11)(z + 11)$ **7.** $(2w - 3)(2w + 3)$ **9.** not factorable **11.** $(3z - 2)(3z + 2)$ **13.** $(x - 1)(x + 1)(x^2 + 1)$ **15.** $(z - 1)(z + 1)(z^2 + 1)(z^4 + 1)$ **17.** $(y - 1)^2$ **19.** $(y + 7)^2$ **21.** $(3z - 2)^2$ **23.** $(2x - 3)^2$ **25.** $2(x - 8)^2$ **27.** $(2x + 1)(x - 4)$ **29.** $4(y - 1)(y + 1)$ **31.** $a(2a + 3)(a + 1)$ **33.** $3(b + 1)(2b + 3)$ **35.** $m(4m + 1)(m + 2)$ **37.** $(7x - 1)(3x + 1)$ **39.** not factorable **41.** $2(z - 3)^2$ **43.** $3x^2(x + 5)^2$ **45.** $2xy(x - y)(x + y)$

Problem Set 5.5, Pages 123–124

1. $-5, 3$ **3.** $-\frac{3}{2}, 3$ **5.** $-9, 12$ **7.** $2, 3$ **9.** $-\frac{7}{3}, -1$ **11.** $0, 14$ **13.** $-6, 11$ **15.** $-\frac{1}{2}, \frac{5}{3}$ **17.** $-4, 4$ **19.** $\frac{3}{7}, \frac{5}{2}$ **21.** $\frac{2}{3}, \frac{5}{3}$ **23.** $-\frac{1}{3}, \frac{1}{7}$ **25.** $-\frac{5}{3}, 2$ **27.** $-2, -5$ **29.** $-\frac{3}{2}, \frac{3}{2}$ **31.** The dimensions are 5 ft by 9 ft. **33.** The dimensions are 3 in. by 8 in. **35.** The strip is 10 yd wide.

5.6 Review Self-Test, Pages 124–125

1. a. 2^6 **b.** $2^3 \cdot 5^3$ **c.** $2^2 \cdot 7^2$ **d.** $2^8 \cdot 5^3$ **2. a.** $5(xy + 1)$ **b.** $2h\left(\frac{1}{3}k + 1\right)$ **c.** $5(x^2 + 3x + 4)$ **d.** $y^2(3y^2 - 5y + 7)$ **3. a.** $(x - 5)(x - 1)$ **b.** $(x - 5)(x + 1)$ **c.** not factorable **d.** $(x - 4)(x - 1)$ **4. a.** $(x - 6)(x + 1)$ **b.** $(x + 6)(x - 1)$ **c.** $(x + 2)(x + 3)$ **d.** $(x - 2)(x - 3)$ **5. a.** $(2x - 5)(x - 3)$ **b.** $(3x - y)(x + 2y)$ **c.** $3(2x + 5)(3x - 4)$ **6. a.** $(2x - 1)(x + 4)$ **b.** $(x - 6y)(x - 4y)$ **c.** $(4x + 1)(3x + 1)$ **7. a.** $(x - 7)(x + 7)$ **b.** not factorable **c.** $(x - 11)^2$ **d.** $(x - 6y)^2$ **8. a.** $3(x - 3)(x + 3)$ **b.** $8(x + 2)(x + 1)$ **c.** $2(2x + 3)^2$ **d.** $x^2(3x + 7)(x - 2)$ **9. a.** -5 **b.** $-13, 13$ **c.** $5, -\frac{7}{2}$ **d.** $\frac{4}{3}, -\frac{7}{5}$ **10.** The dimensions are 11 ft by 16 ft **11.** $3r^2(16 - \pi)$ **12.**

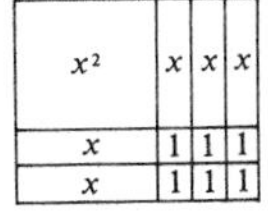

Review Problems, Pages 125–126

1. $3(x+6)$ **3.** $4(5x+1)$ **5.** $x(3x-2)$ **7.** $xy(3y^2+5x)$ **9.** $5xy(xy-1)(xy+1)$
11. $(x+2)(x+5)$ **13.** $(x-6)(x-1)$ **15.** $(x-3)(x-5)$ **17.** $(x-2)(x+5)$
19. $(2x+1)(3x+4)$ **21.** $(5x+2)(x+3)$ **23.** $(4x-1)(3x+2)$ **25.** $(6x-1)(x+2)$
27. $(12x-5)(2x+1)$ **29.** $x(x-1)(x+2)$ **31.** $x^2(x-1)(x-3)$ **33.** $x^2(x+4)(x-2)$
35. $4(x-2)(x+3)$ **37.** $2(x+4)(x-3)$ **39.** $3(x+2)(x-5)$ **41.** $2(x+5)(x-3)$
43. $(x-6)(x+6)$ **45.** $(x-1)(x+1)$ **47.** $(x-12)(x+12)$ **49.** $5(x-1)(x+1)$
51. $2(x-3)(x+3)$ **53.** $4(2-x)(2+x)$ **55.** $(2x-3)(2x+3)$ **57.** $(5x-7)(5x+7)$
59. $(x-4)^2$ **61.** $(2x+1)^2$ **63.** $(4x-5)^2$ **65.** $x^2(x-3)^2$ **67.** $3(x+4)^2$ **69.** $5(x-1)^2$
71. $x(3x+2)(x+1)$ **73.** $3(x-4)(2x+1)$ **75.** $2x(2x-3)(3x+1)$ **77.** $-3, -5$ **79.** $2, 7$
81. $-2, 1$ **83.** $-5, 5$ **85.** $-\frac{1}{6}, \frac{1}{6}$ **87.** $-\frac{5}{3}, \frac{5}{3}$ **89.** 3 **91.** $\frac{2}{3}$ **93.** $-3, 1$
95. $-3, 2$ **97.** $-2, 6$ **99.** 2

Problem Set 6.1, Pages 132–133

1. not equal **3.** equal **5.** equal **7.** equal **9.** equal **11.** $\frac{1}{2}$ **13.** $\frac{1}{2}$ **15.** $\frac{2}{3}$
17. $\frac{3}{4}$ **19.** $\frac{2}{3}$ **21.** $\frac{3}{7}$ **23.** $\frac{12}{13}$ **25.** $\frac{2}{3}$ **27.** $\frac{3}{4}$ **29.** $\frac{a}{b}$ **31.** $\frac{2}{35}$ **33.** $\frac{2}{3}$
35. $\frac{28}{15}$ or $1\frac{13}{15}$ **37.** $\frac{5}{6}$ **39.** 4 **41.** $\frac{7}{9}$ **43.** $\frac{5}{7}$ **45.** $\frac{5}{4}$ or $1\frac{1}{4}$ **47.** $\frac{7}{12}$ **49.** $\frac{35}{24}$ or $1\frac{11}{24}$
51. $\frac{25}{36}$ **53.** $\frac{-1}{72}$ **55.** 1 **57.** $\frac{1}{3}$ **59.** $\frac{-1}{5}$ **61.** $\frac{-1}{2}$ **63.** $\frac{3}{5}$ **65.** $\frac{8}{13}$ **67.** $\frac{-21}{34}$
69. $\frac{-55}{89}$

Problem Set 6.2, Pages 137–138

1. $\frac{x}{2}$ **3.** $\frac{1}{2s}$ **5.** $\frac{2ab}{3}$ **7.** $\frac{3b}{4}$ **9.** $\frac{2}{3rs}$ **11.** $\frac{3st}{7r}$ **13.** $\frac{12x^3}{13z}$ **15.** $\frac{2a^2b}{3}$ **17.** $\frac{4a^2}{3bc}$
19. $\frac{3b^4}{5c}$ **21.** $\frac{-y}{3}$ **23.** $\frac{-2a}{b}, b \neq 0$ **25.** $\frac{x}{y+1}, y \neq -1$ **27.** $\frac{-x}{2-x}$ or $\frac{x}{x-2}, x \neq 2$
29. $\frac{-(a-b)}{a+b}$ or $\frac{b-a}{a+b}, a \neq -b$ **31.** -1 **33.** -1 **35.** -5 **37.** -1 **39.** $m-n$
41. $x-y$ **43.** $\frac{x}{3}$ **45.** $\frac{-2}{3m}$ **47.** $\frac{m-3}{m+3}$ **49.** $\frac{x-6}{x+3}$ **51.** $\frac{3-m}{m+1}$ **53.** $\frac{-3}{r+3}$
55. -1 **57.** $\frac{z+3}{3z}$ **59.** $\frac{-(n+5)}{n+2)}$

Problem Set 6.3, Pages 140–141

1. $\frac{3a}{2}$ **3.** $\frac{6}{y}$ **5.** $\frac{6}{y}$ **7.** $\frac{y}{2}$ **9.** $\frac{18}{35a^2}$ **11.** $\frac{1}{4a}$ **13.** $\frac{-9x^2}{y^2}$ **15.** $\frac{m}{n^2}$ **17.** $3ab^2d$
19. $\frac{3}{2(a-b)}$ **21.** $\frac{2y}{3x}$ **23.** $\frac{-m}{n}$ **25.** $\frac{4(x-3)}{9(x-2)}$ **27.** $\frac{-9}{10}$ **29.** 2 **31.** $\frac{x-2}{x+1}$
33. $3-m$ **35.** $\frac{-r(r+2)}{r+1}$ **37.** $(x-4)(3x+1)$ **39.** 1 **41.** $2x+1$ **43.** $\frac{1}{3x+1}$
45. $\frac{2x+1}{x-5}$ **47.** $\frac{x+1}{x-2}$ **49.** $\frac{x+3}{x-2}$

Problem Set 6.4, Pages 144–145

1. 42 **3.** 54 **5.** 72 **7.** $12x^2y$ **9.** $18a^2b$ **11.** a^4 **13.** $18xy$ **15.** $12m^2n$
17. $(x-3)(x+3)$ **19.** $2x(x-2)$ **21.** $2x(3x+2)$ **23.** $2(n-2)(n+2)$
25. $(x-2)(x+2)(x-3)$ **27.** $(a+2)(a+3)(a-4)$ **29.** $(x-1)(x+4)(x-5)$ **31.** $\frac{3x+2y}{42}$
33. $\frac{7m+5n}{70}$ **35.** $\frac{15-4y}{5y}$ **37.** $\frac{a^2+1}{a}$ **39.** $\frac{1-c^2}{c}$ **41.** $\frac{21r+2s}{6rs}$ **43.** $\frac{20y+3x}{12x^2y}$
45. $\frac{4a+15b}{18a^2b}$ **47.** $\frac{a^3+a+2}{a^4}$ **49.** $\frac{x}{(x-2)(x+2)}$ **51.** $\frac{x-2}{2(x-1)}$
53. $\frac{-(x^2+4)}{2x(2x-3)}$ or $\frac{x^2+4}{2x(3-2x)}$ **55.** $\frac{(x+2)^2}{x(x-1)(x+1)}$ **57.** $\frac{-3a^2+9a+4}{2(a-3)(a+3)}$ **59.** $\frac{3x^2-x+4}{6(x-1)(x+2)}$
61. $\frac{9x^2-50}{15x(3x-5)}$ **63.** $\frac{3x-2}{2x(2x-1)}$ **65.** $\frac{6x+5y}{(x-y)(x+y)}$ **67.** $\frac{5x^2-39x+30}{6(x-3)(x+3)}$ **69.** $\frac{7x-6}{6(x+2)(x-3)}$

Problem Set 6.5, Pages 147–148

1. $x=9$ **3.** $z=10$ **5.** $y=1$ **7.** $x=1$ **9.** $z=1$ **11.** $y=-1$ **13.** $x=2$
15. $a=-1$ **17.** $m=0$ **19.** no solution $\left(x=\frac{1}{2}\text{ is extraneous}\right)$ **21.** $x=2$ **23.** $x=8$
25. $x=-1$ ($x=1$ is extraneous) **27.** $x=3$ **29.** $x=-\frac{1}{2}$ **31.** $x=6$ **33.** $x=-2$
35. $a=-8$ **37.** $x=-1$ ($x=3$ is extraneous) **39.** $x=\frac{1}{2}$ or 3

Problem Set 6.6, Pages 152–153

1. 1 **3.** $\frac{1}{2}$ **5.** $\frac{1}{3^2}$ **7.** 3^2 **9.** 5^{10} **11.** 7^{12} **13.** $2^3\cdot 3^3$ or 6^3 **15.** 2^6 **17.** 2^4
19. $\frac{1}{5^2}$ **21.** $\frac{3}{2}$ **23.** $\frac{2^2}{3^2}$ **25.** 3^4 **27.** 5^7 **29.** 1 **31.** $\frac{1}{2^{10}}$ **33.** $\frac{1}{5^{16}}$ **35.** 3^3
37. 1 **39.** $\frac{1}{5^5}$ **41.** $6x^3$ **43.** $2z^7$ **45.** $8y^6$ **47.** x^2 **49.** $\frac{1}{z^2}$ **51.** $\frac{y^9}{z^6}$ **53.** $\frac{y^6}{x^4}$
55. $\frac{x^9}{z^6}$ **57.** $\frac{b^8}{25c^6}$ **59.** 1 **61.** $\frac{1}{x^5}$ **63.** $\frac{a^2}{b^2}$ **65.** $\frac{5xy^4}{2}$ **67.** $3m^3n^2$ **69.** $-x$
71. 4.3×10^7 **73.** 6.81×10^{-4} **75.** 6.831×10^{-2} **77.** 3.666×10^9 **79.** 1.166×10^{-5}
81. 331.46 **83.** 57,870,000 **85.** 4,200,000,000,000 **87.** 270,000,000
89. 0.00000 00000 00000 00000 003

Problem Set 6.7, Pages 155–156

1. $\frac{3}{10}$ **3.** $\frac{28}{45}$ **5.** $\frac{3x}{2}$ **7.** $\frac{1}{xy}$ **9.** $\frac{2b}{a}$ **11.** $\frac{ab^2}{6}$ **13.** $\frac{5}{13}$ **15.** $\frac{13}{4}$ **17.** $\frac{x+3}{x+1}$
19. $\frac{x+2}{x-4}$ **21.** $\frac{(a+1)b}{a(b-1)}$ **23.** $\frac{3(x^2+1)}{x(3x-1)}$ **25.** $\frac{1+x}{1-x}$ **27.** $\frac{a^2+1}{a(1-2a)}$ **29.** $\frac{(a+b)^2}{ab}$
31. $\frac{x^2-1}{x^2+1}$ **33.** $\frac{(a+b)^2}{ab}$ **35.** $\frac{x^2-1}{x^2+1}$ **37.** $2(x-3)$ **39.** 1

6.8 Review Self-Test, Page 157

1. a. $\frac{-35}{72}$ **b.** $\frac{-3}{7}$ **2. a.** $\frac{3y^2}{4xz}$ **b.** $\frac{2}{a-2}$ **3. a.** $\frac{x}{75yz}$ **b.** $\frac{1}{x}$ **4. a.** $\frac{x-3}{(x-5)(2x+1)}$

b. 0 **5. a.** $\frac{x^2 - x + 1}{x}$ **b.** $\frac{-7}{6(x + y)}$ **6. a.** $\frac{b^6}{a^9}$ **b.** $\frac{3y^7}{5x^7}$ **7. a.** $\frac{7}{m - n}$ **b.** $\frac{b(4b - a)}{5b^2 - a^2}$

8. a. $b \cdot b \cdot b \cdot b \cdots b$ (n factors) **b.** 1 **c.** $\frac{1}{b^n}$ **d.** b^{m+n} **e.** b^{nm} **f.** b^{m-n}

9. a. i. 1.53×10^7 **ii.** 6.13×10^{-3} **iii.** 1×10^7 **iv.** 3×10^{-21} **b. i.** 0.00301 **ii.** 529,000 **iii.** 0.1 **iv.** 1,370,000,000 **10. a.** $x = \frac{35}{4}$ **b.** $x = \frac{48}{5}$ **c.** $y = 51$ **d.** $x = 3$

Review Problems, Pages 157–160

1. $\frac{1}{3}$ **3.** $\frac{1}{4}$ **5.** $\frac{25}{36}$ **7.** $-\frac{2}{5}$ **9.** $\frac{2}{3}$ **11.** $\frac{3y}{2z}$ **13.** $\frac{4x}{5y}$ **15.** $\frac{3}{a + 1}$ **17.** $\frac{3}{2}$ **19.** $\frac{x}{x + 1}$ **21.** $\frac{2a}{3}$ **23.** $\frac{3x}{y^2}$ **25.** $\frac{3}{20xyz}$ **27.** $\frac{x^2y}{10z}$ **29.** $\frac{a}{b}$ **31.** $\frac{4}{9}$ **33.** $\frac{1}{2x}$ **35.** $-x$ **37.** $\frac{3x}{2}$ **39.** $\frac{x - 1}{x}$ **41.** $\frac{2x + 3y}{12}$ **43.** $\frac{xy + 24}{6y}$ **45.** $\frac{xy + 4}{y}$ **47.** $\frac{x^2 + 6}{6x}$ **49.** $\frac{y + 2x}{xy}$ **51.** $\frac{x}{x^2 - 1}$ **53.** $\frac{1}{6(x - 2)}$ **55.** $\frac{13}{6(y + 2)}$ **57.** -8 **59.** 9 **61.** $\frac{1}{9}$ **63.** $\frac{-2}{3}$ **65.** $\frac{1}{5^7}$ **67.** $\frac{1}{9}$ **69.** $\frac{1}{b^2}$ **71.** $\frac{1}{x^6y^2}$ **73.** $\frac{x^6}{y^2}$ **75.** $\frac{b^2}{a^2}$ **77.** $\frac{x^5}{y^4}$ **79.** $\frac{y^3}{x^9}$ **81.** 1 **83.** 2 **85.** 16 **87.** 4 **89.** -5 **91.** 1 **93.** 3 **95.** 3 **97.** no solution (2 is extraneous) **99.** no solution (2 is extraneous)

Problem Set 7.1, Pages 166–167

1. Telegraph Hill **3.** Embarcadero **5.** Ferry Building **7.** Old Mint Building **9.** Coit Tower **11.** Niels Sovndal **13.** Clint Stevens **15.** Missy Smith **17.**

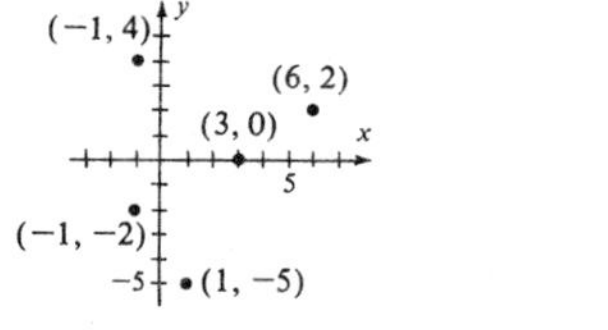

19. **21.**

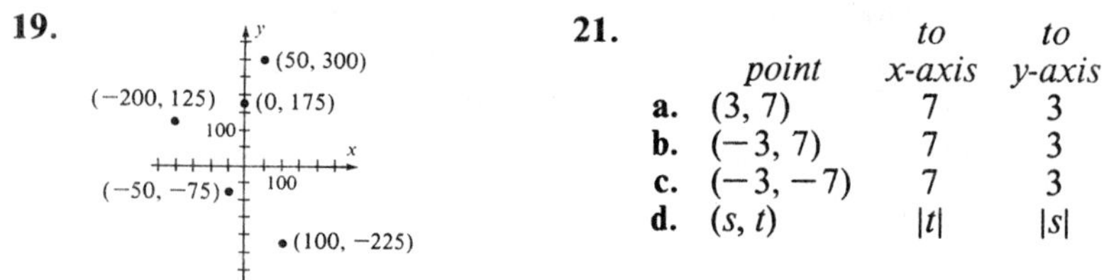

	point	to x-axis	to y-axis
a.	(3, 7)	7	3
b.	(−3, 7)	7	3
c.	(−3, −7)	7	3
d.	(s, t)	|t|	|s|

23. **25.**

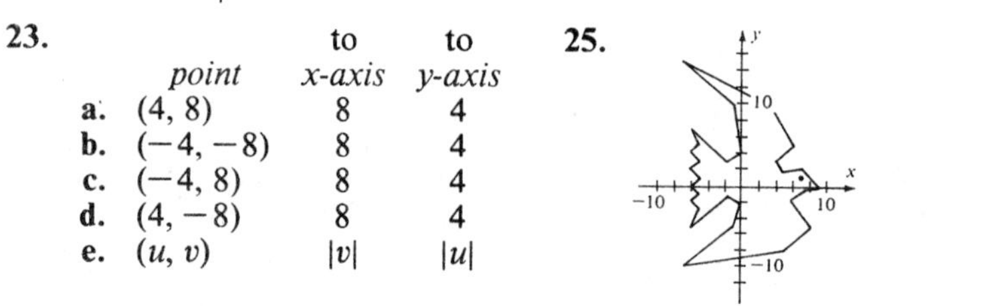

	point	to x-axis	to y-axis
a.	(4, 8)	8	4
b.	(−4, −8)	8	4
c.	(−4, 8)	8	4
d.	(4, −8)	8	4
e.	(u, v)	|v|	|u|

27. They all lie on the same line. **29.**

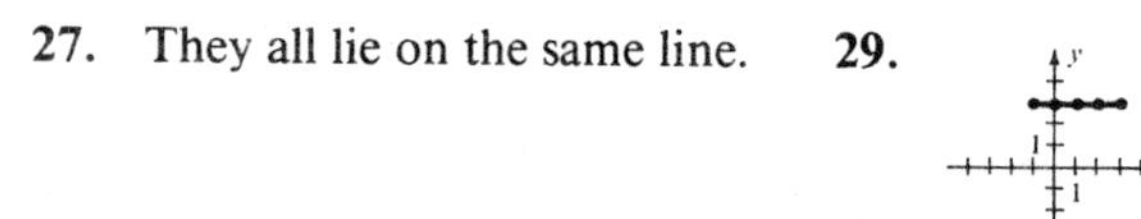

Problem Set 7.2, Pages 172–173

1. (0, 3), (1, 4), (2, 5), (3, 6), (4, 7) **3.** (0, 0), (1, 4), (2, 8), (3, 12), (4, 16) **5.** (0, 0), (1, π), (2, 2π), (3, 3π), (4, 4π) Answers in Problems 7–17 vary. **19.**

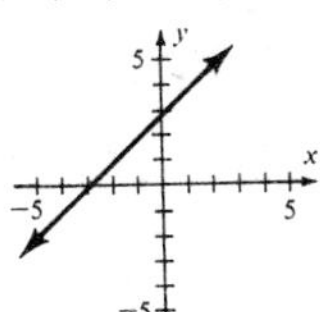

21.

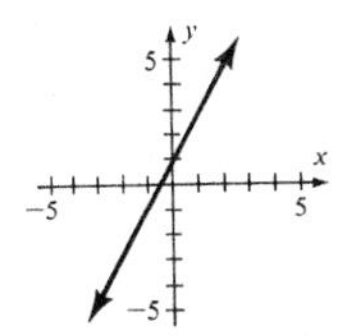

23.

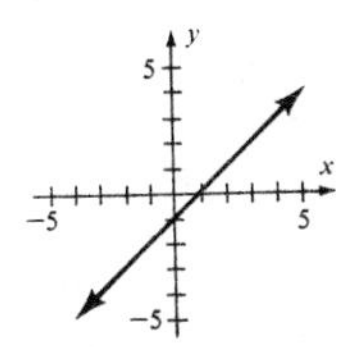

25.

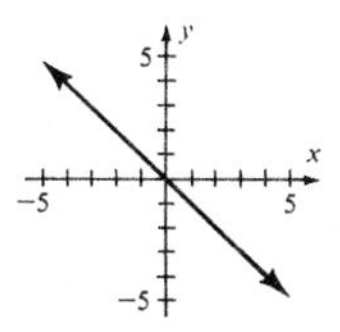

27.

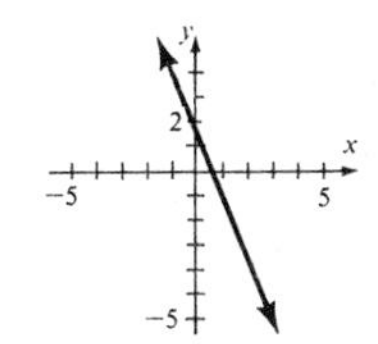

29.

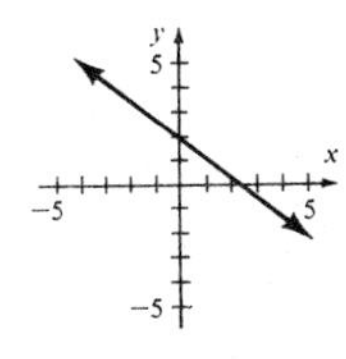

31.

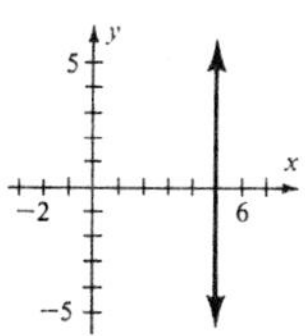

33.

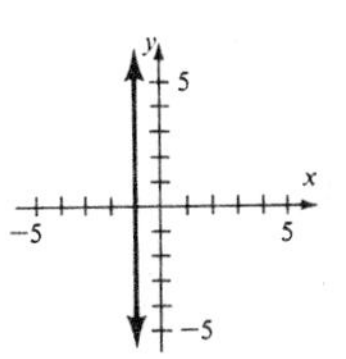

35.

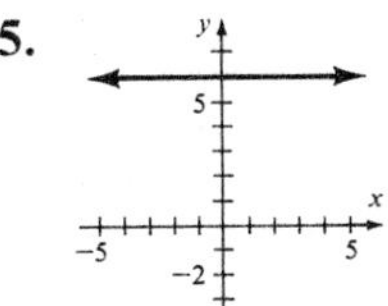

37.

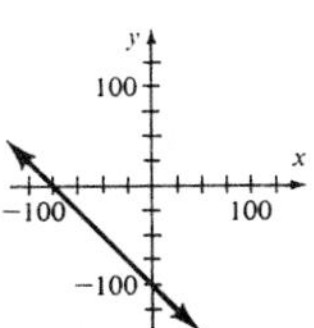

39.

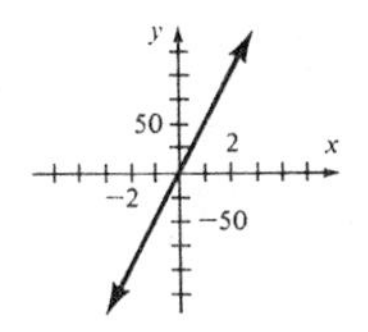

41. (125, 400), (200, 325), (247, 278) **43.** (0, 0), (3, 96), (6, 192)

Problem Set 7.3, Pages 179–180

1. $m = \frac{1}{2}$ **3.** $m = \frac{-2}{3}$ **5.** $m = \frac{7}{1}$ **7.** $m = \frac{-11}{1}$ **9.** $m = \frac{3}{2}$ **11.** $m = \frac{-1}{1}$

13. $m = \frac{-3}{1}$ **15.** $m = \frac{3}{2}$ **17.** $m = \frac{5}{9}$ **19.** $m = \frac{-3}{4}$ **21.**

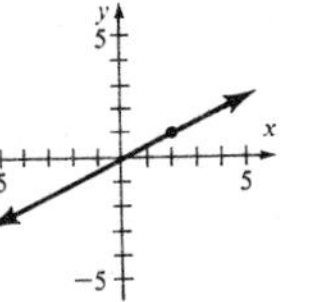

23.

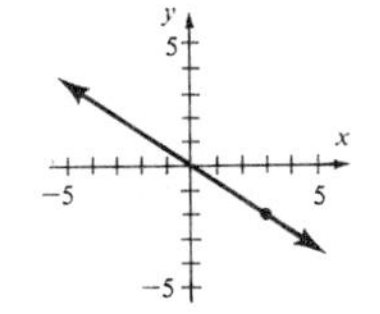

25.

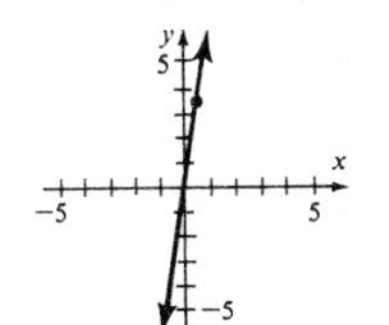

27.

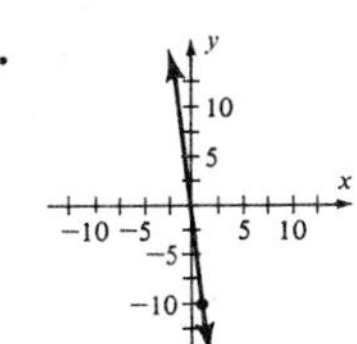

29. 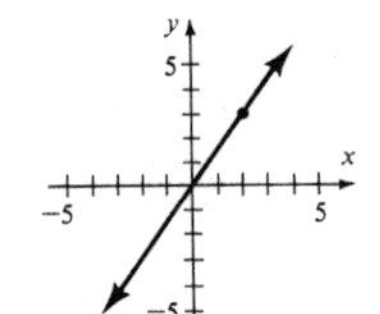

31. 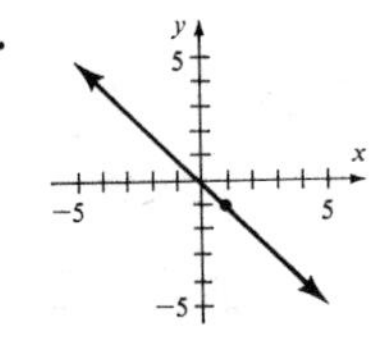

33. 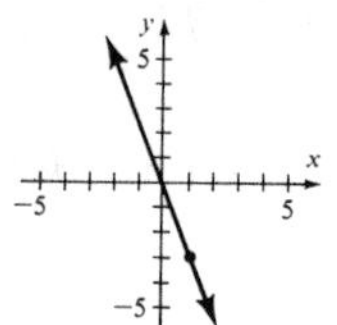

35. 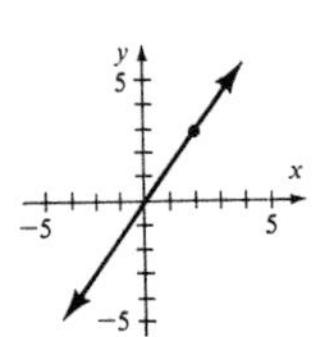

37. 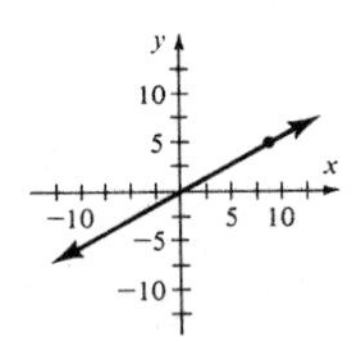

39. 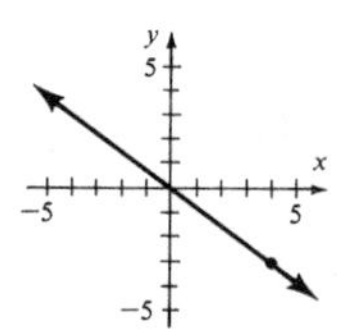

41. 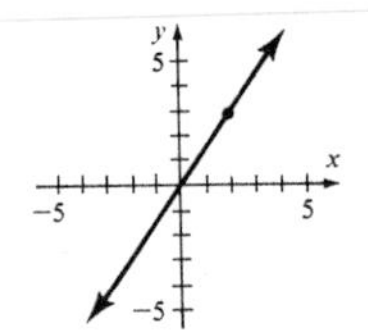

43. 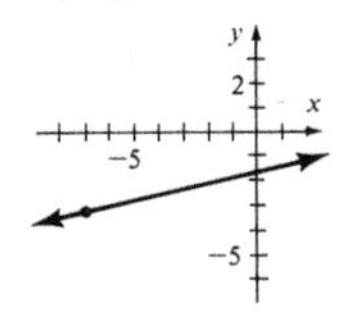

45. 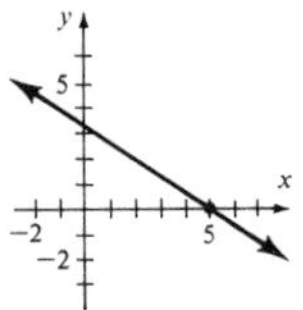

47. 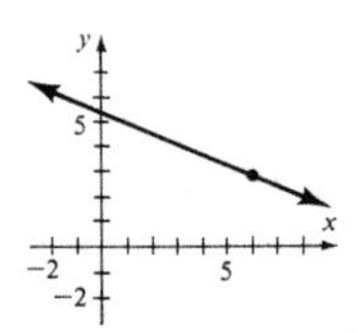

49. 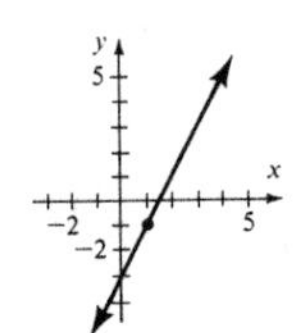

51. 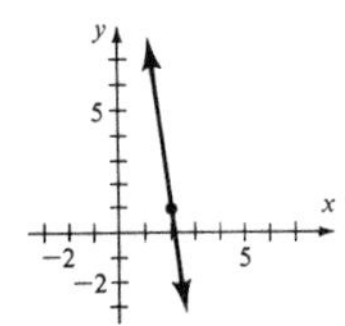

53. 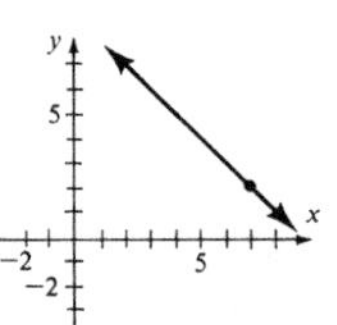

55. 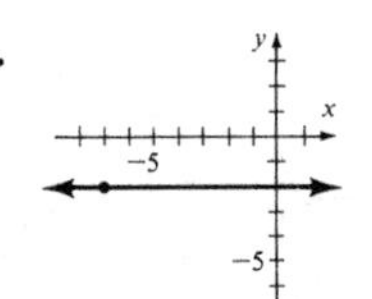

57. 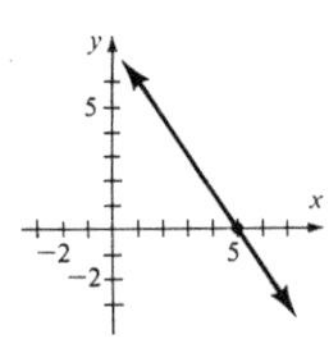

59. 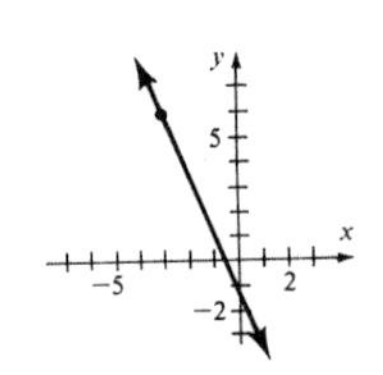

61. $m = \frac{-1}{8}$ **63.** $m = \frac{1}{2}$ **65.** $m = \frac{2}{5}$ **67.** $m = 0$

Problem Set 7.4, Pages 185–186

1. $m = 3, b = 2$ **3.** $m = 1, b = -4$ **5.** $m = -3, b = -2$ **7.** $m = \frac{1}{2}, b = -\frac{3}{2}$
9. $m = \frac{2}{3}, b = -\frac{1}{3}$ **11.** $m = \frac{-2}{3}, b = \frac{1}{6}$ **13.** $m = \frac{4}{5}, b = 8$ **15.** $m = \frac{3}{4}, b = -12$
17. no slope; no y-intercept **19.** $m = 0, b = -3$ **21.** $m = 0, b = 0$

23. 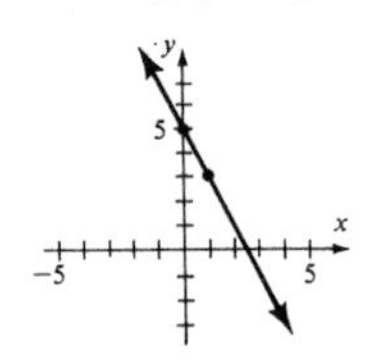

25. 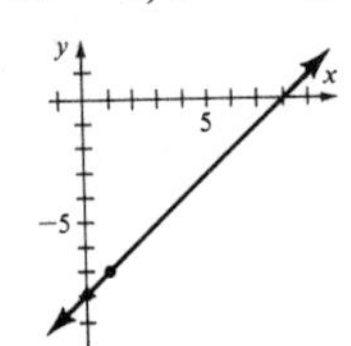

27. 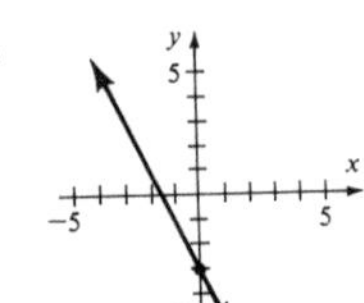

29.

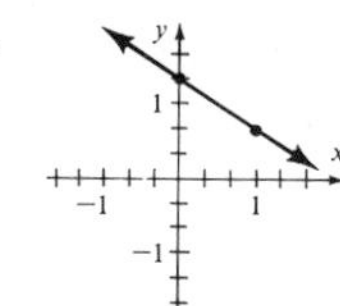

31.

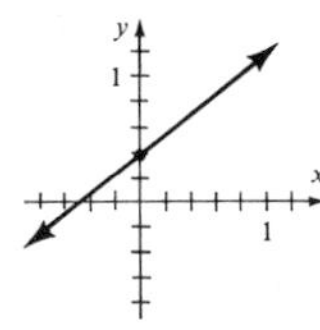

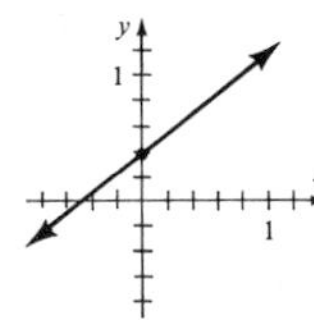

33.

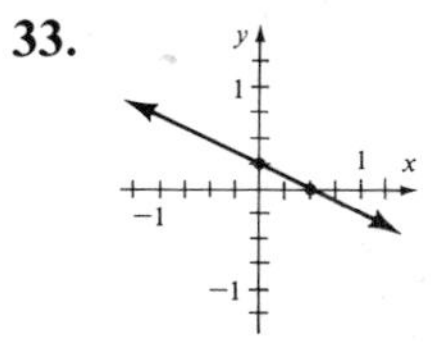

35.

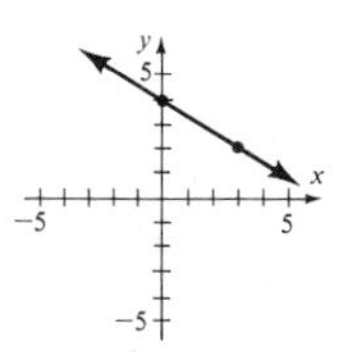

37.

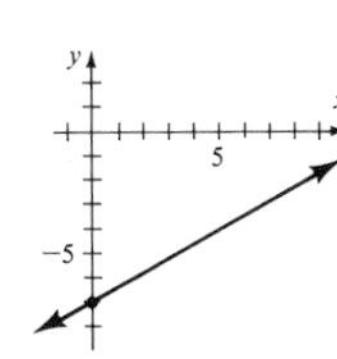

39.

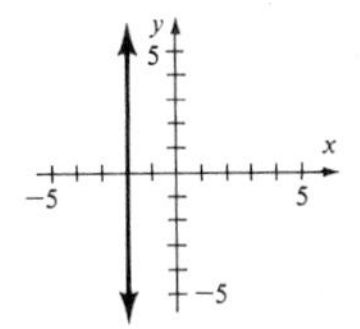

41.

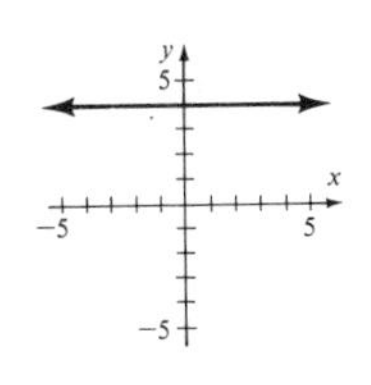

43. $y = -2x - 4$

45. $y = 4x + 5$

47. $3y = -2x - 6$

$$y = -\frac{2}{3}x - 2$$

49. $y = \frac{3}{2}x + \frac{5}{2}$

51. $4y = 5x - 8$

$$y = \frac{5}{4}x - 2$$

53.

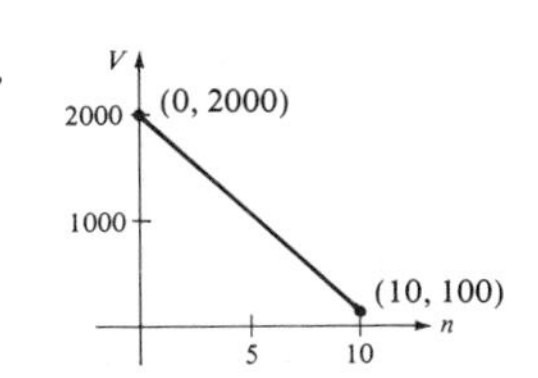

Problem Set 7.5, Pages 188–189

1.

3.

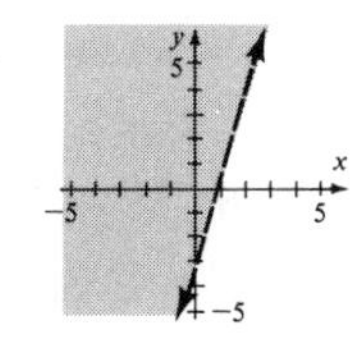

5.

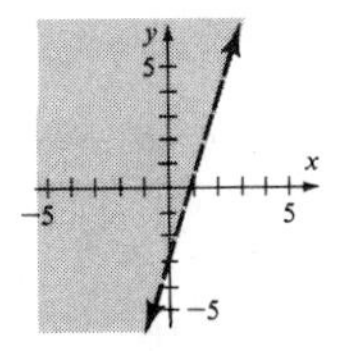

7.

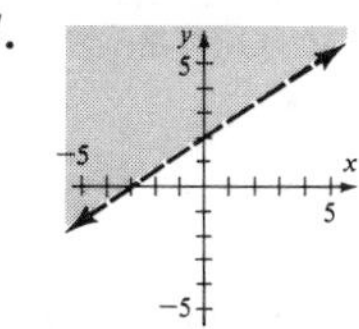

9.

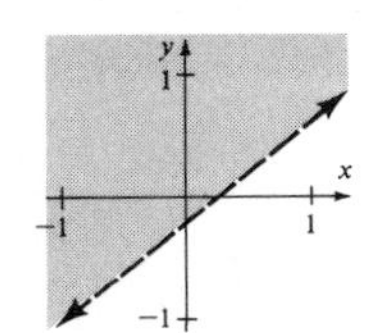

11.

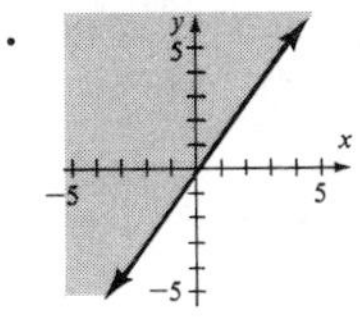

13. 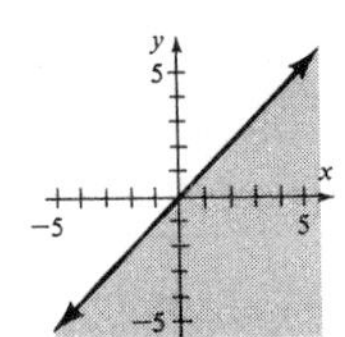

15. 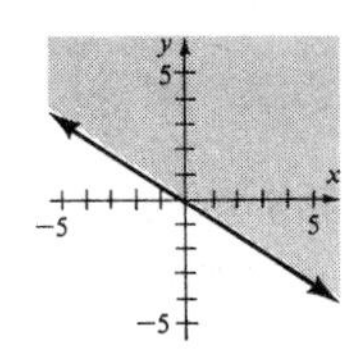

17. 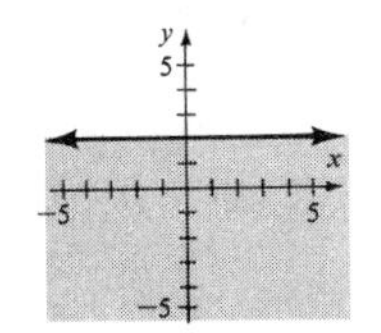

19. 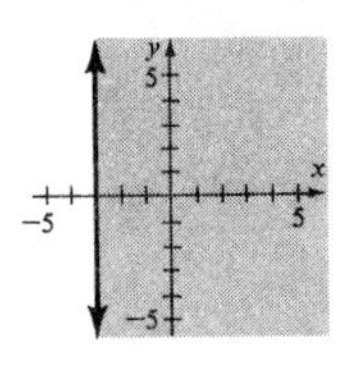

21. 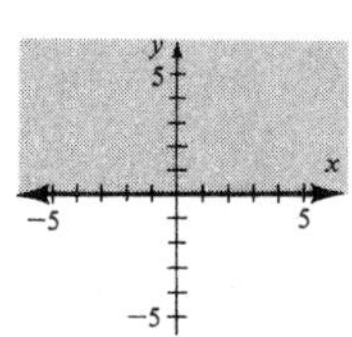

23. $2y \geq -4x - 1$
$y \geq -2x - \frac{1}{2}$

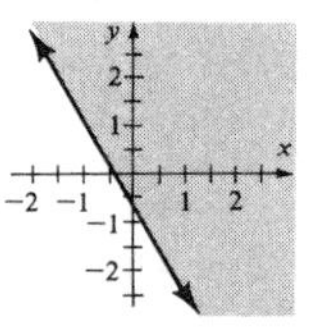

25. $-3y \geq -x + 9$
$y \leq \frac{1}{3}x - 3$

27. $2y > -3x + 1$
$y > -\frac{3}{2}x + \frac{1}{2}$

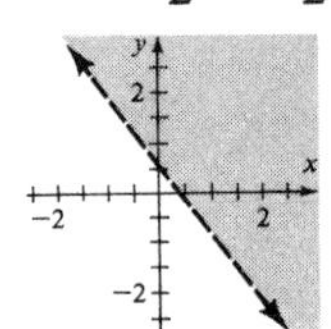

29. $-7y > -6x + 1$
$y < \frac{6}{7}x - \frac{1}{7}$

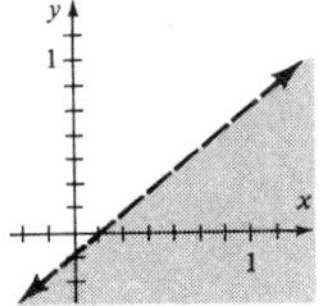

7.6 Review Self-Test, Page 189

1.

	point	*to x-axis*	*to y-axis*
a.	$(-2, 4)$	4	2
b.	$(6, -8)$	8	6
c.	$(-5, -3)$	3	5
d.	(m, n)	$\lvert n \rvert$	$\lvert m \rvert$

2.

3. $m = \frac{1}{3}$

4. **a.** 0 **b.** no slope

5. 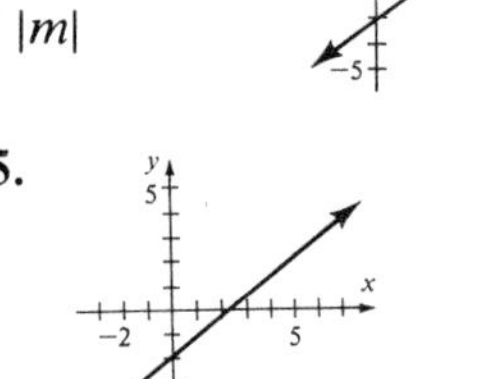

6. 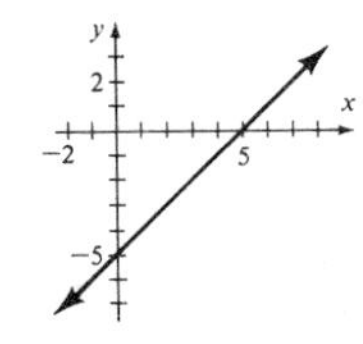

7. 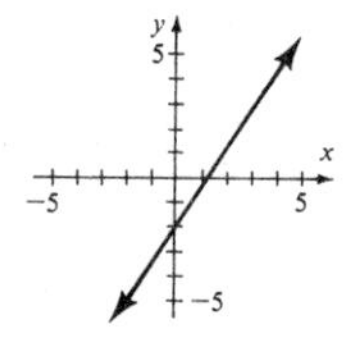

8. 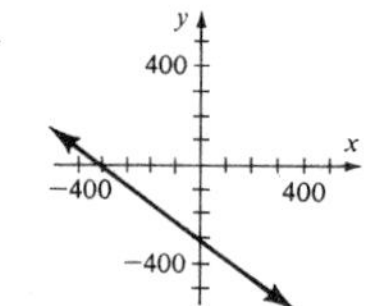

9. 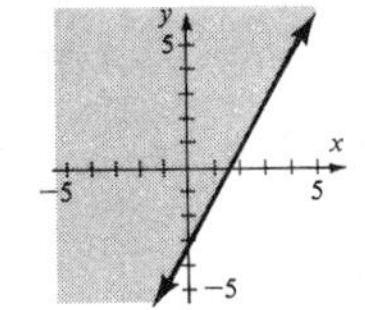

10. 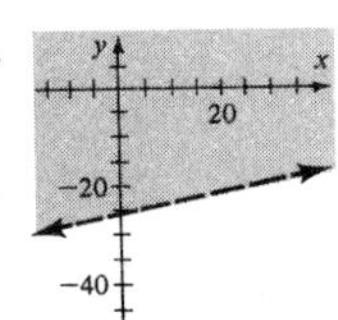

Review Problems, Pages 189 – 190

1. 3 **3.** $\frac{4}{5}$ **5.** $\frac{9}{5}$ **7.** $\frac{2}{3}$ **9.** $\frac{4}{9}$ **11.** $\frac{-2}{3}$ **13.** 0 **15.** 0 **17.** no slope
19. $\frac{d-b}{c-a}$ **21.** $\frac{z-x}{y-w}$ **23.** $m=2, b=-5$ **25.** $m=\frac{1}{2}, b=-5$ **27.** $m=\frac{1}{3}, b=-1$
29. $m=-3, b=0$ **31.** $m=-1, b=5$ **33.** $m=-3, b=1$ **35.** $m=2, b=3$ **37.** $m=3, b=5$ **39.** $m=3, b=-2$ **41.** $m=\frac{3}{2}, b=2$ **43.** $m=-3, b=0$ **45.** $m=\frac{1}{2}, b=0$
47. $m=-\frac{2}{3}, m=0$ **49.** $m=\frac{5}{3}, b=0$ **51.** no slope, no y-intercept **53.** $m=0, b=-3$

55.

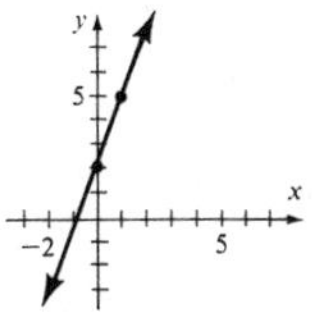

57.

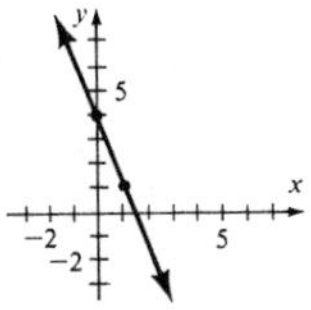

59.

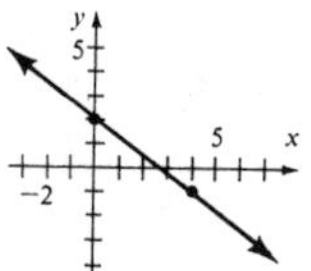

61.

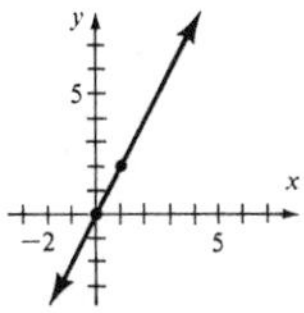

63.

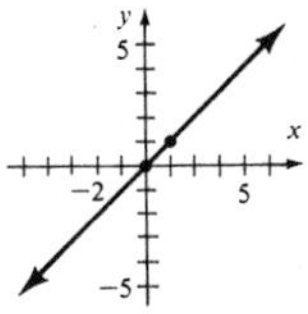

65.

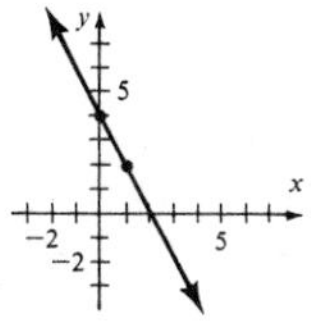

67.

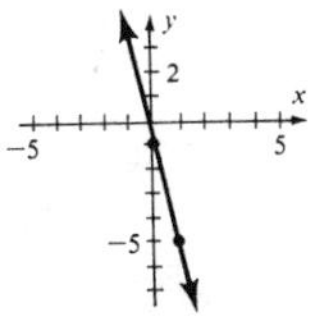

69.

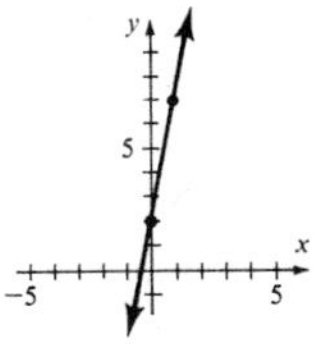

71.

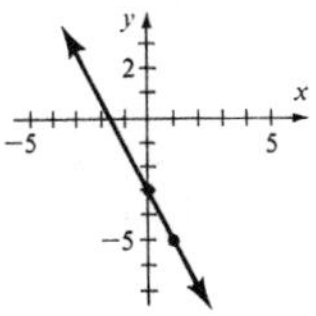

73.

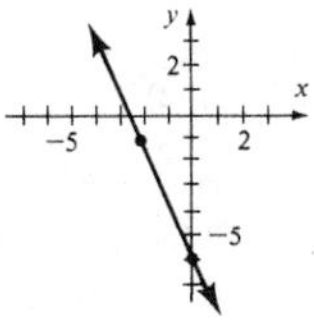

75.

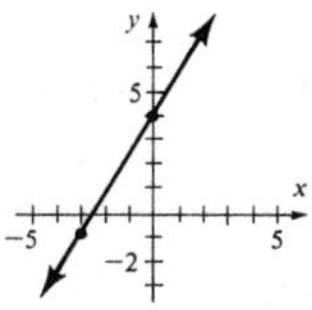

77.

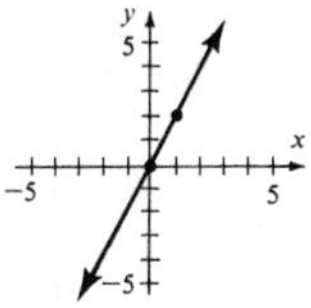

79.

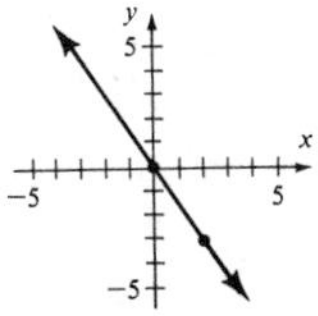

81.

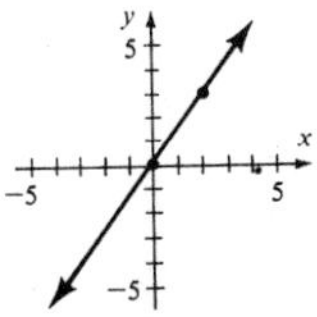

83.

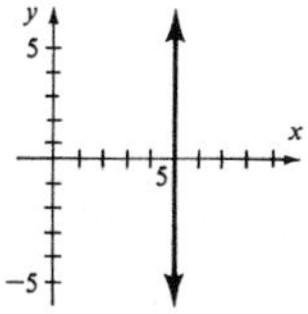

85.

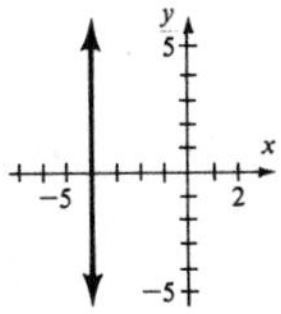

87.

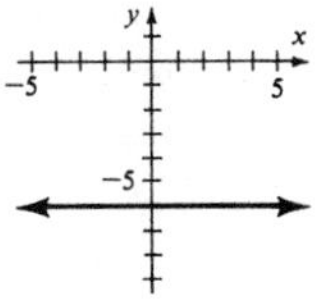

89. 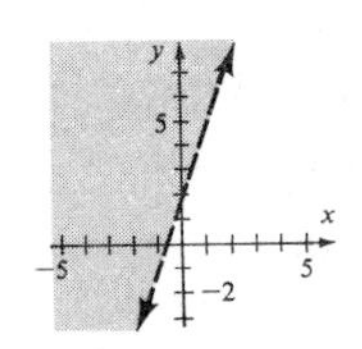

91. 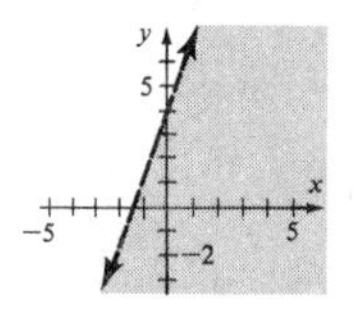

93. 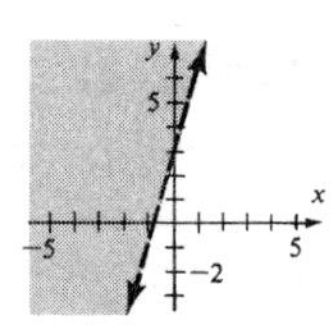

95. 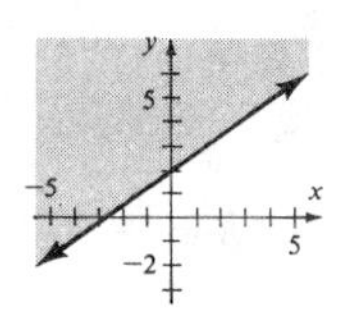

97. 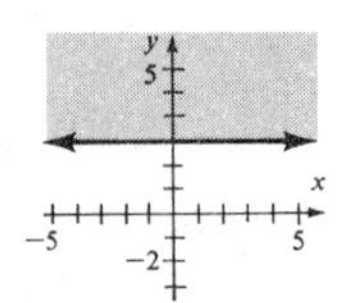

99. 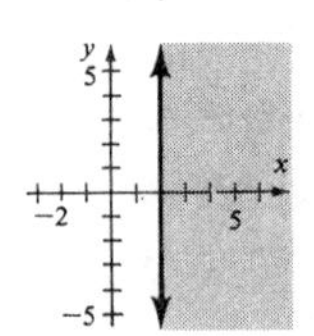

Problem Set 8.1, Pages 196–197

1. yes **3.** no **5.** yes **7.** no **9.** yes **11.** no

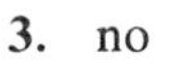

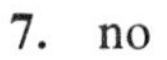

13. 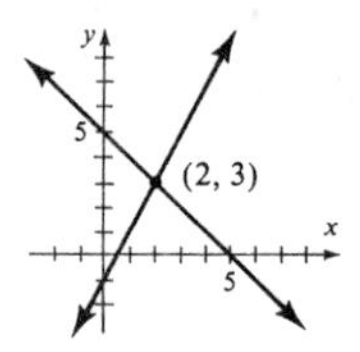

15. 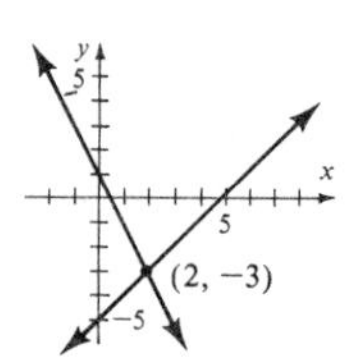

17. 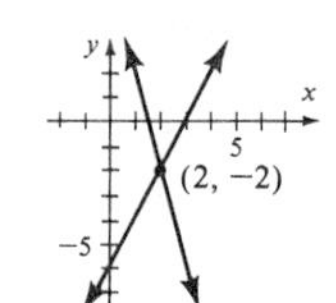

19. 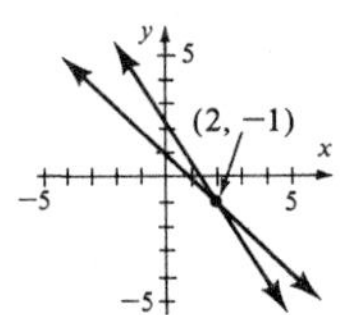

21. 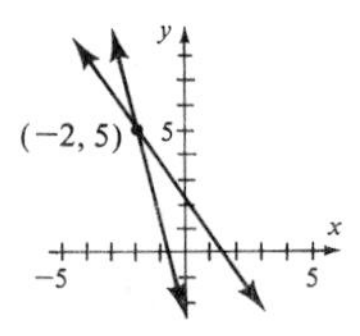

23. 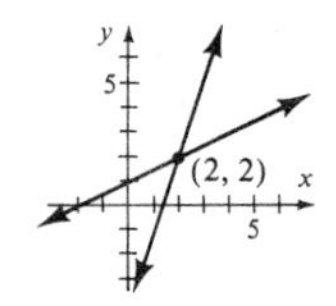

25. Demand: (6, 200) and (2, 800); Supply: (7, 800), (1, 200) **27.** Demand: (1, 5000) and (7, 1000); Supply: (2, 2000), (8, 5000) **29.** Demand: (10, 180) and (40, 20); Supply: (10, 20) and (40, 180)

31. A #4 price maximizes the profit. **33.** (4, 3000)

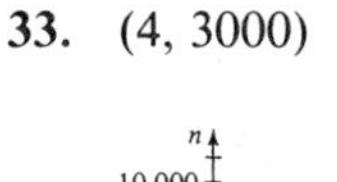

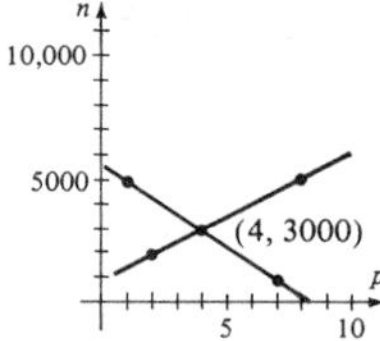

35. The optimum price is $25

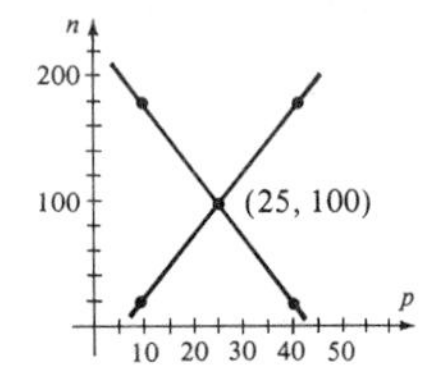

Problem Set 8.2, Pages 201–202

1. (19, 17) **3.** (5, −13) **5.** $(m, n) = (10, 4)$ **7.** $\left(6, \frac{1}{2}\right)$ **9.** (5, 1)

11. $(r, s) = (4, -1)$ **13.** (1, 0) **15.** (6, −4) **17.** $\left(0, \frac{1}{2}\right)$ **19.** $(p, q) = (1, -1)$

21. $(r, s) = (-1, 4)$ **23.** (−4, 3) **25.** (3, 2) **27.** (1, 1) **29.** $(F, S) = (3, 4)$

31. The numbers are 54 and 46. **33.** The numbers are 24 and 7. **35.** The smaller number is 57.

37.
$$\begin{cases} \begin{pmatrix}\text{RATE OF}\\ \text{BOAT}\end{pmatrix} + \begin{pmatrix}\text{RATE OF}\\ \text{CURRENT}\end{pmatrix} = \begin{pmatrix}\text{RATE WITH}\\ \text{CURRENT}\end{pmatrix} \\ \begin{pmatrix}\text{RATE OF}\\ \text{BOAT}\end{pmatrix} - \begin{pmatrix}\text{RATE OF}\\ \text{CURRENT}\end{pmatrix} = \begin{pmatrix}\text{RATE AGAINST}\\ \text{CURRENT}\end{pmatrix} \end{cases}$$

Let B = rate of boat
C = rate of current

$$-1\begin{cases} B + C = 30 \\ B - C = 26 \end{cases}$$

$$+\begin{cases} B + C = 30 \\ -B + C = -26 \end{cases}$$

$$2C = 4$$
$$C = 2$$

The current's rate is 2 mph.

39.
$$\begin{cases} \begin{pmatrix}\text{CANOE}\\ \text{RATE}\end{pmatrix} + \begin{pmatrix}\text{CURRENT}\\ \text{RATE}\end{pmatrix} = \begin{pmatrix}\text{TRIP RATE}\\ \text{DOWNSTREAM}\end{pmatrix} \\ \begin{pmatrix}\text{CANOE}\\ \text{RATE}\end{pmatrix} - \begin{pmatrix}\text{CURRENT}\\ \text{RATE}\end{pmatrix} = \begin{pmatrix}\text{TRIP RATE}\\ \text{UPSTREAM}\end{pmatrix} \end{cases}$$

$$\begin{cases} \begin{pmatrix}\text{CANOE}\\ \text{RATE}\end{pmatrix} + \begin{pmatrix}\text{CURRENT}\\ \text{RATE}\end{pmatrix} = \frac{9}{1.5} \\ \begin{pmatrix}\text{CANOE}\\ \text{RATE}\end{pmatrix} - \begin{pmatrix}\text{CURRENT}\\ \text{RATE}\end{pmatrix} = \frac{9}{3} \end{cases}$$

The rate of the current is 1.5 mph.

Problem Set 8.3, Pages 206–207

1. (−1, 4) **3.** (−3, 9) **5.** (2, −5) **7.** (−2, 0) **9.** (3, 5) **11.** (5, 4) **13.** (−12, 37) **15.** (5, −3) **17.** (3, −1) **19.** (9, −4) **21.** (5, 1) **23.** (1, 0) **25.** $3 **27.** $.85 **29.** Value of dimes + Value of quarters = 375 **31.** Value of nickels + Value of dimes + Value of pennies = 1214 **33.** Number of dimes = 2(number of nickels) − 1 **35.** Number of dimes = 2(number of pennies) − 6

37.
$$\begin{pmatrix}\text{VALUE}\\ \text{OF}\\ \text{QUARTERS}\end{pmatrix} + \begin{pmatrix}\text{VALUE}\\ \text{OF}\\ \text{DIMES}\end{pmatrix} = \begin{pmatrix}\text{VALUE}\\ \text{OF}\\ \text{COINS}\end{pmatrix}$$

$$25\begin{pmatrix}\text{NO. OF}\\ \text{QUARTERS}\end{pmatrix} + 10\begin{pmatrix}\text{NO. OF}\\ \text{DIMES}\end{pmatrix} = 840$$

$$25\left[2\cdot\begin{pmatrix}\text{NO. OF}\\ \text{DIMES}\end{pmatrix}\right] + 10\begin{pmatrix}\text{NO. OF}\\ \text{DIMES}\end{pmatrix} = 840$$

Let d = NO. OF DIMES
$25(2d) + 10d = 840$
$d = 14$

There are 14 dimes and 28 quarters.

39.
$$(\text{VALUE OF NICKELS}) + (\text{VALUE OF DIMES}) + (\text{VALUE OF PENNIES}) = \text{TOTAL VALUE}$$
$$5(\text{NO. OF NICKELS}) + 10(\text{NO. OF DIMES}) + (\text{NO. OF PENNIES}) = \text{VALUE IN CENTS}$$
$$5(\text{NO. OF DIMES}) + 10(\text{NO. OF DIMES}) + (\text{NO. OF PENNIES}) = 840$$
$$5(2\cdot\text{NO. OF PENNIES} - 6) + 10(2\cdot\text{NO. OF PENNIES} - 6) + (\text{NO. OF PENNIES}) = 840$$
Let p = NO. OF PENNIES
$$5(2p - 6) + 10(2p - 6) + p = 840$$
$$p = 30$$
$$2p - 6 = 54$$
The box contains 30 pennies, 54 dimes, and 54 nickels.

Problem Set 8.4, Pages 208–210

1. (5, 3) **3.** (−2, 3) **5.** (2, 0) **7.** (4, 2) **9.** (1, 4) **11.** inconsistent system **13.** (1, 1) **15.** dependent system **17.** (4, −2) **19.** $\left(6, \frac{1}{2}\right)$ **21.** (1, −5) **23.** (9, −4) **25.** (−6, 2) **27.** (3, 4) **29.** (4, −5)

31.
$$\begin{cases} \begin{pmatrix}\text{SPEED}\\\text{OF}\\\text{PLANE}\end{pmatrix} + \begin{pmatrix}\text{SPEED}\\\text{OF}\\\text{WIND}\end{pmatrix} = \begin{pmatrix}\text{SPEED}\\\text{WITH}\\\text{WIND}\end{pmatrix} \\ \begin{pmatrix}\text{SPEED}\\\text{OF}\\\text{PLANE}\end{pmatrix} - \begin{pmatrix}\text{SPEED}\\\text{OF}\\\text{WIND}\end{pmatrix} = \begin{pmatrix}\text{SPEED}\\\text{AGAINST}\\\text{WIND}\end{pmatrix} \end{cases}$$
$$\begin{cases} p + w = 400 \\ p - w = 100 \end{cases}$$
The rate of the plane in still air is 250 mph.

33.
$$\begin{pmatrix}\text{VALUE}\\\text{OF}\\\text{QUARTERS}\end{pmatrix} + \begin{pmatrix}\text{VALUE}\\\text{OF}\\\text{DIMES}\end{pmatrix} = \begin{pmatrix}\text{TOTAL}\\\text{VALUE}\end{pmatrix}$$
$$25\begin{pmatrix}\text{NO. OF}\\\text{QUARTERS}\end{pmatrix} + 10\begin{pmatrix}\text{NO. OF}\\\text{DIMES}\end{pmatrix} = 2400$$
$$25\begin{pmatrix}\text{NO. OF}\\\text{QUARTERS}\end{pmatrix} + 10\left(147 - \begin{matrix}\text{NO. OF}\\\text{QUARTERS}\end{matrix}\right) = 2400$$
$$25q + 10(147 - q) = 2400$$
There are 62 quarters.

35.

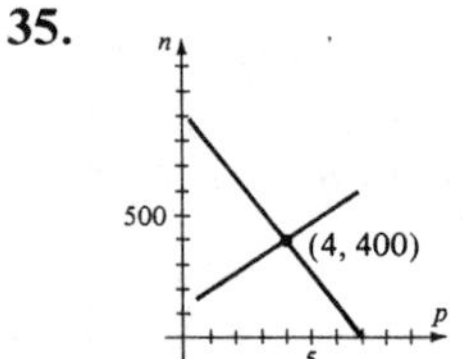

The price is $4.

Problem Set 8.5, Pages 216–218

1.
$$\begin{cases} \begin{pmatrix}\text{DOM DELUISE'S}\\\text{BIRTH YEAR}\end{pmatrix} - \begin{pmatrix}\text{MEL BROOKS'S}\\\text{BIRTH YEAR}\end{pmatrix} = 7 \\ \begin{pmatrix}\text{DOM DELUISE'S}\\\text{BIRTH YEAR}\end{pmatrix} + \begin{pmatrix}\text{MEL BROOKS'S}\\\text{BIRTH YEAR}\end{pmatrix} = 3859 \end{cases}$$
Dom DeLuise was born in 1933.

3.
$$\begin{cases} \begin{pmatrix}\text{FELICIANO'S}\\ \text{BIRTH YEAR}\end{pmatrix} - \begin{pmatrix}\text{CHARLES'S}\\ \text{BIRTH YEAR}\end{pmatrix} = 15 \\ \begin{pmatrix}\text{FELICIANO'S}\\ \text{BIRTH YEAR}\end{pmatrix} + \begin{pmatrix}\text{CHARLES'S}\\ \text{BIRTH YEAR}\end{pmatrix} = 3875 \end{cases}$$

Jose Feliciano was born in 1945.

5.
$$\begin{cases} \begin{pmatrix}\text{DISTANCE}\\ \text{VIPER}\\ \text{TRAVELS}\end{pmatrix} = \begin{pmatrix}\text{DISTANCE}\\ \text{CYLON}\\ \text{TRAVELS}\end{pmatrix} + 4500 \\ \begin{pmatrix}\text{VIPER'S}\\ \text{RATE}\end{pmatrix}(\text{TIME}) = \begin{pmatrix}\text{CYLON'S}\\ \text{RATE}\end{pmatrix}(\text{TIME}) + 4500 \end{cases}$$

$$15{,}000(\text{TIME}) = 12{,}000(\text{TIME}) + 4500$$

It will take $1\frac{1}{2}$ hours.

7.
$$\begin{cases} \text{COMMUTER DISTANCE} = 55 \\ \text{COMMUTER TIME} = 1.5 \end{cases}$$

$$\begin{cases} \begin{pmatrix}\text{TRAIN}\\ \text{DISTANCE}\end{pmatrix} + \begin{pmatrix}\text{TAXI}\\ \text{DISTANCE}\end{pmatrix} = 55 \\ \begin{pmatrix}\text{TRAIN}\\ \text{TIME}\end{pmatrix} + \begin{pmatrix}\text{TAXI}\\ \text{TIME}\end{pmatrix} = 1.5 \end{cases}$$

$$\begin{cases} \begin{pmatrix}\text{TRAIN}\\ \text{RATE}\end{pmatrix}\begin{pmatrix}\text{TRAIN}\\ \text{TIME}\end{pmatrix} + \begin{pmatrix}\text{TAXI}\\ \text{RATE}\end{pmatrix}\begin{pmatrix}\text{TAXI}\\ \text{TIME}\end{pmatrix} = 55 \\ \begin{pmatrix}\text{TRAIN}\\ \text{TIME}\end{pmatrix} + \begin{pmatrix}\text{TAXI}\\ \text{TIME}\end{pmatrix} = 1.5 \end{cases}$$

$$\begin{cases} 45\begin{pmatrix}\text{TRAIN}\\ \text{TIME}\end{pmatrix} + 20\begin{pmatrix}\text{TAXI}\\ \text{TIME}\end{pmatrix} = 55 \\ \begin{pmatrix}\text{TRAIN}\\ \text{TIME}\end{pmatrix} + \begin{pmatrix}\text{TAXI}\\ \text{TIME}\end{pmatrix} = 1.5 \end{cases}$$

Let T be the time spent on the train and S be the time spent in the taxi.

$$\begin{cases} 45T + 20S = 55 \\ T + S = 1.5 \end{cases}$$

She spends $\frac{1}{2}$ hour in a cab.

9.
$$\begin{cases} \text{FLIGHT DISTANCE} = 1100 \\ \text{FLIGHT TIME} = 7 \end{cases}$$

$$\begin{cases} \begin{pmatrix}\text{FIRST LEG}\\ \text{DISTANCE}\end{pmatrix} + \begin{pmatrix}\text{SECOND LEG}\\ \text{DISTANCE}\end{pmatrix} = 1100 \\ \begin{pmatrix}\text{FIRST LEG}\\ \text{DISTANCE}\end{pmatrix} + \begin{pmatrix}\text{SECOND LEG}\\ \text{DISTANCE}\end{pmatrix} = 7 \end{cases}$$

$$\begin{cases} \begin{pmatrix}\text{FIRST}\\ \text{RATE}\end{pmatrix}\begin{pmatrix}\text{FIRST}\\ \text{TIME}\end{pmatrix} + \begin{pmatrix}\text{SECOND}\\ \text{RATE}\end{pmatrix}\begin{pmatrix}\text{SECOND}\\ \text{TIME}\end{pmatrix} = 1100 \\ \begin{pmatrix}\text{FIRST}\\ \text{TIME}\end{pmatrix} + \begin{pmatrix}\text{SECOND}\\ \text{TIME}\end{pmatrix} = 7 \end{cases}$$

$$\begin{cases} 180\binom{\text{FIRST}}{\text{TIME}} + 140\binom{\text{SECOND}}{\text{TIME}} = 1100 \\ \binom{\text{FIRST}}{\text{TIME}} + \binom{\text{SECOND}}{\text{TIME}} = 7 \end{cases}$$

Let F be the time of first leg and S be the time on second leg.

$$\begin{cases} 180F + 140S = 1100 \\ F + S = 7 \end{cases}$$ The legs were 540 miles and 560 miles.

11.
$$\begin{cases} (.50)(6) = .20\binom{\text{AMT}}{\text{AFTER}} \\ 6 + \binom{\text{AMT}}{\text{WATER}} = \binom{\text{AMT}}{\text{AFTER}} \end{cases}$$

9 oz of water must be added.

13.
$$\begin{cases} \binom{\text{AMT}}{\text{MILK}} + \binom{\text{AMT}}{\text{CREAM}} = 180 \\ .20\binom{\text{AMT}}{\text{MILK}} + .60\binom{\text{AMT}}{\text{CREAM}} = .50(180) \end{cases}$$

45 gal of milk must be mixed with 135 gal of cream.

15.
$$\binom{\text{AMT}}{\text{MILK}} + \binom{\text{AMT}}{\text{CREAM}} = 30$$
$$.03\binom{\text{AMT}}{\text{MILK}} + .23\binom{\text{AMT}}{\text{CREAM}} = .04(30)$$

28.5 gal of milk must be mixed with 1.5 gal of cream.

17. Texas is 262,134 sq mi and Florida is 54,090 sq mi. **19.** The plane's flight is 2 hours. **21.** The Sears Tower is 1454 ft tall and the Standard Oil Building is 1136 ft. **23.** *My Fair Lady* was performed 2717 times. **25.** The Verrazano-Narrows Bridge is 4260 ft long. **27.** Replace 9 quarts. **29.** Equilibrium at \$3.50 for 4000 items.

Problem Set 8.6, Pages 220–221

1. 
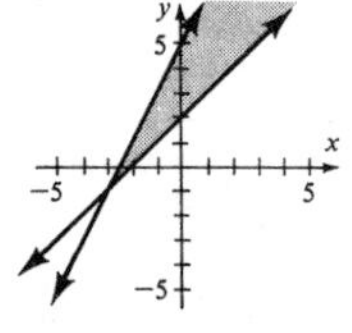

3.
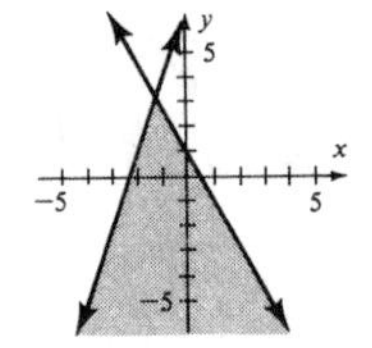

5.
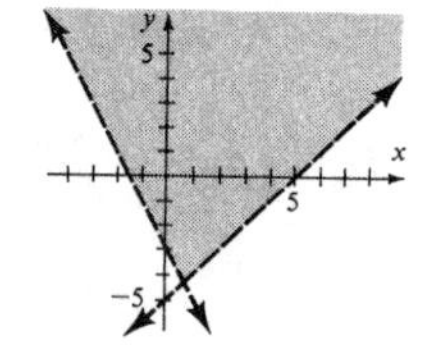

7.
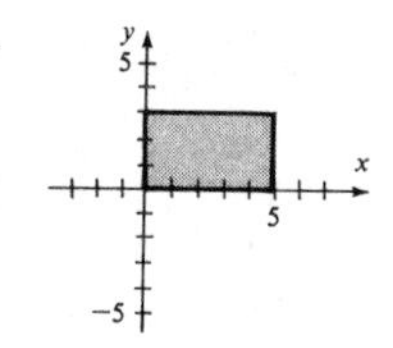

9.
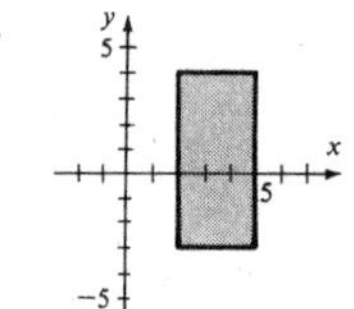

11.
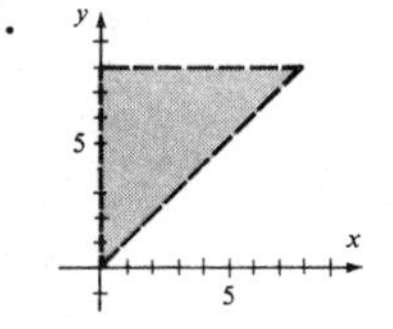

13. 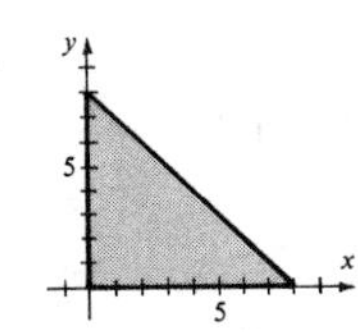

15. 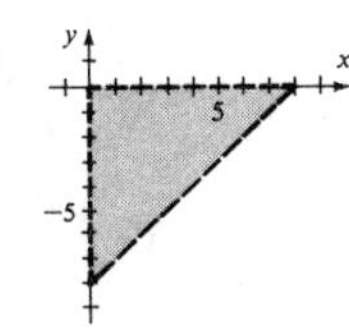

17. 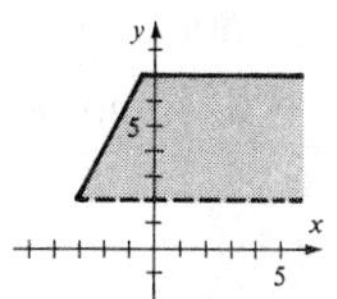

19. 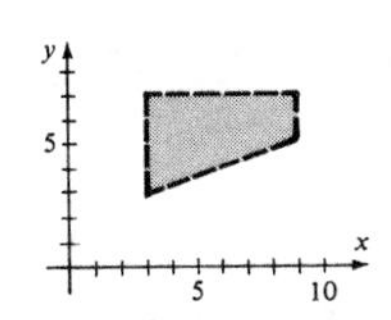

21. 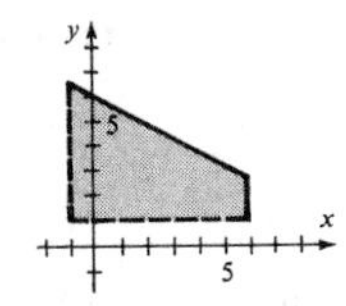

23. 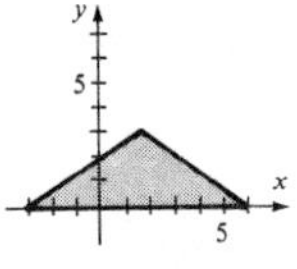

25. 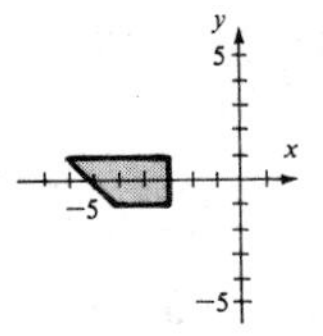

27. 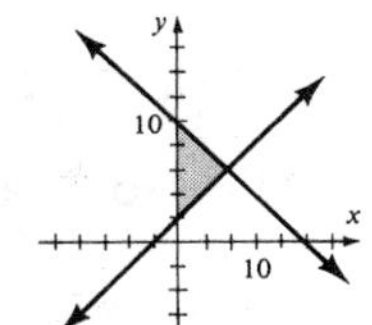

29. 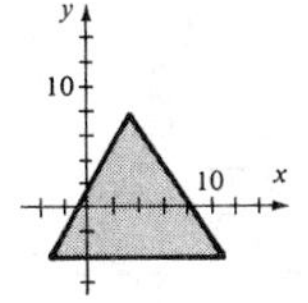

8.7 Review Self-Test, Pages 221–222

1. **a.** (5, 0) **b.** (5, 9) **2.** a. (2, −1) **b.** $\left(2, \frac{1}{2}\right)$ **3.** **a.** (5, −3) **b.** (2, 3)
4. **a.** (2, 3) **b.** inconsistent system **5.** **a.** inconsistent system **b.** (2, 0) **6.** (2, 40000)
7. The current's speed is 8 mph. **8.** There were 717 children admitted. **9.** 40 ml of 12.5% solution and 20 ml of 65% solution **10.** **a.**

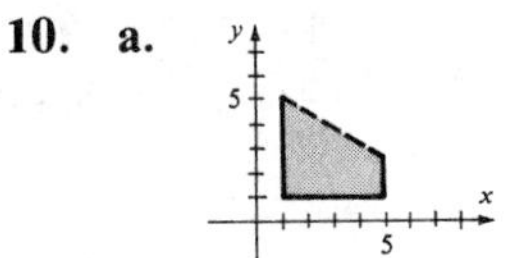

b.

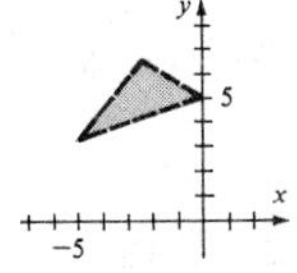

Review Problems, Pages 222–223

1. (3, 5) **3.** (−2, 1) **5.** (−3, −1) **7.** (2, −1) **9.** (−1, −1) **11.** (−2, −3) **13.** (0, 3)
15. (−2, 0) **17.** (1, 3) **19.** (5, 1) **21.** (−5, 1) **23.** (−2, −3) **25.** (1, 8) **27.** (−2, 4)
29. (3, 13) **31.** (1, 4) **33.** (−1, 2) **35.** (−2, 1) **37.** (1, 6) **39.** (2, −3) **41.** (6, 6)
43. (2, 3) **45.** (3, 4) **47.** (2, 1) **49.** (−6, 5)

Problem Set 9.1, Pages 231–232

1. right triangle **3.** right triangle **5.** right triangle **7.** right triangle **9.** right triangle

11.
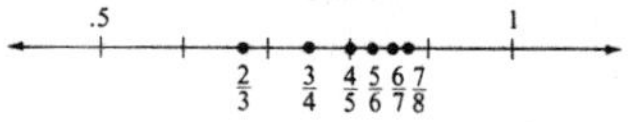

13.
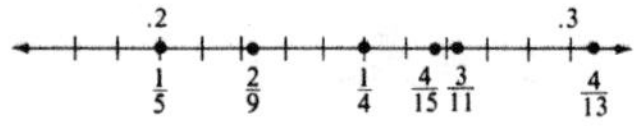

15.
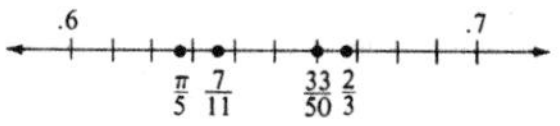

17.
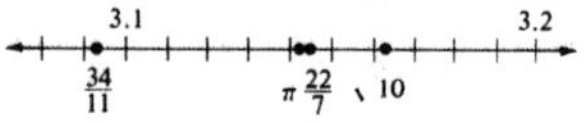

19.
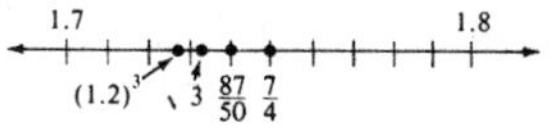

21. 5 **23.** 8 **25.** not real **27.** $\frac{1}{3}$

29. $\frac{1}{4}$ **31.** $\frac{7}{8}$ **33.** 12 **35.** 11 **37.** irrational **39.** -1.2 **41.** false **43.** true
45. false **47.** true, although they are also equal **49.** true

Problem Set 9.2, Pages 234–235

1. $\sqrt{10}$ **3.** $\sqrt{15}$ **5.** $\sqrt{6b}$ **7.** $\sqrt{2xy}$ **9.** $\sqrt{5xz}$ **11.** $\sqrt{14mn}$ **13.** $\sqrt{3}$ **15.** 2 **17.** 3
19. $\sqrt{2}$ **21.** $\sqrt{5c}$ **23.** 3 **25.** 2 **27.** $2\sqrt{3}$ **29.** $\sqrt{7c}$ **31.** $3\sqrt{3}$ **33.** $3\sqrt{2}$
35. $-5\sqrt{2}$ **37.** $\frac{\sqrt{2}}{2}$ **39.** $\frac{-\sqrt{5}}{5}$ **41.** $\frac{\sqrt{15}}{5}$ **43.** $4\sqrt{3}$ **45.** $-5\sqrt{3}$ **47.** $-5\sqrt{6}$ **49.** $3\sqrt{2}$
51. $2\sqrt{7}$ **53.** 24 **55.** $\frac{3\sqrt{5}}{10}$ **57.** $\frac{\sqrt{6}}{15}$ **59.** $2\sqrt{2}$ **61.** $x\sqrt{x}$ **63.** $x^2\sqrt{x}$ **65.** $-3n\sqrt{n}$
67. $\frac{2\sqrt{2x}}{5}$ **69.** $\frac{3\sqrt{30y}}{5y}$ **71.** $4\sqrt{3} \approx 6.93$ miles per second

Problem Set 9.3, Pages 237–238

1. $4\sqrt{5}$ **3.** $\sqrt{2}$ **5.** $7\sqrt{3}$ **7.** $-2\sqrt{2}$ **9.** $3\sqrt{17}$ **11.** $10\sqrt{2}$ **13.** $12\sqrt{3}$ **15.** $\sqrt{5}$
17. $-\sqrt{3}$ **19.** $13\sqrt{13}$ **21.** $3\sqrt{2}+\sqrt{6}$ **23.** 9 **25.** $10+5\sqrt{2}+5\sqrt{3}$ **27.** $-6\sqrt{3}$
29. $6\sqrt{5}$ **31.** $\sqrt{3}-3$ **33.** $\frac{3+\sqrt{2}}{2}$ **35.** $\frac{1+2\sqrt{2}}{2}$ **37.** $\frac{2-\sqrt{2}}{3}$ **39.** $\frac{12+\sqrt{7}}{9}$
41. $\frac{\sqrt{2}+\sqrt{3}}{3}$ **43.** $\frac{3\sqrt{2}+2\sqrt{3}}{3}$ **45.** $\frac{3\sqrt{2}+4\sqrt{6}}{3}$ **47.** $7+\sqrt{3}$ **49.** $\frac{6-2\sqrt{6}}{3}$
51. $2+\sqrt{2}+\sqrt{3}$ **53.** $6\sqrt{5}$ **55.** $2\sqrt{10}$ **57.** 4 **59.** $5+5\sqrt{3}$ **61.** $-1-\sqrt{3}$
63. $8+2\sqrt{7}$ **65.** $-19+5\sqrt{11}$ **67.** $-\sqrt{2}$ **69.** $9+4\sqrt{5}$ **71.** approximately 9 miles

Problem Set 9.4, Pages 240–242

1. -1 **3.** -46 **5.** 1 **7.** -71 **9.** -1 **11.** 6 **13.** 1 **15.** 3 **17.** 3 **19.** 1
21. $25-b$ **23.** x^2-y **25.** n^2-6 **27.** -3 **29.** -276 **31.** $\frac{3+\sqrt{5}}{2}$ **33.** $\frac{15-5\sqrt{6}}{3}$
35. $15+3\sqrt{19}$ **37.** $\frac{-2-3\sqrt{2}}{7}$ **39.** $5-2\sqrt{5}$ **41.** $\frac{14+4\sqrt{7}}{3}$ **43.** $-4+3\sqrt{3}$
45. $\frac{5+3\sqrt{3}}{2}$ **47.** $-1-2\sqrt{2}$ **49.** $\frac{1+t+2\sqrt{t}}{1-t}$ **51.** $\frac{1+s+2\sqrt{s}}{s-1}$ **53.** $5\sqrt{2}+5\sqrt{3}$ **55.** 75
57. 73 **59.** .07 in. **61.** .04 in.

Problem Set 9.5, Pages 244–245

1. $x = 9$ **3.** $z = 16$ **5.** no solution (b = 9 is extraneous) **7.** $x = 22$ **9.** $x = 50$ **11.** $x = 25$ **13.** $x = 9$ **15.** $x = \frac{16}{25}$ **17.** $x = 3$ ($x = -2$ is extraneous) **19.** $x = 5$ **21.** $x = 3$ **23.** $x = \frac{5}{2}$ **25.** $x = 2$ **27.** no solution ($x = -3$ is extraneous) **29.** $x = 3$ **31.** $x = 5$ ($x = 4$ is extraneous) **33.** $x = 2$ ($x = 3$ is extraneous) **35.** $x = 11$ ($x = 3$ is extraneous) **37.** $x = 8$ ($x = 1$ is extraneous) **39.** $x = -1$ or 7 **41.** $x = 1$ or -2 **43.** $x = 3$ $\left(x = \frac{3}{4}\right.$ is extraneous$\left.\right)$ **45.** ($\sqrt{x+5} = 3$) The number is four. **47.** ($x = \sqrt{6x}$) The number is zero or six. **49.** ($x - 2 = \sqrt{x+4}$) The number is five.

9.6 Review Self-Test, Page 246

1. a. 6 **b.** -2 **c.** $\frac{1}{5}$ **d.** $\frac{2}{3}$ **2. a.** $3\sqrt{3}$ **b.** $a\sqrt{a}$ **c.** $3x\sqrt{2}$ **d.** $4x^3\sqrt{2x}$ **3. a.** $\frac{\sqrt{2}}{4}$ **b.** $\frac{2\sqrt{x}}{x}$ **c.** $\sqrt{3}$ **d.** $\frac{\sqrt{15x}}{5}$ **4. a.** $5\sqrt{2}$ **b.** $5\sqrt{2} - 2\sqrt{3}$ **c.** $4\sqrt{3} - 15\sqrt{2}$ **d.** $\frac{5\sqrt{2}}{2}$ **5. a.** $\frac{2+\sqrt{2}}{3}$ **b.** $\frac{3-\sqrt{2}}{4}$ **c.** $\frac{3+\sqrt{6}}{3}$ **d.** $1 + \sqrt{3}$ **6. a.** 1 **b.** $a + 2\sqrt{a} + 1$ **c.** $4 + 2\sqrt{3}$ **d.** 2 **7. a.** $6 + 3\sqrt{3}$ **b.** $\frac{3-\sqrt{5}}{2}$ **c.** $-2\sqrt{2} - \sqrt{3}$ **d.** $\frac{3\sqrt{3}-5}{2}$

8. a. $x = 16$ **b.** no solution ($x = 625$ is extraneous) **c.** $x = 4$ ($x = -1$ is extraneous) **d.** $x = -2$ ($x = -\frac{31}{9}$ is extraneous) **9. a.** $2\sqrt{10} \approx 6.3$ seconds **b.** 45 mph **10.** ($1 + 2n = \sqrt{n+2}$) The number is one-fourth.

Review Problems, Pages 247–248

1. 3 **3.** 9 **5.** $\frac{1}{5}$ **7.** -0.5 **9.** 1.5 **11.** $-5\sqrt{5}$ **13.** $2\sqrt{3a}$ **15.** $3a\sqrt{2}$ **17.** $\frac{-\sqrt{3}}{6}$ **19.** $\frac{-\sqrt{15}}{5}$ **21.** $\frac{\sqrt{6}}{2}$ **23.** $\frac{-\sqrt{10xy}}{5y}$ **25.** $3\sqrt{6}$ **27.** $-5\sqrt{2}$ **29.** $-\sqrt{5}$ **31.** $4\sqrt{3}$ **33.** $5 - 2\sqrt{5}$ **35.** $\frac{-11\sqrt{2}}{2}$ **37.** $\sqrt{5} + 5\sqrt{3}$ **39.** $\sqrt{11}$ **41.** $3 + \sqrt{3} + \sqrt{6}$ **43.** $6 + 2\sqrt{3}$ **45.** $9 + 3\sqrt{3}$ **47.** $2 - \sqrt{3}$ **49.** $3 + 2\sqrt{3}$ **51.** $3\sqrt{3} + 2\sqrt{6}$ **53.** 5 **55.** $4\sqrt{3}$ **57.** $5\sqrt{5}$ **59.** 2 **61.** $-1 - \sqrt{3}$ **63.** $10 + 2\sqrt{5}$ **65.** 11 **67.** $25 - x$ **69.** $9 - 4z$ **71.** $\frac{6+3\sqrt{2}}{2}$ **73.** $\frac{15+5\sqrt{2}}{7}$ **75.** $5 - 2\sqrt{5}$ **77.** $\frac{-6-5\sqrt{3}}{13}$ **79.** $-4 - 3\sqrt{3}$ **81.** $\frac{1-\sqrt{3}}{2}$ **83.** $\frac{1+b+2\sqrt{b}}{b-1}$ **85.** $-11 + 5\sqrt{5}$ **87.** $\frac{-23-7\sqrt{11}}{5}$ **89.** no solution (9 is extraneous) **91.** 13 **93.** no solution (10 is extraneous) **95.** 1 (-7 extraneous) **97.** -2 (-3 is extraneous) **99.** ± 1

Problem Set 10.1, Pages 252–253

1. $x = \pm 6$ **3.** $z = \pm 11$ **5.** $y = \pm\sqrt{10}$ **7.** $x = \pm 3$ **9.** $z = \pm 2$ **11.** $y = \pm\frac{10}{3}$ **13.** $x = \pm\frac{4}{3}$ **15.** $z = \pm\frac{3}{2}$ **17.** $y = -1$ or 5 **19.** $x = 1$ or -7 **21.** $z = -2$ or -6

23. $y = 2$ or $-\frac{6}{5}$ **25.** $x = 0$ or 1 **27.** $z = 0$ or $\frac{2}{5}$ **29.** $y = 0$ or $\frac{4}{5}$ **31.** The smaller square is 8 ft on a side. **33.** The larger square is 10 cm on a side. **35.** The number is $-5 + \sqrt{3}$ or $-5 - \sqrt{3}$. **37.** The number is $-2 + \sqrt{5}$ or $-2 - \sqrt{5}$. **39.** The number is 3 or $-\frac{5}{3}$. **41.** $x = 2 \pm \sqrt{5}$ (-0.236 or 4.236) **43.** $x = \frac{1 \pm 2\sqrt{2}}{2}$ (-0.914 or 1.914) **45.** $x = \frac{3 \pm \sqrt{6}}{5}$ (0.110 or 1.090)

Problem Set 10.2, Pages 256–257

1. $16; (x + 4)^2$ **3.** $1; (z + 1)^2$ **5.** $25; (y - 5)^2$ **7.** $144; (x + 12)^2$ **9.** $196; (z + 14)^2$ **11.** $625; (y + 25)^2$ **13.** $\frac{9}{4}; \left(x + \frac{3}{2}\right)^2$ **15.** $\frac{49}{4}; \left(z + \frac{7}{2}\right)^2$ **17.** $\frac{1}{4}; \left(y - \frac{1}{2}\right)^2$ **19.** $\frac{121}{4}; \left(x - \frac{11}{2}\right)^2$ **21.** $x = -2$ or -6 **23.** $x = -1$ or 5 **25.** $x = -6$ or 30 **27.** $x = -1$ or -2 **29.** $y = -1$ or 2 **31.** $x = 1 \pm \sqrt{3}$ **33.** $x = 2 \pm \sqrt{3}$ **35.** $y = \frac{3}{2}$ or -2 **37.** $x = \frac{1}{2}$ or 1 **39.** $x = \frac{-4 \pm \sqrt{13}}{2}$ **41.** $x = \frac{4 \pm \sqrt{7}}{9}$ (0.150 or 0.738) **43.** $h = -\frac{7}{4}$ or 3 (-1.750 or 3.000) **45.** $m = 1$ or $\frac{7}{3}$ (1.000 or 2.333) **47.** $x = \frac{3 \pm \sqrt{17}}{2}$ (-0.562 or 3.562) **49.** $x = -\sqrt{5} \approx -2.236$

51. **53.**

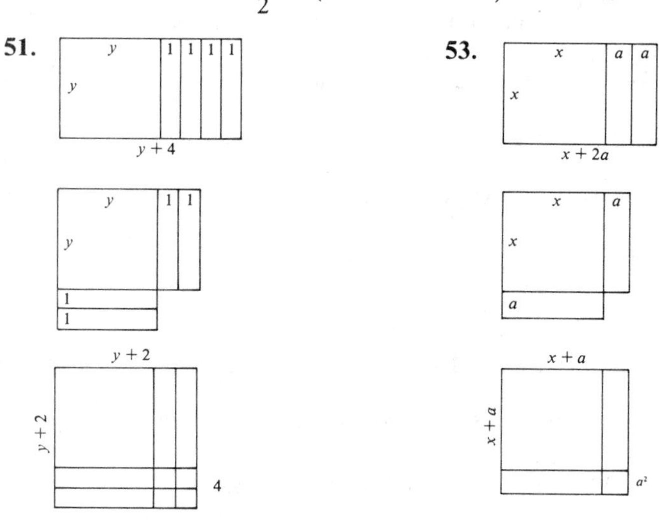

Problem Set 10.3, Pages 259–260

1. $a = 1, b = 2, c = -15$ **3.** $a = 6, b = -7, c = -3$ **5.** $a = 1, b = -6, c = 4$ **7.** $a = 4, b = -4, c = -2$ **9.** $a = 9, b = 24, c = 1$ **11.** $x = 2$ or -7 **13.** $x = 2$ or -8 **15.** $x = \frac{1}{3}$ or 1 **17.** $c = -\frac{1}{2}$ or $-\frac{2}{3}$ **19.** $x = -\frac{1}{3}$ or $-\frac{1}{2}$ **21.** $x = 2$ **23.** $x = 2 \pm \sqrt{3}$ **25.** $x = 5 \pm \sqrt{2}$ **27.** $x = -4 \pm \sqrt{15}$ **29.** $x = \frac{1 \pm \sqrt{2}}{2}$ **31.** $x = \frac{1 \pm \sqrt{5}}{2}$ **33.** $x = \frac{1 \pm \sqrt{3}}{3}$ (-0.244 or 0.911) **35.** $x = \frac{5 \pm \sqrt{11}}{2}$ (4.158 or 0.842) **37.** no real solution **39.** $x = \frac{1 \pm 2\sqrt{2}}{3}$ (-0.610 or 1.276)

41.
$$
\begin{aligned}
ax^2 + bx + c &= 0 \\
ax^2 + bx &= \quad -c \\
4a^2x^2 + 4abx &= \quad -4ac \\
4a^2x^2 + 4abx + b^2 &= b^2 - 4ac \\
(2ax + b)^2 &= b^2 - 4ac \\
2ax + b &= \pm\sqrt{b^2 - 4ac} \\
2ax &= -b \pm \sqrt{b^2 - 4ac} \\
x &= \frac{-b \pm \sqrt{b^2 - 4ac}}{2a}
\end{aligned}
$$

Comments should vary. However, notice that there are no fractions introduced until the very last step.

Problem Set 10.4, Page 264

1. $x = \pm 3$ **3.** $x = \pm\sqrt{5}$ **5.** $x = \pm 4\sqrt{3}$ **7.** $x = 0$ or 9 **9.** $x = 0$ or $\frac{3}{5}$ **11.** $x = 0$ or $-\frac{4}{3}$ **13.** $x = 2$ or -10 **15.** $x = \frac{-7 \pm \sqrt{41}}{2}$ **17.** no real solution **19.** $x = 1$ or 9 **21.** $x = 1$ or -7 **23.** $x = -2$ or $\frac{10}{3}$ **25.** $x = 1 \pm \sqrt{7}$ **27.** $x = 4 \pm \sqrt{13}$ **29.** $x = \frac{4 \pm \sqrt{11}}{5}$ **31.** The width is 6 cm and the length is 19 cm. **33.** The side of the square is 18 mm. **35.** The sides of the rectangle are 5 m and 12 m. **37.** The sides of the triangle are $3\sqrt{10}$, $9\sqrt{10}$, and 30 cm. **39.** The vertical rise is $\frac{100}{13}$ ft or approximately 7.7 ft or 7 ft 8 in. to the nearest inch.

Problem Set 10.5, Pages 269–270

1.
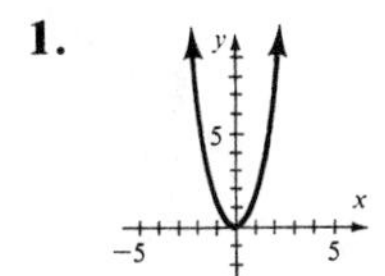

3.
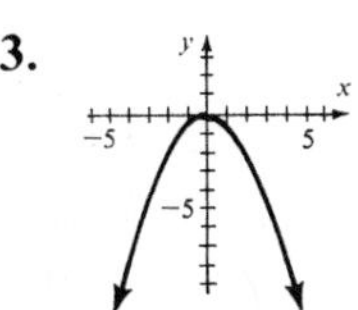

5.
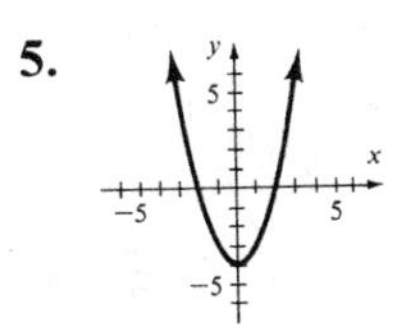

7.
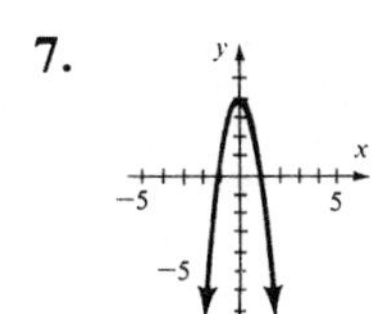

9.
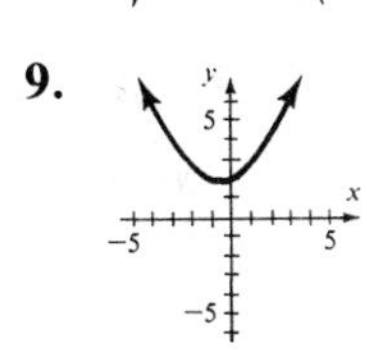

11. 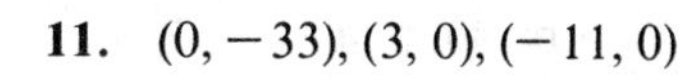
$(0, -33), (3, 0), (-11, 0)$

13. $(0, -1), \left(\frac{-5 \pm \sqrt{33}}{4}, 0\right)$ **15.** $(0, -3), (\sqrt{3}, 0), (-\sqrt{3}, 0)$ **17.** $(0, -1), (-2 \pm \sqrt{5}, 0)$

19. $(0, -3), \left(\frac{2 \pm \sqrt{7}}{2}, 0\right)$

21.
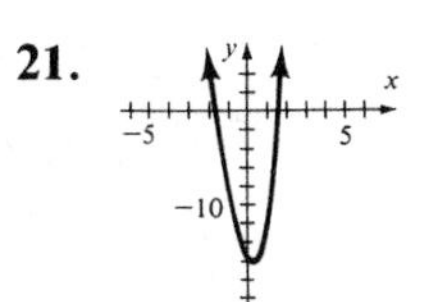

23.
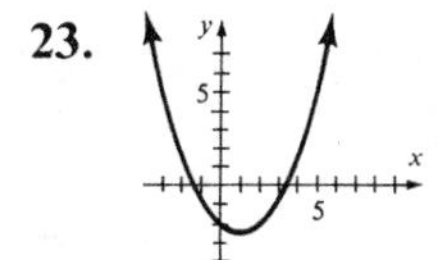

25.
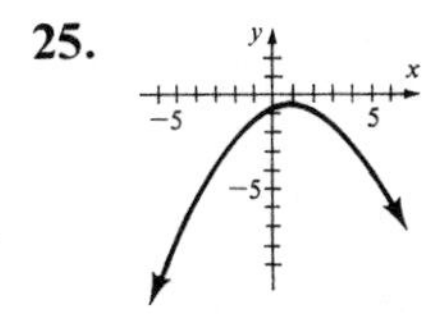

27.
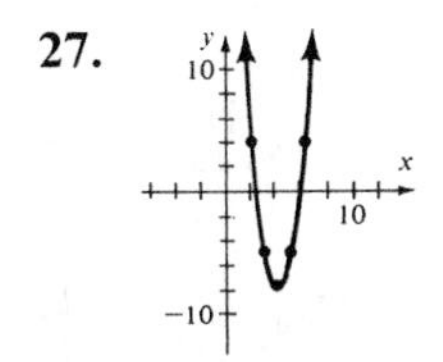

29.
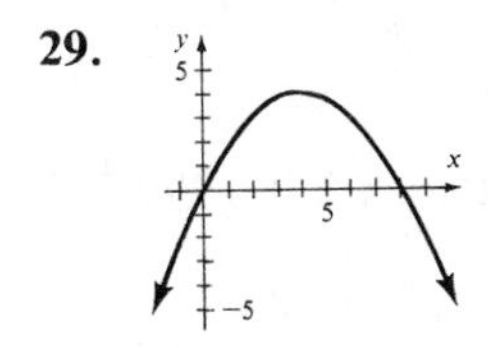

31.

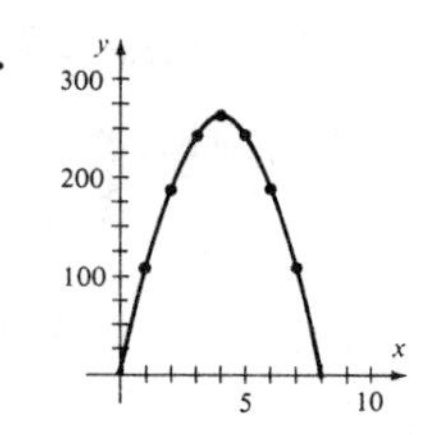

33.

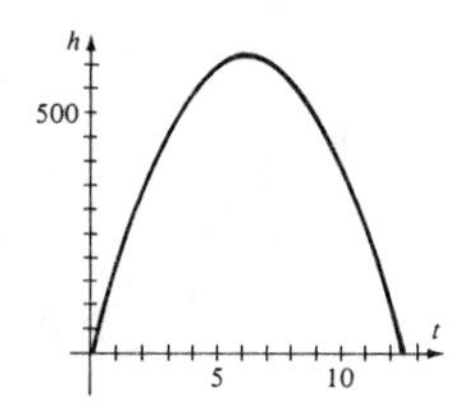

35.

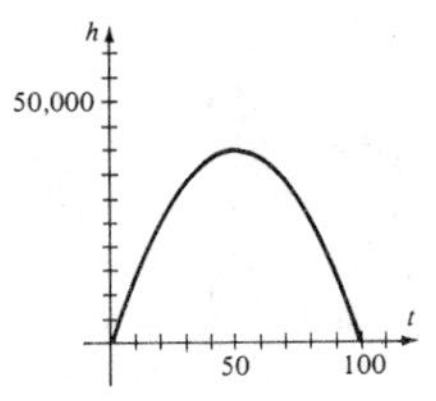

37. At approx. 3 and 7 sec.

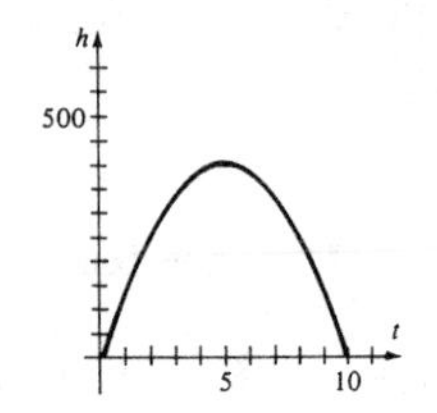

39. Set zero at water level and use the scale shown.

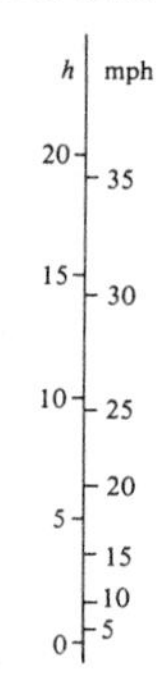

10.6 Review Self-Test, Page 271

1. **a.** $13, -13$ **b.** $3\sqrt{3}, -3\sqrt{3}$ **c.** $14, -8$ **d.** $-3, 2$ **2.** **a.** 25 **b.** 4 **c.** $\frac{121}{4}$ **d.** $\frac{1}{4}$ **3.** **a.** $3, -1$ **b.** $\frac{1}{3}, -1$ **c.** $2 \pm \sqrt{3}$ **d.** $\frac{1 \pm \sqrt{5}}{3}$ **4.** Same answers as #3.

5. **a.** $\frac{\pm\sqrt{5}}{5}$; ± 0.45 **b.** $\frac{5}{3}, -\frac{2}{7}$; $1.67, -0.29$ **c.** $2 \pm \sqrt{7}$; $4.65, -0.65$ **d.** $\frac{-2 \pm \sqrt{7}}{3}$; $0.22, -1.55$

6. The field is 9 by 40 meters. **7.** **a.** $(0, -24), (-3, 0), (8, 0)$ **b.** $(0, -10), \left(\frac{2}{3}, 0\right), \left(-\frac{5}{2}, 0\right)$, **c.** $(0, -1), (2 + \sqrt{5}, 0), (2 - \sqrt{5}, 0)$ **d.** $(0, -2), \left(\frac{3 + \sqrt{17}}{2}, 0\right), \left(\frac{3 - \sqrt{17}}{2}, 0\right)$

8. **a.**

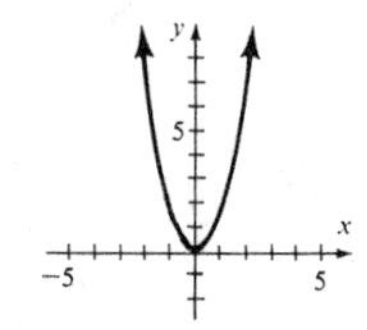

b.

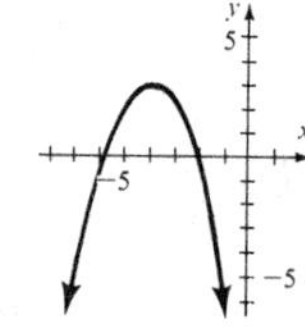

c.

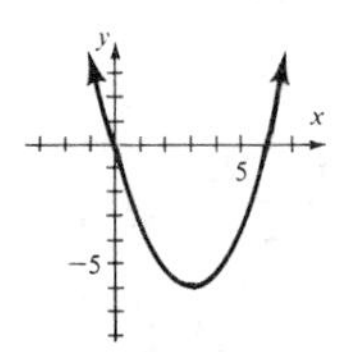

d.

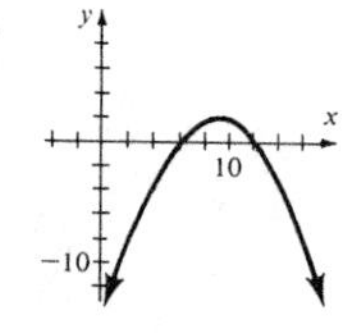

9.

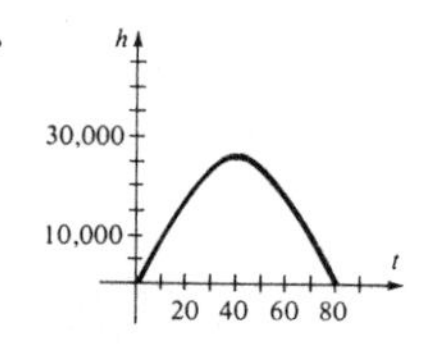

10. It will take less than 10 seconds. (Actually less than 9.)

Review Problems, Pages 271–272

1. ± 7 **3.** ± 14 **5.** $\pm 3\sqrt{2}$ **7.** $\pm\frac{2}{3}$ **9.** $\pm\frac{8}{5}$ **11.** $0, 4$ **13.** $\frac{1}{2}, \frac{5}{2}$ **15.** $\frac{-1 \pm \sqrt{3}}{4}$

17. $1, -5$ **19.** $3, 4$ **21.** $2, -12$ **23.** $3 \pm \sqrt{5}$ **25.** $-3, \frac{1}{2}$ **27.** $1, \frac{1}{3}$ **29.** no real solution **31.** $2, -5$ **33.** $3, -5$ **35.** $1 \pm \sqrt{3}$ **37.** $\frac{1}{3}, \frac{2}{3}$ **39.** $-\frac{1}{3}, \frac{1}{2}$ **41.** $\frac{-2 \pm \sqrt{3}}{2}$ **43.** $\frac{4 \pm \sqrt{17}}{3}$ **45.** $\frac{-4 \pm \sqrt{17}}{3}$ **47.** $\pm 2\sqrt{2}$ or ± 2.83 **49.** $\pm 4\sqrt{2}$ or ± 5.66 **51.** $\pm 4\sqrt{6}$ or ± 9.80 **53.** $0, \frac{3}{2}$ **55.** $2, -7$ **57.** $1, -9$ **59.** $-2 \pm \sqrt{7}$ or $0.65, -4.65$ **61.** $-2, 8$ **63.** $4, 6$ **65.** $\frac{1}{5}, -\frac{1}{2}$ **67.** $\frac{3 \pm \sqrt{5}}{2}$ or $0.38, 2.62$ **69.** $\frac{3 \pm \sqrt{7}}{5}$ or $0.07, 1.13$ **71.** $\frac{-1 \pm \sqrt{2}}{3}$ or $0.14, -0.80$ **73.** $\frac{-2 \pm \sqrt{11}}{3}$ or $0.44, -1.77$ **75.** $\frac{5 \pm \sqrt{37}}{2}$ or $-0.54, 5.54$ **77.** $(3, 0), (-3, 0), (0, -9)$ **79.** $(-3, 0), (7, 0), (0, -21)$ **81.** $(2.09, 0), (-9.09, 0), (0, -19)$ **83.** $(-0.32, 0), (6.32, 0), (0, -2)$ **85.** $(0.57, 0)\ (-1.5, 0), (0, -12)$

87.

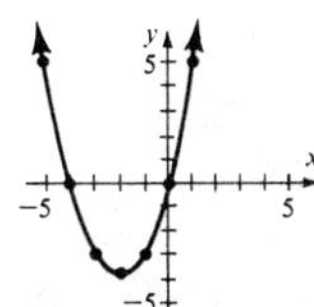

89.

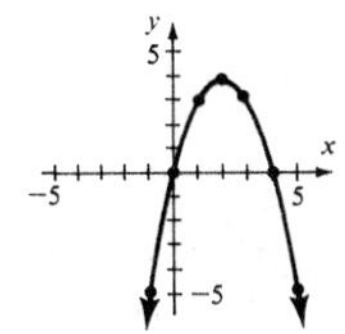

91.

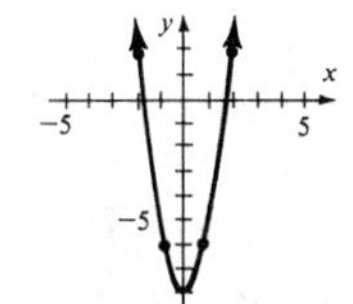

93.

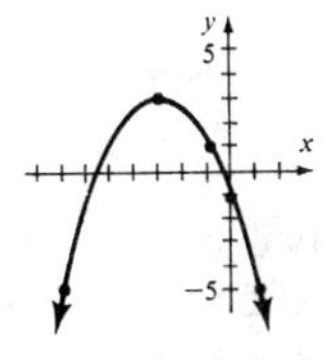

95.

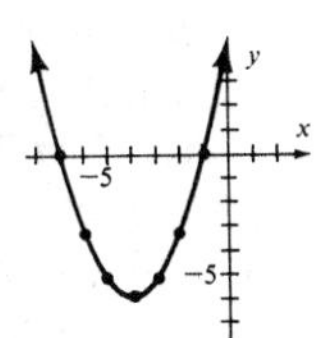

97.

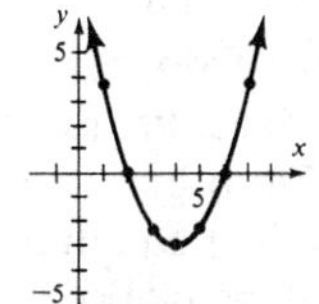

99.

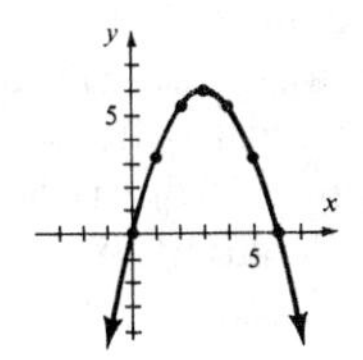

Glossary

Abscissa (7.1). The horizontal coordinates in a two-dimensional system of rectangular coordinates; usually denoted by x.

Absolute value (2.1). The absolute value of n is its undirected distance from zero. We use the symbol $|n|$.

Addition property of equality (3.7). If a number is added to or subtracted from both sides of an equation, the result is an equivalent equation.

Addition property of inequality (3.6). If $x < y$, then $x + c < y + c$. The property also holds for $\leq$, $\geq$ and $>$.

Associative property (1.5). A property of grouping; if a, b, and c are real numbers, then $(a + b) + c = a + (b + c)$ and $(ab)c = a(bc)$.

Axis, axes (7.1). *See* Cartesian coordinate system.

Base (1.3, 4.2). *See* Exponent.

Binomial (4.1). *See* Polynomial.

Braces (1.2). The symbols, { }, used to enclose elements of a set. Also used as grouping symbols if parentheses and brackets are also used, as in

$$2\{6[4 + (-3)] + 5\}$$

See Grouping symbols.

Brackets (1.2). The grouping symbols, [ ], that are usually used when a set of parentheses is also used, as in

$$6 + [(-2) + (-3)] \text{ or } 6[4(8 + 2) + 5]$$

See Grouping symbols.

Cartesian coordinate system (7.1). Two intersecting lines, called *axes*, used to locate points in a plane. If the intersecting lines are perpendicular, the system is also called a *rectangular coordinate system*.

Circle (1.4). The set of all points in a plane that are a fixed distance from a fixed point. The fixed point is called the *center*, the fixed distance the *radius*. The *diameter* is twice the radius.

Circumference (1.4). The length of a complete circular arc. $C = 2\pi r$ and $C = \pi D$ for a circle of radius r and a diameter D.

Coefficient (2.5). In a product of two or more factors, any collection of factors is said to be the coefficient of the rest of the factors. For example, in 5xyz, 5x is the coefficient of yz; $5z$ is the coefficient of xy; $5xy$ is the coefficient of z. Most commonly used is the term *numerical coefficient*, which refers to the constant factor. The numerical coefficient of $5xyz$ is 5. If no constant factor appears, it is understood to be 1. For example, x, a^2, t, and $(x + y)$ all have a numerical coefficient of 1.

Common factor (5.1, 5.4). A number (or expression) that two or more numbers (or expressions) share as a factor.

Common monomial factor (5.1). A term that several terms share as a factor. For example,

$$3a + 3b + 3c = 3(a + b + c)$$

so 3 is a common monomial factor;

$$2x^2y + 3xy^2 = x(2xy + 3y^2)$$

so x is a common monomial factor;

$$5m^2n + 10mn^2 + 5mn = 5mn(m + 2n + 1)$$

so $5mn$ is a common monomial factor.

Commutative property (1.5). A property of order; if a

and b are real numbers, then $a + b = b + a$ and $ab = ba$.

Comparison property (3.7). For any two numbers x and y, exactly one of the following is true: (1) $x = y$, (2) $x > y$, or (3) $x < y$.

Completely factored (5.1, 5.2). A factored form of a polynomial is said to be completely factored if none of its terms can be factored further. For example, 5, $(x - 2)(x + 2)$, $3(x + y)$, $x^2 + y^2$, and $3x^2 + 2x + 1$ are completely factored.

Completing the square (10.2). A process of adding the necessary constant to make a complete square. For example, in $x^2 + px = q$, the square is completed by adding $(p/2)^2$ to both sides. Notice that the coefficient of x^2 must be 1.

Complex fraction (6.7). If a or b in a/b is a fraction, then a/b is called a complex fraction.

Conjugate (5.4, 9.4). The binomials $a + b$ and $a - b$ are conjugates. The product of conjugates is the difference of squares: $(a + b)(a - b) = a^2 - b^2$.

Constant (1.1). If the domain of a variable has only one member, then that variable is called a constant. A constant is also a single number, such as 5, 47, or 623.

Counting numbers (1.1, 1.5). Numbers belonging to the set $\{1,2,3 \ldots\}$. Also called *natural numbers.*

Degree (4.1). The degree of a term is the number of variable factors in the term. The degree of a polynomial is the largest degree of any of its terms. A first-degree polynomial is called *linear,* and a second-degree polynomial is called *quadratic.* For example,

zero-degree polynomials:

$$5,\ 19,\ \pi + 2$$

first-degree polynomials:

$$5x,\ 3x + 2y,\ x + 1$$

second-degree polynomials:

$$5x^2 + 2x + 1,\ xy,\ 3x^2 + 4y^2$$

third-degree polynomials:

$$5x^3 + 2x^2 + 5x + 1,\ x^2y,\ 5x^3 + 2xy^2 + 2x + 4$$

Dependent variable (7.2). *See* Ordered pair.

Diameter of a circle (1.4). *See* Circle.

Difference (1.1). The result obtained from a subtraction problem.

Difference of squares (4.4, 5.4, 9.4). The special binomial product: $(x + a)(x - a) = x^2 - a^2$.

Directed numbers (2.1). *See* Signed numbers.

Discriminant (10.3). *See* Quadratic formula.

Distance formula (9.5). *See* Distance.

Distributive property for multiplication over addition (1.5). If a, b, and c are real numbers, then $a(b + c) = ab + ac$.

Division property of radicals (9.2). $\sqrt{a}/\sqrt{b} = \sqrt{a/b}$, $b \neq 0$

Domain of a variable (1.1). A set of numbers from which the values of a variable are chosen.

Element of a set (1.5). Any member of a set. We write $x \in S$ to mean x is an element of set S and $\notin S$ to mean x is not an element of set S.

Equation (3.1). A statement of equality.

Equation of a graph (7.2). (1) Every point on the graph has coordinates that satisfy the equation, and (2) every ordered pair satisfying the equation has coordinates that lie on the graph.

Equivalent equations (3.2). Two equations are equivalent if they have the same solution set.

Exponent (1.3, 4.2, 6.6). (1) x^n, where x is any nonzero real number and n is any natural number, is defined as

$$x^n = \underbrace{x \cdot x \cdots x}_{n \text{ factors}}$$

x is called the *base,* n is called the exponent, and x^n is called a *power.* (2) In addition, $x^0 = 1$, $x^{-n} = 1/x^n$.

Expression (1.4). A very general term used to designate any symbolic mathematical form. See also numerical expression and variable expression.

Extraneous root (6.5, 9.5). A number obtained in the process of solving an equation that is not a root of the original equation.

Factor (1.1, 4.2, 4.5). Numbers being multiplied are called factors.

Factoring (1.3, 5.1–5.5, 6.1, 6.3). Resolving numbers or expression into factors.

First component (7.1). *See* Ordered pair.

First-degree equation (3.4, 4.1). An equation for which the exponent on the variable is one. For example, $2x + 1 = 0$ and $2x + 4(x - 3) = 5x + 2$ are first-degree equations.

Formula (1.4). A name given to a variable expression that has some special meaning.

Fractions, equality of (6.1). $a/b = c/d$ if and only if $ad = bc$, $d \neq 0$

Graph of an equation (7.2). (1) Every point on the graph has coordinates that satisfy the equation, and (2) every ordered pair satisfying the equation has coordinates that lie on the graph.

Grouping symbols (1.2). *Parentheses,* ( ), *brackets,* [ ], and *braces,* { }. These symbols indicate the order of operations and are also sometimes used to indicate multiplication, as in (2)(3) = 6.

Half-plane (7.5). The part of a plane that lies on one side of a line in the plane.

Horizontal line (7.3, 7.4). A line parallel to the x-axis; its equation has the form $y = k$, where k is any real number. This line has zero slope.

Hypotenuse (9.1). In a right triangle, the side opposite the right angle.

Independent variable (7.2). *See* Ordered pair.

Integers (2.1). The set of natural numbers, the opposites of the natural numbers, and zero, denoted by $I = \{\ldots, -3, -2, -1, 0, 1, 2, 3, \ldots\}$

Interest (1.4). The amount of money earned by investing the principal. The formula for simple interest is $I = prt$, where I is the interest, p the principal, r the rate, and t the time.

Intersection (8.1). The intersection of two sets consists of the elements in both sets. The symbol for intersection is $\cap$.

Irrational numbers (9.1). Numbers that are not rational. Irrational numbers do not have *repeating decimal representations.* Some examples of irrational numbers are $\sqrt{2}$, $\sqrt{3}$, $\sqrt{5}$, π, $\pi + 2$, $3\sqrt{7} + 6$, and $\sqrt{3} + \pi - 5$.

Laws of exponents (4.2, 6.6). For real x and y, if a and b are natural numbers:

1. $x^a x^b = x^{a+b}$
2. $(xy)^a = x^a y^a$
3. $(x^a)^b = x^{ab}$
4. $\left(\dfrac{x}{y}\right)^a = \dfrac{x^a}{y^a},\ y \neq 0$
5. $\dfrac{x^a}{x^b} = x^{a-b},\ x \neq 0$

Least common denominator, LCD (6.4). The smallest number that is exactly divisible by each of the given numbers. To find the LCD:

1. completely factor each of the given expressions, and write the factorization using exponents, and
2. find the product of the representatives of each factor, where the representative chosen is the one with the largest exponent.

For example,

$$\begin{aligned} 12x^2y &= 2^2 \cdot 3 \cdot \quad x^2 \cdot y \\ 15xy^3 &= \quad\ \ 3 \cdot 5 \cdot x \cdot y^3 \\ \text{LCD:}\quad & 2^2 \cdot 3 \cdot 5 \cdot x^2 \cdot y^3 = 60x^2y^3 \end{aligned}$$

Also,

$$\begin{aligned} 2x + 2y &= 2 \quad (x + y) \\ 3x + 3y &= \quad\ 3 \cdot (x + y) \\ \text{LCD:}\quad & 2 \cdot 3 \cdot (x + y) = 6x + 6y \end{aligned}$$

Linear equation (3.1, 7.4). A first-degree equation. For example, $2x + 3y + 5 = 0$, $Ax + By + C = 0$, $y = \dfrac{2}{3}x + 2$, and $y = mx + b$.

Linear system (8.1). System of first-degree (linear) equations. The equations may represent (1) two intersecting lines with a single point of intersection, (2) two parallel lines with no point of intersection, or (3) the same line with an unlimited number of solutions. Also refers to a system of linear inequalities, which is the intersection of half-planes.

Literal equation (3.5). An equation in which constants are represented by letters. For example, $ax + b = c$ is a literal equation, whereas $2x + 3 = 4$ is not.

Lowest terms (6.1). If 1 is the only common factor of the numerator and denominator of a fraction, it is said to be reduced to lowest terms.

Monomial (4.1). *See* Polynomial.

Multiplication property of equality (3.3). If both sides of an equation are multiplied or divided by a nonzero number, the result is an equivalent equation.

Multiplication property of inequality (3.7). If $x < y$ and c is positive, then $xc < yc$. If c is negative, then $xc > yc$. The property also holds for $\leq$, $>$, and $\geq$. Remember, the inequality is reversed if both sides are multiplied or divided by a negative number.

Multiplication property of square roots (9.2). $\sqrt{a}\sqrt{b} = \sqrt{ab}$ for nonnegative a and b.

Natural numbers (1.1, 1.5). Numbers belonging to the set {1,2,3, . . . }; also called *counting numbers.*

Negative (2.1). A number less than zero is called negative.

Numerical coefficient (2.5, 4.1). *See* Coefficient.

Numerical expression (1.2, 2.5). A phrase containing only constants and at least one defined operation.

Opposites (2.1). The pair of numbers n and $-n$ are called opposites. $(-1)n = -n$ and $n + (-n) = 0$

Ordered pair (4.1). A pair of numbers written (x,y), where the *order* in which they are named is important. That is, $(4,5) \neq (5,4)$. The first number listed is called the *first component* or *independent variable,* and the other is called the *second component* or *dependent variable.*

Order-of-operations convention (1.2, 1.3). An agreement that tells the order in which an expression is to be simplified:

1. first perform any operations enclosed by parentheses;

2. next, perform any operations involving raising to a power;
3. next, perform multiplications and divisions by working from left to right;
4. finally, perform additions and subtractions by working from left to right.

Ordinates (7.1). The vertical coordinates in a two-dimensional system of rectangular coordinates; usually denoted by y.

Origin (2.1, 7.1). The zero point on a number line, or the (0,0) point in the Cartesian coordinate system.

Parabola (10.5). The graph of $y = ax^2 + bx + c$, $a \neq 0$. As $|a|$ increases, the parabola is narrower, and, as $|a|$ decreases, the curve is wider.

Parentheses (1.1, 1.2). The symbol, ( ), used to indicate multiplication, as in 4(3), or order of operations, as in $6(4 + 3)$. *See* Grouping symbols.

Perfect square (4.4, 5.4, 10.2). (1) The special binomial product: $(x + a)^2 = x^2 + 2ax + a^2$. (2) The squares of the integers: $1^2 = 1$, $2^2 = 4$, $3^2 = 9$, $4^2 = 16$, $5^2 = 25$, $6^2 = 36, \ldots$.

Perimeter (1.4). The sum of the lengths of the sides of a polygon. For a square with side s the perimeter is $4s$; for a rectangle whose sides have lengths l and w the perimeter is $2l + 2w$.

Polynomial (4.1). A term or the sum or difference of two or more terms. *Monomial, binomial,* and *trinomial* are polynomials with one, two, and three terms, respectively. Examples of polynomials are 6, $5x$, $4y + 2$, $3x^2 + 2x - 3$, and $6x^2 + y - 3xy - 2x - 4y + 15$. Examples of expressions that are not polynomials are $1/x$, x/y, *and* $(x + 2)/(x - 3)$.

Positive (2.1). A number greater than zero is called positive.

Power (1.3). *See* Exponent.

Prime number (5.1). A number that has exactly two divisors. The primes less than 50 are: 2, 3, 5, 7, 11, 13, 17, 19, 23, 29, 31, 37, 41, 43, and 47.

Principal (1.4). Invested money. *See also* Interest.

Product (1.1). The result obtained from a multiplication problem.

Profit (1.3, 5.1). The selling price of an item less the cost.

Pythagorean Theorem (9.1). The sum of the squares of the legs of a right triangle is equal to the square of the hypotenuse. That is, $a^2 + b^2 = c^2$ for hypotenuse c and legs a and b.

Quadratic (4.1). *See* Degree.

Quadratic equation (5.5, 10.1–10.4). A second-degree polynomial equation with the general form $ax^2 + bx + c = 0$.

Quadratic formula (10.3, 10.4). If $ax^2 + bx + c = 0$, $a \neq 0$, then

$$x = \frac{-b \pm \sqrt{b^2 - 4ac}}{2a}$$

The number $b^2 - 4ac$ is the *discriminant* of the quadratic. If $b^2 - 4ac < 0$, there are *no* real solutions. If $b^2 - 4ac = 0$, there is *one* real solutuion. If $b^2 - 4ac > 0$, there are *two* solutions.

Quotient (1.1). The result obtained from a division problem.

Radical equation (6.5). An equation involving the square root of a variable.

Radical symbol (9.1). The symbol $\sqrt{\ }$. It denotes the square root.

Radicand (9.1). The number inside the radical symbol.

Radius of a circle (1.4). *See* Circle.

Rational equation (6.5). An equation that has at least one variable in the denominator.

Rational expression (6.2). A polynomial divided by a nonzero polynomial.

Rationalizing the denominator (9.2, 9.3). The process whereby radicals are eliminated from denominators.

Rational numbers (6.1, 9.1). The set, denoted by Q, of quotients of integers (except division by zero). Every rational number has a *repeating decimal representation.* For example, 6, -2, $\frac{1}{2}$, $\frac{3}{4}$, .2, .333..., .123123123... are rational numbers. See Irrational numbers for examples of numbers that are not rational.

Real numbers (9.1). The set, denoted by R, containing all the rationals and all the irrationals. See Rational numbers and Irrational numbers for examples of real numbers.

Reciprocal (6.1). $1/n$ is called the reciprocal of n; m/n is the reciprocal of n/m. Here are some examples

Number	Reciprocal
5	$\frac{1}{5}$
$\frac{2}{3}$	$\frac{3}{2}$
.1	10

Sometimes called *multiplicative inverse.*

Rectangular coordinate system (7.1). *See* Cartesian coordinate system.

Region (8.6). *See* System of inequalities.

Relation symbol (1.1). A symbol used to express the relationship between quantities, such as equality, =, less than, <, less than or equal to, ≤, greater than, >, and greater than or equal to, ≥.

Repeating decimal representation (9.1). A rational number whose decimal representation has a finite block of digits that repeats indefinitely such as $\frac{1}{3} = .3333\ldots$ or $\frac{1}{7} = .142857142857\ldots$ or $\frac{29}{22} = 1.3181818181\ldots$. *See* Rational numbers.

Root of an equation (3.1). A number that makes an equation true. We say that a root *satisfies* the equation.

Satisfy an equation (3.1, 8.1). A set of values of the variables that make an equation or equations true are said to satisfy the equation or equations.

Scientific notation (6.6). A number written in the form of the product of a number between 1 and 10 and a power of 10, or a power of 10. For example,

$$93{,}000{,}000 = 9.3 \cdot 10^7$$
$$5{,}001{,}000 = 5.001 \cdot 10^6$$
$$.0000347 = 3.47 \cdot 10^{-5}$$

Second component (7.1). *See* Ordered pair.

Signed numbers (2.1). Positive and negative numbers are called *signed* or *directed numbers.*

Similar terms (2.5, 4.2). Terms that contain the same power or powers of the variables, sometimes called *like terms.* When simplifying expressions you add (or subtract) similar terms. For example,

similar terms

$$5x + 3y + 4 + 6x - 2y - 5$$

and

similar terms

$$6x^2 + 3xy + 4y^2 + 2x - 3y + 4xy + 3y^2$$

Simplest form of square roots (9.2, 9.4). A square root is in simplified form if it contains no:

1. square factors in the radicand (if the radicand is completely factored, there is no factor raised to a power greater than 1); for example, if x is positive, $\sqrt{x^2} = x$, $\sqrt{x^3} = \sqrt{x^2}\sqrt{x} = x\sqrt{x}$, $\sqrt{x^4} = \sqrt{(x^2)^2} = x^2$, $\sqrt{x^5} = \sqrt{x^4}\sqrt{x} = x^2\sqrt{x}$, $\sqrt{27} = \sqrt{3^3} = 3\sqrt{3}$
2. radical as a denominator; for example, if x is positive,

$$\frac{a}{\sqrt{x}} = \frac{a}{\sqrt{x}} \cdot \frac{\sqrt{x}}{\sqrt{x}} = \frac{a\sqrt{x}}{x}$$
$$\frac{a}{\sqrt{x+2}} = \frac{a}{\sqrt{x+2}} \cdot \frac{\sqrt{x+2}}{\sqrt{x+2}} = \frac{a\sqrt{x+2}}{x+2}$$
$$\frac{a}{\sqrt{x}+2} = \frac{a}{\sqrt{x}+2} \cdot \frac{\sqrt{x}-2}{\sqrt{x}-2} = \frac{a(\sqrt{x}-2)}{x-4}$$

3. fraction in the radicand; for example, if x is positive,

$$\sqrt{\frac{1}{x}}, = \frac{\sqrt{1}}{\sqrt{x}} = \frac{1}{\sqrt{x}} \cdot \frac{\sqrt{x}}{\sqrt{x}} = \frac{\sqrt{x}}{x}$$
$$\sqrt{2^{-1}} = \sqrt{\frac{1}{2}} = \frac{1}{\sqrt{2}} \cdot \frac{\sqrt{2}}{\sqrt{2}} = \frac{\sqrt{2}}{2}$$

Simplify (2.5). The process of reducing an expression or a statement to a briefer form. Simplified form can mean (1) the briefest, least complex form or (2) the form best adapted to the next step to be taken in the process of seeking a certain result. One of the most indefinite terms used in mathematics, its meaning depends on the context in which it occurs.

Simultaneous (8.1). *See* System of equations and System of inequalities.

Slope of a line (7.3). The steepness of a line. For the line $y = mx + b$, the slope is m, and

$$m = \frac{\text{rise}}{\text{run}} = \frac{\text{change in vertical distance}}{\text{change in horizontal distance}} = \frac{y_2 - y_1}{x_2 - x_1}$$

where (x_1, y_1) and $(x_2 y_2)$ are two points on a line. For example, the slope of $y = -2x + 5$ is -2; for $2x - 5y + 6 = 0$

$$-5y = -2x - 6$$
$$y = \frac{2}{5}x + \frac{6}{5}, \text{ so the slope is } \frac{2}{5}$$

for the line passing through $(-6, 4)(5, -1)$ the slope is

$$m = \frac{-1-4}{5-(-6)} = \frac{-5}{11}$$

Slope-intercept form of a line (7.4). $y = mx + b$, where b is the y-intercept and m is the slope. For example, the slope-intercept form of the line $2x - 5y + 6 = 0$ is

$$y = \frac{2}{5}x + \frac{6}{5}$$

Solution of an equation (3.1). A number (or numbers) that makes an equation true. Also called a *root.* We say that a root satisfies the equation.

Solution of a system of equations (7.2). *See* System of equations.

Solution of a system of inequalities (5.5). *See* System of inequalities.

Solve an equation (3.3). Find all possible roots.

Square root (9.1). The square root of a positive number a is denoted by $\sqrt{a}$ and $\sqrt{a} = b$ only if $a = b^2$. For example, $\sqrt{1} = 1$, $\sqrt{4} = 2$, $\sqrt{9} = 3$, and $\sqrt{75} = 5\sqrt{3}$, since $(5\sqrt{3})^2 = (5\sqrt{3})(5\sqrt{3}) = 25 \cdot 3 = 75$.

Square-root method (10.1). If $x^2 = a$, then $x = \sqrt{a}$ or $x = -\sqrt{a}$ or both.

Standard form for a fraction (6.1). If p and q are positive numbers,

$$\frac{p}{q} \text{ and } \frac{-p}{q}.$$

Standard form of the equation of a line (7.3). $Ax + By + C = 0$, where A, B, and C are numbers, not all zero, and (x, y) is any point on the line.

Subtraction (2.3). To subtract, add the opposite of the number to be subtracted: $a - b = a + (-b)$.

Sum (1.1). The result obtained from an addition problem.

System of equations (8.1). Two (or more) equations with two (or more) variables. The *solution* of the system is all of the values of the variables that satisfy each of the equations. We say that the solutions satisfy the equations *simultaneously.*

System of inequalities (8.6). Two (or more) inequalities with two (or more) variables. The *solution* is the *region* determined by the *intersection* of the individual inequalities. A point that satisfies the system *simultaneously* satisfies each of the inequalities in the system.

Term (1.1, 2.5, 4.1). (1) Numbers being added or subtracted. (2) A single collection of factors—that is, an indicated product of numbers and variables. (3) A monomial. *See also* Polynomial.

Trichotomy property (3.7). *See* Comparison property.

Trinomial (4.1). *See* Polynomial.

Union (3.7). The *union* of two sets consists of all elements in both of the sets. The symbol for union is $\cup$.

Variable (1.1). A symbol used to represent an unspecified member of some set, called the *domain.* A variable is a "placeholder" for the name of some member in the domain.

Variable expression (1.2). A phrase containing one or more variables and at least one defined operation.

Vertical line (7.3, 7.4). A line parallel to the y-axis; its equation has the form $x = k$, where k is any real number. This line has no slope.

***x*-axis** (7.1). The horizontal axis in a Cartesian coordinate system.

***y*-axis** (7.1). The vertical axis in a Cartesian coordinate system.

***y*-intercept of a line** (7.4). The point at which a line intersects the y-axis. For the line $y = mx + b$, the y-intercept is $(0, b)$.

Index

Definition of Exponent	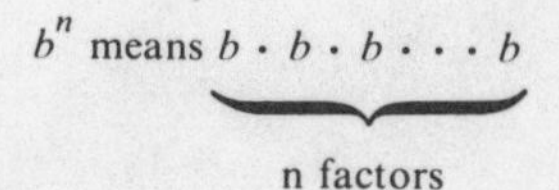 b^n means $\underbrace{b \cdot b \cdot b \cdots b}_{\text{n factors}}$ b is called the *base* and n is called the *exponent*. This expression is called "b to the nth power," or simply "b to the nth."
Zero Exponent	$b^0 = 1$ for all nonzero b.
Negative Exponents	$b^{-n} = \frac{1}{b^n}$ for all nonzero b and natural numbers n.
First Law of Exponents	$b^n \cdot b^m = b^{n+m}$ To multiply two numbers with like bases, add the exponents.
Second Law of Exponents	$(ab)^n = a^n b^n$ where n is a positive integer. To raise a product to a power, raise each factor to that power.
Third Law of Exponents	$(b^m)^n = b^{mn}$ where m and n are positive integers. To raise a power to a power, multiply the exponents.
Fourth Law of Exponents	$\frac{b^m}{b^n} = b^{m-n}$ for all nonzero b and integers m and n. To divide two numbers with like bases, subtract the exponents.
Fifth Law of Exponents	$\left(\frac{a}{b}\right)^m = \frac{a^m}{b^m}$ for all nonzero b To raise a quotient to a power, raise both the numerator and the denominator to that power.
Binominal Products	$(x + a)(x + b) = x^2 + (a + b)x + ab$ $\begin{pmatrix}\text{product of}\\ \text{FIRST}\\ \text{terms}\end{pmatrix} + \begin{pmatrix}\text{product of}\\ \text{OUTER}\\ \text{terms}\end{pmatrix} + \begin{pmatrix}\text{product of}\\ \text{INNER}\\ \text{terms}\end{pmatrix} + \begin{pmatrix}\text{product of}\\ \text{LAST}\\ \text{terms}\end{pmatrix}$
Difference of Squares	$(x + a)(x - a) = x^2 - a^2$ $\begin{pmatrix}\text{SUM OF}\\ \text{TERMS}\end{pmatrix}\begin{pmatrix}\text{DIFFERENCE}\\ \text{OF TERMS}\end{pmatrix} = \begin{pmatrix}\text{FIRST}\\ \text{TERM}\end{pmatrix}^2 - \begin{pmatrix}\text{SECOND}\\ \text{TERM}\end{pmatrix}^2$
The Perfect Square	$(x + a)^2 = (x + a)(x + a) = x^2 + 2ax + a^2$ $\begin{pmatrix}\text{FIRST}\\ \text{TERM}\end{pmatrix} + \begin{pmatrix}\text{SECOND}\\ \text{TERM}\end{pmatrix}^2 = \begin{pmatrix}\text{FIRST}\\ \text{TERM}\end{pmatrix}^2 + 2\begin{pmatrix}\text{FIRST}\\ \text{TERM}\end{pmatrix}\begin{pmatrix}\text{SECOND}\\ \text{TERM}\end{pmatrix} + \begin{pmatrix}\text{SECOND}\\ \text{TERM}\end{pmatrix}^2$